METHODS IN MOLECULAR BIOLOGY

For further volumes:
http://www.springer.com/series/7651

For over 35 years, biological scientists have come to rely on the research protocols and methodologies in the critically acclaimed Methods in Molecular Biology series. The series was the first to introduce the step-by-step protocols approach that has become the standard in all biomedical protocol publishing. Each protocol is provided in readily-reproducible step-by step fashion, opening with an introductory overview, a list of the materials and reagents needed to complete the experiment, and followed by a detailed procedure that is supported with a helpful notes section offering tips and tricks of the trade as well as troubleshooting advice. These hallmark features were introduced by series editor Dr. John Walker and constitute the key ingredient in each and every volume of the Methods in Molecular Biology series. Tested and trusted, comprehensive and reliable, all protocols from the series are indexed in PubMed.

Manuscript Instructions https://drive.google.com/file/d/1K4Az5vpAFVzYadJ3jKx2-QSo7Dqa_SNA/view?usp=sharing

Bioconjugation Methods

Edited by

Alexander Deiters

Department of Chemistry, University of Pittsburgh, Pittsburgh, PA, USA

Humana Press

Editor
Alexander Deiters
Department of Chemistry
University of Pittsburgh
Pittsburgh, PA, USA

ISSN 1064-3745 ISSN 1940-6029 (electronic)
Methods in Molecular Biology
ISBN 978-1-0716-5111-7 ISBN 978-1-0716-5112-4 (eBook)
https://doi.org/10.1007/978-1-0716-5112-4

This Humana imprint is published by the registered company Springer Science+Business Media, LLC, part of Springer Nature.
The registered company address is: 1 New York Plaza, New York, NY 10004, U.S.A.

Preface

Modern biological inquiry increasingly depends upon precise, modular, and quantitative methods to manipulate, label, and track biomolecules. Bioconjugation methodologies enable the covalent modification of biological macromolecules, such as proteins and nucleic acids. The utilized approaches often include chemical reactions that provide high selectivity, are compatible with an aqueous environment, and are bioorthogonal to chemical functionalities that are present due to the complexity of the macromolecules and the biological environment they reside in. Applications of bioconjugation reactions include the assembly of therapeutic agents, such as antibody drug conjugates; the detection and imaging of peptides, proteins, and nucleic acids; and the targeting of viruses and cells toward disease biomarkers. The protocols presented in this volume leverage cutting-edge chemical and molecular strategies—including bioorthogonal cycloadditions, use of electrophiles, self-labeling systems, and noncanonical amino acids—to enable powerful, site-selective bioconjugation across proteins, nucleic acids, cells, and viral assemblies. Each chapter adheres to the longstanding *Methods in Molecular Biology* tradition: an introductory overview, detailed and reproducible protocols, and, most importantly, notes that provide guidance and tips designed to support researchers to successfully implement these methods.

This volume brings together expert contributions from leading laboratories, each with a focus on functionality and clarity. The *Chatterjee* lab presents a site-specific, post-translational conjugation platform for AAV capsids. Using noncanonical amino acid mutagenesis and ultra-fast click chemistry, this method enables precise protein attachment, including full-length antibodies, to viral capsids, preserving infectivity while enabling customizable receptor targeting. The *Wang* lab presents metabolic labeling of enveloped virions using azido-modified sugars and lipids, enabling click labeling of viral envelopes for fluorescent visualization. The chapter gives detailed protocols for virus propagation, purification, and "click chemistry" labeling to support imaging and tracking studies. The *Cropp* lab focuses on activity-based di-ubiquitin probes synthesized via a noncanonical amino acid. The chapter encompasses the synthetic route to the amino acid and the assembly of diUb probes with site-specific isopeptide bonds, facilitating the study of ubiquitin conjugating/deconjugating enzyme specificity. The *Liu* lab details the synthesis of the fluorogenic deubiquitinase (DUB) substrate Ub-coumarin using activated cysteine-directed protein ligation. The method supports high-yield, high-purity production of Ub-coumarin from recombinantly expressed Ub-G76C, enabling sensitive assays of DUB activity. The *Pratt* lab delivers protocols for native chemical ligation (NCL) and derivative strategies to assemble full-length proteins with site-specific modifications, combining recombinant and synthetic fragments with chemoselective ligation. The *Deiters* lab describes a quantitative, in-cell bioconjugation assay using inverse electron demand Diels-Alder chemistry and HaloTag-mediated gel-shift analysis. This adaptable method enables quantification of labeling efficiency and is compatible with diverse proteins and noncanonical amino acid systems using widely available laboratory instruments. The *Howarth* lab presents their versatile SpyVLP platform, based on the broadly adapted SpyTag/SpyCatcher system, for multivalent antigen display. The chapter covers expression and purification of SpyCatcher003-mi3 nanoparticles, antigen coupling, validation, and immunogen design, emphasizing modularity and downstream applications in vaccinology, diagnostics, and catalysis. As another self-labeling enzyme technology, the *Lampkin* lab introduces the BenzoTag, an optimized protein evolved from HaloTag7 with enhanced reaction kinetics toward the fluorogenic Bz1 dye, enabling wash-free, orthogonal live-cell imaging alongside HaloTag7 systems. The *Lohmueller* lab describes SNAP-tag-based cell surface engineering, involving conjugation of antibodies with benzylguanine motifs and attachment to SNAP-CAR on T cells. Protocols span retroviral transduction, antibody modification, validation of surface conjugation, and functional assays of CAR activation and target killing. The *Spring* lab introduces protocols to use divinyl pyrimidine (DVP) reagents for irreversible, site-specific conjugation of native antibodies to generate homogeneous antibody-drug conjugates (ADCs) with controlled drug-antibody ratio. Protocols include conjugation chemistry and ADC characterization. The *Ploegh* lab explores sortase A-mediated transpeptidation for site-selective nanobody functionalization. The method leverages the LPxTG tag and oligoglycine nucleophiles to introduce diverse functional moieties with precision and efficiency via sortagging. The *Fox* lab introduces an affinity bioorthogonal chemistry tag strategy, employing a new tetrazine as both a bioorthogonal handle and an affinity purification tag. This dual functionality simplifies purification of site-selective protein conjugates and ensures clean separation from unreacted or mis-reacted species. The *Hamachi* lab offers protocols to implement a reactive peptide tag/probe system using hexahistidine tags paired with a $BisNi^{2+}$-NTA/NASA electrophile. Demonstrated via covalent labeling of a nanobody targeting GluD2 in live-cell imaging, this strategy affords selective lysine targeting and robust in-cell functionality. In a related approach, the *Jia* lab describes a histidine-specific labeling protocol via thiophosphorodichloridate reagents, forming stable thiophosphoramidate bonds. This multi-step method includes reagent synthesis, protein modification, purification via ion-exchange chromatography, and further diversification via copper-catalyzed azide-alkyne cycloaddition. The *Hocek* lab details the creation of reactive DNA probes via enzymatic incorporation of modified nucleotides, including their chemical synthesis, enabling covalent, proximity-directed cross-linking to proteins. Based on mild reaction conditions and high specificity, this method facilitates mapping of protein–DNA interactions. The *Lin* lab showcases the hydrazonyl sultone-BCN (bicyclononyne) ligation, featuring fast kinetics and aqueous stability. The chapter provides protocols for encoding the noncanonical amino acid BCNK into proteins (e.g., nanobodies and GPCRs), performing in vitro and live-cell labeling, and analyzing reaction selectivity, experiments that are particularly valuable for receptor dynamics studies.

Taken together, the bioconjugation protocols here span a broad range of techniques and applications and will be of use to researchers in both academia and industry.

Pittsburgh, PA, USA ***Alexander Deiters***

Contents

Contributors

T. Ashton Cropp • *Department of Chemistry, Virginia Commonwealth University, Richmond, VA, USA*

Gabriella Belsito • *Department of Chemistry, University of Pittsburgh, Pittsburgh, PA, USA*

Matthew T. Buhl • *Department of Chemistry, Virginia Commonwealth University, Richmond, VA, USA*

Saibal Chanda • *Department of Chemistry, Texas A&M University, College Station, TX, USA*

Abhishek Chatterjee • *Department of Chemistry, Boston College, Chestnut Hill, MA, USA*

Alexander Deiters • *Department of Chemistry, University of Pittsburgh, Pittsburgh, PA, USA; UPMC Hillman Cancer Center, University of Pittsburgh, Pittsburgh, PA, USA*

Ming Fang • *Department of Chemistry, State University of New York at Buffalo, Buffalo, NY, USA*

Joseph M. Fox • *Department of Chemistry and Biochemistry, University of Delaware, Newark, DE, USA; Department of Material Science and Engineering, University of Delaware, Newark, DE, USA*

Jake Glicksman • *Department of Chemistry, Boston College, Chestnut Hill, MA, USA*

Denise-Liu' Gracias Leone • *KU Leuven - Department of Cellular and Molecular Medicine, Leuven, Belgium*

Itaru Hamachi • *Department of Synthetic Chemistry and Biological Chemistry, Graduate School of Engineering, Kyoto University, Katsura, Nishikyo-ku, Kyoto, Japan*

Eldon R. Hard • *Department of Chemistry, University of Southern California, Los Angeles, CA, USA*

Rory A. Hills • *Department of Biochemistry, University of Oxford, Oxford, UK; Department of Pharmacology, University of Cambridge, Cambridge, UK*

Michal Hocek • *Institute of Organic Chemistry and Biochemistry of the Czech Academy of Sciences, Praha 6, Czech Republic*

Mark R. Howarth • *Department of Pharmacology, University of Cambridge, Cambridge, UK*

Shan Hu • *Chemistry and Biochemistry Department, University of South Carolina, Columbia, SC, USA*

Afraah Javed • *Department of Chemistry, University of Southern California, Los Angeles, CA, USA*

Shang Jia • *Department of Chemistry, Rutgers University-Newark, Newark, USA*

Monchupa Kingsak • *Chemistry and Biochemistry Department, University of South Carolina, Columbia, SC, USA*

Sona Krajcovicova • *Yusuf Hamied Department of Chemistry, Cambridge, UK*

Bryan J. Lampkin • *Department of Pharmaceutical Sciences, University of Illinois Chicago, Chicago, IL, USA*

Xinyuan Li • *Department of Chemistry, Rutgers University-Newark, Newark, NJ, USA*

QING LIN • *Department of Chemistry, State University of New York at Buffalo, Buffalo, NY, USA*
WENSHE RAY LIU • *Department of Chemistry, Texas A&M University, College Station, TX, USA*
JASON LOHMUELLER • *Department of Immunology, University of Pittsburgh, Pittsburgh, PA, USA; Division of Surgical Oncology, Department of Surgery, University of Pittsburgh, Pittsburgh, PA, USA; UPMC Hillman Cancer Center, University of Pittsburgh, Pittsburgh, PA, USA*
THONGPON MEETHONG • *Chemistry and Biochemistry Department, University of South Carolina, Columbia, SC, USA*
AVANI PARIKH • *Department of Immunology, University of Pittsburgh, Pittsburgh, PA, USA; Division of Surgical Oncology, Department of Surgery, University of Pittsburgh, Pittsburgh, PA, USA; UPMC Hillman Cancer Center, University of Pittsburgh, Pittsburgh, PA, USA*
QUAN PHAM • *Department of Chemistry, Boston College, Chestnut Hill, MA, USA*
HIDDE L. PLOEGH • *Program in Cellular and Molecular Medicine, Boston Children's Hospital, Boston, MA, USA*
MATTHEW R. PRATT • *Department of Chemistry, University of Southern California, Los Angeles, CA, USA*
PARAMESH K. RAMARAJ • *Department of Chemistry and Biochemistry, University of Delaware, Newark, DE, USA*
ELISA RUFFO • *Department of Immunology, University of Pittsburgh, Pittsburgh, PA, USA; Division of Surgical Oncology, Department of Surgery, University of Pittsburgh, Pittsburgh, PA, USA; UPMC Hillman Cancer Center, University of Pittsburgh, Pittsburgh, PA, USA*
SAMUEL L. SCINTO • *Department of Chemistry and Biochemistry, University of Delaware, Newark, DE, USA*
OLIVIA SHADE • *Department of Chemistry, University of Pittsburgh, Pittsburgh, PA, USA*
AMBRA SPAMPINATO • *Institute of Organic Chemistry and Biochemistry of the Czech Academy of Sciences, Praha 6, Czech Republic*
DAVID R. SPRING • *Yusuf Hamied Department of Chemistry, Cambridge, UK*
TOMONORI TAMURA • *Department of Synthetic Chemistry and Biological Chemistry, Graduate School of Engineering, Kyoto University, Katsura, Nishikyo-ku, Kyoto, Japan*
FUNG KIT TANG • *Department of Chemistry, Rutgers University-Newark, Newark, NJ, USA*
VIKRAM THIMARADKA • *Department of Synthetic Chemistry and Biological Chemistry, Graduate School of Engineering, Kyoto University, Katsura, Nishikyo-ku, Kyoto, Japan*
FLORIS J. VAN DALEN • *Program in Cellular and Molecular Medicine, Boston Children's Hospital, Boston, MA, USA*
QIAN WANG • *Chemistry and Biochemistry Department, University of South Carolina, Columbia, SC, USA*
THOMAS WHARTON • *Yusuf Hamied Department of Chemistry, Cambridge, UK*

Chapter 1

Site-Specific Chemical Conjugation of Recombinant Proteins onto the AAV Capsid

Quan Pham, Jake Glicksman and Abhishek Chatterjee

Abstract

Adeno-associated virus (AAV) is the leading vector for developing in vivo gene therapies. However, native AAV capsids exhibit limited control over tissue tropism, often necessitating high doses that lead to off-target effects and reduced therapeutic efficacy. Here, we present a robust and generalizable strategy to posttranslationally conjugate recombinant proteins, including full-length antibodies, onto the AAV capsid with precise control over the site and stoichiometry of attachment. By genetically encoding a bioorthogonal chemical handle into select capsid proteins using noncanonical amino acid (ncAA) mutagenesis, we enable selective labeling of the virus capsid with recombinant proteins using bifunctional linkers. This modular approach allows for subsequent rapid conjugation of site-specifically modified recombinant proteins through ultrafast click chemistry. Our method preserves AAV integrity and infectivity, and enables programmable retargeting of AAV tropism toward disease-relevant receptors. This platform provides a powerful approach for engineering next-generation AAV vectors with customizable targeting profiles for diverse therapeutic applications.

Keywords Adeno-associated virus, Recombinant proteins, Antibodies, Bioconjugation, Genetic code expansion, Noncanonical amino acid, AAV-protein conjugates, Retargeting, Click chemistry, Targeted gene delivery

1 Introduction

Adeno-associated virus (AAV) has become the vector of choice for in vivo gene therapy, with several FDA-approved treatments—such as Luxturna, Zolgensma, Hemgenix, Roctavian, Elevidys, Beqvez, and Kebilidi—highlighting its clinical success [1–3]. AAV vectors are favored for their nonpathogenic profile, low immunogenicity, and capacity to enable long-term gene expression [4–6]. However, natural AAV serotypes exhibit limited tropism, often requiring high doses to achieve therapeutic benefit. This leads to off-target transduction, increased toxicity, and higher treatment costs. To overcome these limitations, new strategies are needed to precisely reprogram AAV tropism for targeted gene delivery [1, 4, 7, 8].

Alexander Deiters (ed.), *Bioconjugation Methods*, Methods in Molecular Biology, vol. 3009,
https://doi.org/10.1007/978-1-0716-5112-4_1,

Traditionally, AAV capsid engineering has relied on either directed evolution or genetic modification approaches like peptide insertions or fusion to enhance specificity [1, 7]. While directed evolution can yield improved variants, it is laborious, time-intensive, and can produce unpredictable outcomes [9]. On the other hand, direct genetic fusion of targeting elements to the capsid often disrupts capsid integrity, compromising viral assembly, packaging efficiency, and infectivity [10]. To address these challenges, postpackaging chemical bioconjugation has emerged as a promising alternative, enabling modification of the capsid without affecting virus production or function [11, 12].

Among these strategies, site-specific incorporation of noncanonical amino acids (ncAAs) into the AAV capsid offers unique advantages [13–15]. Through genetic code expansion (GCE), bioorthogonal chemical handles can be precisely installed at defined sites on the capsid, enabling homogeneous and controlled labeling (Fig. 1). This approach allows conjugation of a wide range of functional groups—including small molecules, peptides, and full-length proteins—without compromising viral infectivity or stability. Additionally, it allows tunable control over the stoichiometry of modification, facilitating systematic optimization for enhanced targeting.

Here, we describe a versatile and efficient strategy to construct AAV-protein conjugates with precise control over both the modification site and degree of labeling (Fig. 2). Using GCE, we incorporate an ncAA bearing an azide group into specific positions on the AAV capsid and the protein of interest. We then apply a two-step post-translational labeling strategy: first, the azide-functionalized AAV or protein is prelabeled with a bifunctional linker (bearing tetrazine or TCO) via SPAAC; second, rapid and selective crosslinking is achieved through inverse-electron demand Diels–Alder (IEDDA) chemistry [16, 17]. This approach offers several key advantages: (A) AzK incorporation in mammalian cells is more efficient compared to bulkier ncAAs, minimizing yield loss; (B) AzK is well-tolerated at many positions on the AAV capsid; and (C) the use of bifunctional

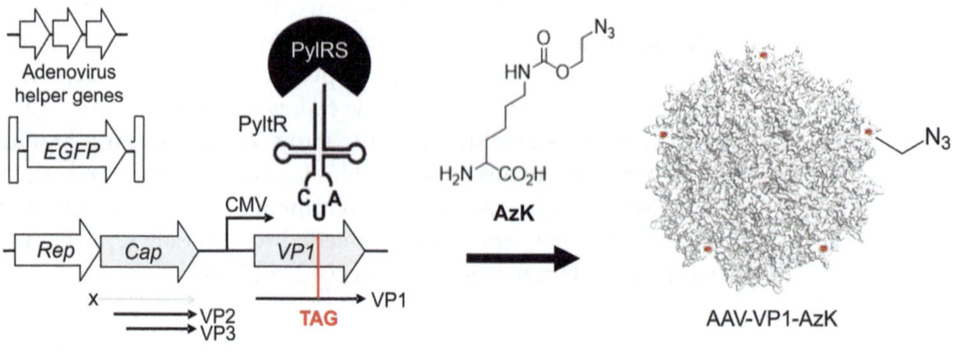

Fig. 1 Schematic description of genetic elements to produce AAV with ncAA mutagenesis at selected capsid protein

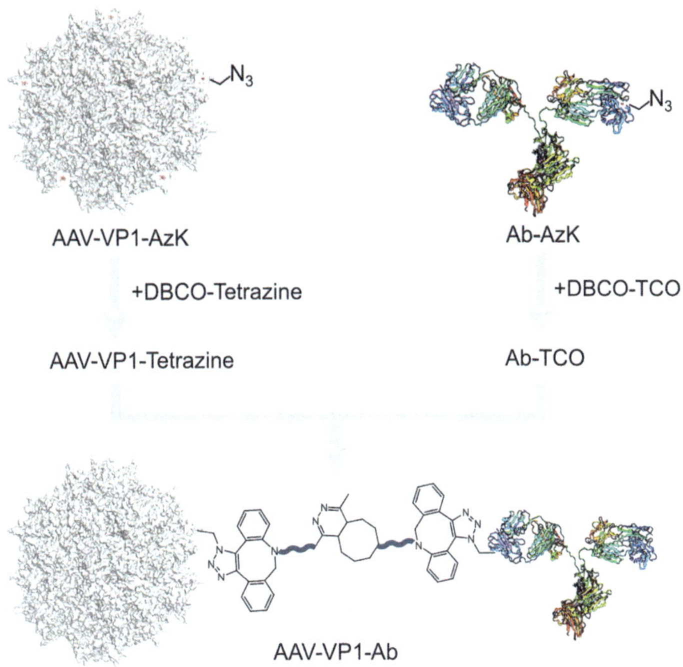

Fig. 2 Schematic description of bioconjugation strategy to crosslink AAV capsid with an antibody

linkers allows for additional tuning of linker properties to optimize conjugate behavior. This platform enables modular and programmable redirection of AAV tropism toward user-defined cellular targets, unlocking new possibilities for the development of next-generation, tissue-specific gene therapies.

2 Materials

2.1 Cell Culture

1. HEK293T cells (ATCC) for the production of AAV vectors (*see* **Note 1**).
2. SK-BR-3 cells (ATCC) for assaying the retargeting efficiency of AAV-antibody conjugates.
3. Expi293 cells (Thermo Fisher Scientific) for the production of antibodies (*see* **Note 2**).
4. Dulbecco's Modified Eagle Medium (DMEM; Thermo Fisher Scientific) supplemented with 10% Fetal bovine serum (FBS; Corning) and 1× antibiotic-antimycotic (Thermo Fisher) for culturing HEK293T cells and SK-BR-3 cells.
5. Expi293™ expression medium (Gibco) supplemented with 0.25× antibiotic-antimycotic (Thermo Fisher) for culturing Expi293 cells.

6. Standard cultureware includes BioLite™ cell culture-treated flasks (T75, T175; Thermo Scientific), cell culture-treated 15 cm dishes (Thermo Scientific), and Clear TC-treated 24-well plates (Corning).
7. Sterile single-use Erlenmeyer flasks with plain bottoms (Thermo Scientific) are used for suspension cultures.
8. Trypsin protease (Cytiva) is used for cell detachment (*see* **Note 3**).

2.2 Production of ncAA Containing AAV by Transient Transfection

1. HEK293T cells.
2. Plasmids: Three plasmids are required for producing AAV containing ncAA mutants via transient transfection.
   (i) pHelper (Cell Biolabs)
   (ii) One of the following pRep-Cap plasmids: pIDTsmart-RC2-TAG,pIDTsmart-RC2-ΔVP1-CMV-VP1,pIDTsmart-RC2-Δ VP2-CMV-VP2, and pIDTsmart-RC2-ΔVP12-CMV-VP12.
   (iii) pGOI-aaRS/tRNA: pIDTsmart-MbPylRS-4xPyOtR-ITR-GFP (*see* **Notes 4 and 5**).
3. Polyethyleneimine (PEI) transfection reagent: 20× stock (20 mg/mL): Prepare by dissolving branched PEI (average MW 25,000 g/mol; Sigma-Aldrich) in sterile H2O to a final concentration of 20 mg/mL. Mix gently and filter through a 0.22 μm hydrophilic PVDF syringe filter. Aliquot and store at −80°C. Working stock (1 mg/mL): Dilute the 20× stock in sterile H_2O and store at −20°C.
4. Opti-MEM I reduced serum medium (Gibco): Use directly without adding FBS or antibiotics.
5. Noncanonical amino acid (ncAA): Prepare a 100 mM stock solution of N6-((2-azidoethoxy)carbonyl)-L-lysine (AzK; Iris Biotech) by dissolving the solid in sterile H_2O. Adjust the pH to 7 with sodium hydroxide and filter through a 0.22 μm hydrophilic PVDF syringe filter.

2.3 Harvesting Virus and PEG Precipitation

1. Cell lifter: Used for harvesting cells from 15 cm dishes.
2. 1× Phosphate-buffered saline (PBS; 137 mM NaCl, 2.7 mM KCl, 10 mM Na_2HPO_4, and 1.8 mM KH_2PO_4; without calcium or magnesium): Prepare by diluting 10× PBS (Cytiva) with sterile H_2O at a 1:9 ratio. Filter through a 0.22 μm hydrophilic PVDF syringe filter.
3. Salty PBS for cell lysis: 1× PBS supplemented with an additional 200 mM NaCl (*see* **Note 6**).
4. Dry ice/ethanol bath: A vessel containing >95% ethanol, cooled with dry ice.
5. 37°C water bath: Used for freeze–thaw cycles.

6. Fresh 40% (w/v) polyethylene glycol (PEG) solution: To prepare 10 mL of 40% PEG solution, dissolve 4 g of PEG 8000 (Fisher Scientific) in 7 mL of sterile H_2O. Shake vigorously to mix, then heat to 55°C in a water bath until completely dissolved.

2.4 Purification of AAV Vectors by AVB Affinity Column

1. 1× PBS: Used to wash the AVB resin column, prepared as described in Sect. 2.3.
2. Elution buffer: 10× elution buffer stock (1 M glycine, pH 2.8): Prepare by dissolving glycine in sterile H_2O. Adjust the pH to 2.8 using HCl, filter through a 0.22 μm hydrophilic PVDF syringe filter, and store at 4°C. 1× elution buffer: Prepare by diluting the 10× stock in sterile H_2O.
3. Column storage buffer (1× PBS supplemented with 0.05% sodium azide): Prepare by adding 5 mL of 5% (w/v) sodium azide solution (Fisher Scientific) into 500 mL of 1× PBS as described in Sect. 2.3.
4. AAV buffer (1× PBS supplemented with 0.001% Pluronic F68 and 0.2 M NaCl): Prepare by mixing 50 mL of 10× PBS (Cytiva), 5.85 g NaCl, and 50 μL of 10% Pluronic F68 solution in 450 mL of H_2O. Filter through a 0.22 μm hydrophilic PVDF syringe filter.
5. 80% glycerol in PBS: Prepare by combining 1 volume of 10× PBS (Cytiva), one volume of H_2O, and eight volumes of glycerol. Filter through a 0.22 μm hydrophilic PVDF syringe filter.
6. AVB sepharose high-performance resin (Cytiva): Used to capture AAV2 capsids.
7. Pierce universal nuclease (Fisher Scientific): Used to digest DNA/RNA from cell lysate.
8. Narrow-bore gravity columns (Thermo Scientific): Disposable plastic columns with top and bottom caps, used to pack AVB resin.
9. Centrifugal filter devices (Amicon 100 kDa MWCO): Used for buffer exchange and concentration of AAV solutions.

2.5 Titration of AAV Vectors by Quantitative PCR

1. A commercially available AAV2-targeted qPCR kit such as the AAVpro® Titration Kit (for real-time PCR) Ver.2 (Takara).
2. Compatible qPCR instrument such as QuantStudio 3 real-time PCR system (Thermofisher).
3. 96-Well semi-skirted PCR plates.
4. Adhesive PCR plate seals.

2.6 Production of Antibodies by Transient Transfection

1. Expi293 cells: Cultured as described in Sect. 2.1.
2. Plasmids: Two plasmids are required for producing antibodies containing ncAA mutants via transient transfection.

   (i) pcDNA-AntiHer2-Ab-TAG

(ii) pIDTsmart-MbPylRS-4xPyOtR (*see* **Note 7**).

3. Polyethyleneimine (PEI) MAX transfection reagent: Prepared as described in Sect. 2.2.
4. Expi293 Expression Medium (Gibco): Use directly without adding antibiotics.
5. ncAA: A 100 mM stock solution of AzK is prepared as described in Sect. 2.2.

2.7 Purification of Antibodies by Affinity Capture Column

1. Wash buffer: Prepare a 20× stock solution of wash buffer (1 M NaOAc, pH 5.5) by dissolving sodium acetate anhydrous (Fisher Scientific) in sterile H_2O. Adjust the pH to 5.5 with acetic acid. Store at 4°C. Prepare 1× wash buffer (50 mM NaOAc, pH 5.5) by diluting the 20× stock in sterile H_2O.
2. Elution buffer (0.1 M Glycine, pH 2.8): Prepare as described in Sect. 2.4.
3. Column storage buffer (1×PBS, supplemented with 0.05% NaN_3): Prepare as described in Sect. 2.4.
4. 1× PBS: Prepare as described in Sect. 2.3.
5. 80% glycerol in PBS: Prepare as described in Sect. 2.4.
6. Pierce Protein G Agarose (Thermo Fisher Scientific) to capture antibodies.
7. Narrow-bore gravity columns: Disposable plastic columns with top and bottom caps, used to pack AVB resin.
8. Centrifugal filter devices (Amicon, 50 kDa MWCO) for buffer exchange and concentration of antibody solutions.

2.8 Bioconjugation of AAV and Antibodies

1. Dimethyl sulfoxide (DMSO; BioReagent Grade, Fisher Scientific): Used to dissolve bifunctional synthetic linkers and to supplement the reaction mixture to enhance the solubility of the linkers in aqueous buffer.
2. 50 mM DBCO–tetrazine bifunctional synthetic linkers: Dissolve commercially available Methyltetrazine-DBCO, Methyltetrazine-PEG12-DBCO, or Methyltetrazine-PEG24-DBCO (BroadPharm) in an appropriate amount of DMSO. Aliquot and store at −80°C.
3. 50 mM DBCO–TCO bifunctional synthetic linkers: Dissolve commercially available TCO-PEG4-DBCO, TCO-PEG12-DBCO, or TCO-PEG24-DBCO (BroadPharm) in DMSO. Aliquot and store at −80°C.
4. Nutating mixer: Used for gentle mixing during bioconjugation reactions.
5. Centrifugal filter devices: Amicon 100 kDa MWCO: For buffer exchange and concentration of AAV solutions. Amicon

10–50 kDa MWCO: For buffer exchange and concentration of antibodies or other proteins.

6. 1× PBS: Used for protein buffer exchange and concentration, as described in Sect. 2.3.
7. Centrifugal filter coating buffer (1× PBS supplemented with 0.1% Pluronic F68): Prepare by adding 500 μL of 10% Pluronic F68 solution (Thermo Fisher Scientific) to 50 mL of 1× PBS as described in Sect. 2.3.
8. AAV buffer (1× PBS supplemented with 0.001% Pluronic F68 and 0.2 M NaCl): Used for AAV buffer exchange and concentration, as described in Sect. 2.4.

2.9 Validation of AAV-Antibody Conjugates by SDS-PAGE

1. 10%–12% polyacrylamide gels.
2. SDS-PAGE running buffer: Prepare a 10× Tris-glycine stock buffer by dissolving 30.3 g Tris base, 114 g glycine, and 10 g SDS in 1 L of H_2O. Prepare 1× Tris-glycine buffer by diluting the 10× stock in H_2O. Use for running gels.
3. 6× SDS-PAGE loading buffer: To prepare 10 mL, combine: 1.5 g sodium dodecyl sulfate (SDS), 3 g DTT, 7 mL of 0.5 M Tris–HCl (pH 6.8), 60 μL 10% bromophenol blue, 2.6 mL glycerol. Aliquot and store at −80°C.
4. PageRuler™ Prestained Protein Ladder (10 to 180 kDa, Fisher Scientific).
5. Compatible electrophoresis chamber.
6. Staining buffer (7.5% acetic acid supplemented with 1× SYPRO Orange Stain): Prepare by diluting acetic acid in H_2O and adding 1/5000 volume of commercial SYPRO Orange stain (5000× concentrate; Thermo Fisher Scientific).

2.10 Assay the Activity of AAV-Antibody Conjugates by Transducing Cell Culture

1. HEK293T and SK-BR-3 cells, cultured as described in Sect. 3.1.
2. 500 mM sodium butyrate solution: Dissolve sodium butyrate (Sigma-Aldrich) in sterile H_2O to a final concentration of 500 mM. Filter through a 0.22 μm sterile filter.
3. Fluorescence microscope equipped with a suitable filter set to image EGFP expression.
4. 0.25% trypsin-EDTA solution (Thermo Fisher Scientific) for cell detachment.
5. 1× PBS, prepared as described in Sect. 2.3.
6. DMEM supplemented with FBS, as described in Sect. 2.1.
7. 100 μm filters or cell strainer tubes (Fisher Scientific).
8. Flow cytometer capable of analyzing EGFP-expressing cell populations.

3 Methods

3.1 Cell Culture

HEK293T cells and SK-BR-3 cells are cultured in DMEM (supplemented with 10% fetal bovine serum and 1× antibiotic-antimycotic) at 37°C with 5% CO_2.

Expi293 cells are cultured at 37°C with 8% CO_2 in Expi293™ Expression Medium (Gibco) supplemented with 0.25× antibiotic-antimycotic on an orbital shaker at 125 rpm.

3.2 Plasmid System for the Production of AAV Vectors with ncAA Mutagensis at the Selected Capsid Proteins

We previously reported the plasmid systems to site-selectively incorporate ncAA throughout the AAV capsid [13, 18]. In brief, along with Adenovirus Helper genes and AAV genes, for the incorporation of ncAA to the AAV capsid, a pair of aminoacyl synthetase (aaRS) and tRNA cassette gene was included. For example, *Methanosarcina barkeri* pyrrolysyl-tRNA-synthetase (MbPylRS) and optimized *Methanosarcina mazei* pyrrolysyl $tRNA_{CUA}$ can be used to incorporate AzK into the capsid protein with high efficiency [19]. The AAV Cap gene contains overlapping coding sequences of VP1, VP2, VP3, which are incorporated into the capsid at a ratio of 1:1:10. To site-selectively incorporate ncAA to the AAV capsid, surface exposed sites at the common region shared among VP1, VP2, VP3 were mutated to TAG, resulting in ncAA incorporation at all 60 capsid proteins.

To exercise better control over the number of ncAAs present per capsid, and thus the degree of capsid modification through bioorthogonal chemistry, we have further developed a "split-cap" system, where the expression of the three capsid proteins is genetically separated (Table 1) [14]. The ability to separately express the capsid proteins allows us to incorporate ncAAs into a selected set of capsid proteins (e.g., only VP1, or only VP2, or both VP1 and VP2). To achieve this, we uncoupled the expression of the minor capsid proteins from the AAV Cap gene by mutating their start codon(s). The uncoupled minor capsid gene was supplied back in *trans*, from a CMV-driven expression cassette, and this cassette was further mutated to abolish expression of VP3. A surface-exposed site at this minor capsid protein-expressing gene was then mutated to TAG for incorporating ncAAs only to the selected capsid subunits (*see* **Note 8**) [14]. Incorporation of ncAAs only at the minor capsid proteins, which are present in fewer numbers, minimizes perturbation of the capsid and affords wild-type-like yield.

3.3 Production of ncAA Containing AAV by Transient Transfection

1. Seed HEK293T cells in 15 cm dishes (~25 million cells per dish) so that cell confluency reaches 70%–80% the next day for transfection (*see* **Note 9**).

Table 1
Summary of mutagenesis in the AAV genome and in the separate minor capsid genes for uncoupling selected capsid proteins from the AAV native capsid genes.

Construct	Deleted start codons in AAV genome	Deleted start codons in CMV-Cap	# of ncAAs incorporated per capsid
pIDTsmart-RC2-Δ VP1-CMV-VP1-TAG	M1L (ATG→CTC)	T138T(ACG→ACC), T197T(ACG→ACC), L202L(CTG→CTC), M203L(ATG→CTC), M211L(ATG→CTC), M235L(ATG→CTC)	~5
pIDTsmart-RC2-Δ VP2-CMV-VP2-TAG	T138T(ACG→ ACC)	L202L(CTG→CTC), M203L(ATG→CTC), M211L(ATG→CTC), M235L(ATG→CTC)	~5
pIDTsmart-RC2-Δ VP1,2-CMV-VP1,2-TAG	M1L (ATG→CTC), T138T(ACG→ ACC)	CMV-VP1: T138T(ACG→ ACC), T197T(ACG→ ACC), L202L(CTG→ CTC), M203L(ATG→ CTC), M211L(ATG→ CTC), M235L(ATG→CTC); CMV-VP2: T197T(ACG→ ACC), L202L(CTG→ CTC), M203L(ATG→ CTC), M211L(ATG→CTC), M235L(ATG→CTC)	~10
pIDTsmart-RC2-TAG	N/A	N/A	60

2. Prepare transfection mixtures in a biosafety cabinet. Add the following components to a 2 mL microcentrifuge tube in this order:
   (a) 1 mL of Opti-MEM™ I reduced serum medium (Gibco).
   (b) A total of 57 μg of plasmid DNA (including the Adenovirus Helper vector, AAV Rep-Cap vector, AAV ITR-transgene and aaRS/tRNA vector). An equimolar ratio of plasmids is a good starting point, but can be further optimized (*see* **Note 10**).
   (c) 200 μL of 1 mg/mL polyethyleneimine (PEI). Use 3.5 μL of PEI for every 1 μg of DNA.
3. Mix the transfection mixture by tapping the side of the tube vigorously after adding PEI. Avoid vortexing.
4. Incubate the mixture at room temperature for 10–15 min.

5. Add the transfection mixture dropwise to the cells using a P1000 pipette, spreading it evenly across the entire dish. Gently mix by rocking the plate back-and-forth a few times (*see* **Note 11**).
6. Add the required volume of ncAA. For incorporation of AzK using the MbPylRS and Pyl tRNA system, supplement the growth media with 1 mM AzK. Because the volume of the ncAA stock is small, tilt the plate a little, add the ncAA to the media, then gently but thoroughly mix by rocking the plate (*see* **Note 12**).
7. Return the transfected plates to the incubator and culture for 3–4 days. (Longer incubation increases AAV titer per mL of media.)
8. Optional: On day 3, collect the media containing AAVs. Then, add fresh growth media supplemented with ncAA and incubate for an additional 2–3 days to further increase overall AAV yield.

3.4 Harvesting and PEG Precipitation of AAV Vectors

After the incubation period of the transfected cell culture, the majority of the virus is secreted in the media, but a fraction remains in the cell pellet. Harvesting virus from both the cell pellet and media will increase the total yield. PEG precipitation can be used to concentrate and semi-purify the virus. The virus should be harvested in a biosafety cabinet.

1. Detach cells gently from the tissue culture plate (15 cm dish) using a cell lifter. Transfer the resuspended cells in culture media (approximately 25 mL) to a 50 mL Falcon tube. Rinse the plate with approximately 5 mL of PBS, and combine the rinse with the resuspended cells.
2. Pellet the cells by centrifugation at 5000× *g* for 5 min at 4°C. Collect the supernatant (the growth media containing AAVs) by gently pouring it into another 50 mL Falcon tube.
3. Resuspend the cell pellet in 1.5 mL of salty PBS and lyse the cells by performing two complete freeze-thaw cycles (snap freeze in dry ice/ethanol followed by thawing in a 37°C water bath).
4. Pellet the cells by centrifugation at 12,000× *g* for 2 min. Collect the cell lysate supernatant and combine it with the collected growth media. Discard the cell pellet.
5. Optional: Resuspend the cell pellet in 0.5 mL of salty PBS and repeat the freeze-thaw step two more times to extract additional virus from the pellet.
6. Add 40% PEG (approximately 11 mL) to the total virus-containing solution (approximately 32 mL) combined from the cell lysate and collected growth media to achieve a final

concentration of 10%. Mix thoroughly by shaking the solution. Let the solution sit for at least 8 h, or overnight, at 4°C.

7. After the incubation period, pellet the virus by centrifugation at 5000× *g* for 30 min at 4°C.
8. Remove the supernatant and resuspend the pelleted virus in 2 mL of AAV buffer. Transfer the resuspended virus to a 2 mL microcentrifuge tube. The virus solution can then be purified by affinity capture. The resuspended virus solution will appear cloudy due to the presence of some insoluble proteins from the cell lysate and growth media.

3.5 Purification of AAV Vectors by AVB Affinity Column

AAV2 can be purified by AVB resin, which contains anti-AAV single-domain antibody fragments crosslinked to agarose matrix. Affinity capture column purification of AAV2 capsids is necessary to obtain a high-purity, concentrated vector solution. This step is also essential for a clear analysis of capsid protein conjugation with a protein using SDS-PAGE. All purification steps should be carried out in a Biosafety cabinet.

1. Centrifuge the virus solution at 12,000× *g* for 5 min to remove insoluble particles and avoid clogging the column. Transfer the supernatant, which contains the virus solution, to a new 5 mL tube.
2. Resuspend the pellet in 0.5 mL AAV buffer (described in Sect. 2.4) to extract any remaining AAV from the insoluble fraction.
3. Centrifuge the virus solution at 12,000× *g* for 5 min to clear the insoluble debris. Combine the supernatant with the supernatant collected from step 1.
4. Treat the virus solution (~2.5 mL) with 2.5 μL of universal nuclease to digest DNA and RNA, which will increase the virus's binding to the resin. Tap the virus solution to mix, and incubate the solution for 10 min at room temperature. Proceed to step 5 during the incubation time.
5. Pack the AVB resin into a narrow-bore gravity column. Use 200 μL of AVB slurry to purify the virus solution harvested from a 15 cm dish.
6. Wash the column with 2 × 1 mL PBS to remove the storage buffer and equilibrate the column.
7. Slowly load the virus solution from step 5 onto the settled resin and let it flow through by gravity. Collect the flow-through.
8. Reload the flow-through at least one more time to increase the binding efficiency of the virus to the affinity resin.

9. Wash the resin bed with 10 × 1 mL PBS. Pressure can be applied to increase the flow rate during washes. Discard the flow-through.
10. Add 200 μL neutralization buffer (1 M Tris, pH 8) to a 4 mL centrifugal filter device (Amicon 4, 100 kDa).
11. Elute the virus from the column quickly with 2 mL of AVB elution buffer (0.1 M glycine, pH 2.8) directly into the Amicon. Place the cap on and mix by inversion immediately (*see* **Notes 13 and 14**).
12. Centrifuge the Amicon centrifugal filter device at 1000 × *g*, 4°C, to reduce the volume to ~500 μL. Discard the flow-through (*see* **Note 15**).
13. Refill the Amicon centrifugal filter device with AAV buffer and mix by inversion. Centrifuge again at 1000 × *g*, 4°C, to reduce the volume to ~500 μL. Discard the flow-through (*see* **Note 16**).
14. Repeat step 13 two more times to ensure efficient buffer exchange of the virus to AAV buffer.
15. Resuspend the concentrated virus solution using a P200 micropipette and transfer the virus to a new 1.5 mL micro-centrifuge tube.
16. Optional: Sterile filter the virus using a 0.22 μm, 4 mm filter.
17. Make a glycerol stock of the AAV solution by adding 1/7th of the virus volume of 80% glycerol in PBS. Mix the solution by tapping (avoid foaming). Aliquot the virus solution, snap-freeze, and store at −80°C.

3.6 Titration of AAV Vectors by Quantitative PCR

Packaged AAV genomes can be titrated using a commercially available qPCR kit like AAVpro® Titration Kit (for Real Time PCR) Ver.2 (Takara) following the manufacturer's protocol.

3.7 Production of ncAA Containing Recombinant Antibodies

This section describes the production of ncAA-containing antibodies for subsequent bioconjugation with AAV capsids, serving as an example of AAV-protein conjugation. It should be noted that our technology is not limited to attaching antibodies to AAV capsids, but represents a general platform that can be used to attach any recombinantly expressed protein. One day prior to transfection, seed Expi293 cells in sterile, single-use Erlenmeyer flasks at a density of 1 million cells/mL. For a 500 mL flask, use 120 million cells in 120 mL of Expi293 expression medium.

1. When the cell density reaches approximately 2 million cells/mL the next day, prepare the transfection mixtures in a cell culture hood as follows. Add the components in the order listed below into a 50 mL Falcon tube:

(a) 10 mL of Expi293 expression medium.

(b) A total of 120 μg of two plasmids (pcDNA-Anti-Her2-Ab-TAG and pIDTsmart-MbPylRS-4xPyOtR). An equimolar ratio of the plasmids is a good starting point but can be further optimized for different plasmid systems.

(c) 720 μL of 1 mg/mL polyethyleneimine (PEI). Use 6 μL of PEI per 1 μg of DNA (*see* **Notes 17 and 18**).

2. Mix the transfection mixture by vigorously tapping the side of the tube after adding PEI. Avoid vortexing.
3. Let the transfection mixture sit at room temperature for 10–15 min.
4. Add the entire transfection mixture to the cells.
5. Add the required volume of noncanonical amino acid (ncAA). For the incorporation of AzK using MbPylRS and Pyl-tRNA, supplement the growth medium with AzK to a final concentration of 1 mM. Return the transfected culture flask to the orbital shaker in the incubator.
6. One day after transfection, add valproic acid to the culture to a final concentration of 2 mM to enhance antibody expression (*see* **Note 19**). Return the flask to the orbital shaker in the incubator and continue incubation for an additional 6–8 days.

3.8 Purification of Antibodies by Affinity Capture Column

1. After 7–8 days of expression, transfer the cell culture to 50 mL Falcon tubes and centrifuge at 5000 × *g* for 10 min at 4°C. Collect the growth medium containing antibodies and discard the cell pellets.
2. Centrifuge the growth medium again at 5000 × *g* for 10 min to remove any insoluble debris.
3. Adjust the pH of the medium to 5.5 using 1 M sodium acetate (NaOAc), pH 5.5 stock solution.
4. Pack Protein G resin into a narrow-bore gravity column. Use 0.5 mL of Pierce™ Protein G resin, which should be sufficient to capture all antibodies from a 120 mL culture.
5. Equilibrate the column with 3 × 1 mL of wash buffer (50 mM NaOAc, pH 5.5).
6. Load the antibody-containing growth medium onto the column via gravity flow.
7. *Optional:* Reloading the flow-through may increase antibody binding efficiency. Discard the flow-through afterward.
8. Wash the column with a volume of wash buffer equal to the amount of loaded medium (~120 mL). Discard the flow-through.

9. Elute the bound antibodies with ten column volumes of elution buffer (~5 mL of 100 mM glycine, pH 2.8).
10. Immediately neutralize the elution fraction by adding 1/10 volume of 1 M Tris, pH 8.0.
11. Transfer the eluted antibodies to Amicon™ centrifugal filter devices (50 kDa MWCO) and centrifuge at 5000× *g*, 4°C, to concentrate the solution to ~200 μL. Discard the flow-through.
12. Refill the Amicon filter with PBS, mix by gentle inversion, and centrifuge again at 5000 × *g*, 4°C, to reduce the volume to ~200 μL. Discard the flow-through.
13. Repeat step 12 two more times to ensure efficient buffer exchange into PBS.
14. Resuspend the concentrated antibody solution using a P200 micropipette and transfer it to a new 1.5 mL Eppendorf tube.
15. *Optional:* Sterile filter the antibody solution using a 0.22 μm, 4 mm syringe filter.
16. Prepare glycerol stocks by adding 1/7 volume of 80% glycerol in PBS to the antibody solution. Mix by tapping (avoid foaming), aliquot, snap-freeze, and store at −80°C.

3.9 Prelabeling Azide Containing AAV Vectors and Antibody with Bifunctional Synthetic Linker

Even though the installation of AzK onto AAV capsid offers advantages in terms of incorporation efficiency and better tolerance, the reaction rate of azides with cyclooctynes via strain-promoted azide–alkyne cycloaddition (SPAAC) is not fast enough to allow efficient protein-protein coupling. Therefore, we functionalize the azide handle to introduce a tetrazine or *trans*-cyclooctene (TCO) group using synthetic bifunctional reagents [15]. This enables subsequent ultra-fast protein conjugation via inverse electron-demand Diels–Alder (IEDDA) reactions and allows for optimization of linker lengths for optimal conjugate properties.

1. In a 0.6 mL microcentrifuge tube, dilute the stock solutions of the protein and synthetic bifunctional linker in PBS to the desired concentrations as indicated below:
   (a) Incubate the purified virus solution containing azide handles at the selected capsid protein concentration (~1×10^9 gc/μL) with 100 μL of DBCO–tetrazine.
   (b) Incubate the purified protein solution containing azide handles (100 μM) with 100 μL of DBCO–TCO (*See* **Notes 20 and 21**).
2. Incubate the reaction mixture overnight at 4°C on a nutating mixer. Protect from light during incubation.

3. Buffer Exchange to Remove Excess Linker: Load each reaction mixture onto an Amicon™ centrifugal filter device and perform buffer exchange in a suitable buffer to remove unreacted synthetic linkers.

   (a) For tetrazine-labeled AAV (*see* **Note 22**):

      (i) Load the tetrazine-labeled AAV solution into the centrifugal filter device.

      (ii) Fill the device to volume with AAV buffer (described in Sect. 2.4), cap, and mix by gentle inversion.

      (iii) Centrifuge at 1000× *g* at 4°C to concentrate the virus solution to approximately 500 μL. Avoid over-concentration. Discard the flow-through.

      (iv) Repeat steps (ii) and (iii) at least three times to efficiently remove excess bifunctional linkers.

   (b) For TCO-labeled proteins:

      (i) Load the TCO-labeled protein solution into a centrifugal filter device.

      (ii) Fill the device with PBS. Cap and mix by gentle inversion.

      (iii) Centrifuge at 14,000× *g* at 4°C to concentrate the protein. Discard the flow-through.

      (iv) Repeat steps (ii) and (iii) at least three times to efficiently remove excess bifunctional linkers.

3.10 Bioconjugation of AAV Capsid and Antibody

The tetrazine-labeled AAV capsids and TCO-labeled proteins from the previous section are now ready for fast protein coupling via IEDDA.

1. Re-titer the virus using quantitative PCR, as the viral titer may slightly change after buffer exchange.
2. Recheck protein concentration using Bradford assay.
3. In a 0.5 mL microcentrifuge tube, combine the tetrazine-labeled virus and the TCO-labeled nanobody to a final concentration of 1×10^9 gc/μL and 1 μM, respectively.
4. Incubate the reaction mixture on a nutating mixer at 4°C for 4 h (*see* **Note 23**).

3.11 Characterizing AAV-Antibody Conjugates by SDS-PAGE

1. The AAV, protein, and AAV–protein conjugate samples were heated in SDS loading buffer for 1 min. Heating is necessary to denature the capsid proteins; however, extended heating may cause protein aggregation and poor resolution on SDS-PAGE.
2. For clear visualization of all capsid protein bands (both minor and major), load approximately 10^{10} genome copies of virus per lane on a 10%–12% SDS-PAGE gel, along with a molecular weight marker.

3. Stain the SDS-PAGE gel using SYPRO orange protein stain (Thermo Fisher Scientific) by incubating it in staining buffer (7.5% acetic acid supplemented with 1× SYPRO Orange stain). Shake gently for 1 h.
4. Remove the staining solution and quickly wash the gel with distilled water.
5. Image the protein bands using a Dylight 540-compatible imager. Wild-type and uncoupled virus samples should display three capsid protein bands: VP1 (~87 kDa), VP2 (~73 kDa), and VP3 (~62 kDa), with relative intensities of approximately 1:1:10. AAV–protein conjugates should show a shift in the band corresponding to the conjugated capsid protein, indicating an increased molecular weight due to protein coupling (Fig. 3).

3.12 Assay the Activity of AAV-Antibody Conjugates by Transducing Cell Culture

This section demonstrates an example of assaying the activity of AAV capsids conjugated to anti-HER2 antibodies. SK-BR-3 cells, which overexpress HER2, are used to assess on-target activity, while HEK293T cells, which express minimal levels of HER2, are used to evaluate off-target activity. Different recombinant proteins or antibodies conjugated to the viral capsid will confer distinct tissue-targeting profiles, and therefore, appropriate target cell lines are required to evaluate the efficiency of each conjugate. The AAV vectors used below carry an EGFP transgene for the analysis of transduction via flow cytometry and fluorescence microscopy.

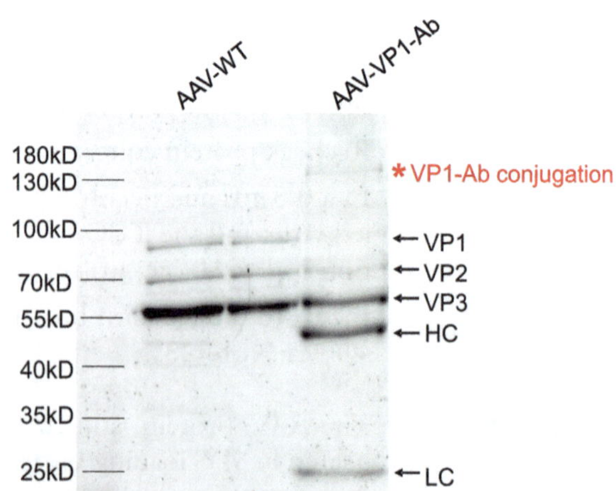

Fig. 3 Representative SDS-PAGE analysis of an AAV–antibody conjugate targeting a specific capsid protein. The AAV–VP1–antibody conjugate (fourth lane) shows a clear shift of the VP1 band to a higher molecular weight, confirming successful conjugation with the antibody

1. Seed SK-BR-3 cells (~0.1 million cells/well) and HEK293T cells (~0.25 million cells/well) in 24-well plates such that they reach >80% confluency the following day.
2. Once the cells have reached the desired confluency (~20–24 h for HEK293T cells, ~30 h for SK-BR-3 cells), transduce them with an equal number of genome copies of the following AAV vectors at a constant MOI:
   - AAV WT
   - AAV-VP1-AzK with additional mutations (R585A R588A throughout the capsid to detarget from heparan sulfate)
   - AAV-VP1-Ab prepared by conjugating AAV-VP1-AzK and anti-HER2 antibody (*see* **Note 24**).
3. Add sodium butyrate to a final concentration of 5 mM to enhance AAV-encoded transgene expression. Mix the plate gently but thoroughly, and return it to the incubator.
4. After 48 h of incubation, visualize cells and EGFP transgene expression using a fluorescence microscope equipped with suitable filter set (excitation 475/40 nm, beamsplitter 500 nm, emission 530/50 nm). The AAV–VP1–Ab conjugate is expected to exhibit significantly enhanced transduction in SKBR3 cells compared to wild-type AAV2 (Fig. 4b), while showing markedly reduced transduction in HEK293T cells.
5. Harvest the cells by removing the media, washing once with PBS, and treating with warm 0.25% trypsin-EDTA (100 μL/well) for 5 min at 37°C.
6. Quench trypsin activity by adding 200 μL of cold DMEM supplemented with 10% FBS to each well.
7. Resuspend the cells and transfer them to 2 mL microcentrifuge tubes. Centrifuge at 2500 × *g* for 5 min at 4°C, then discard the supernatant.
8. Resuspend the cell pellets in 200 μL of cold PBS and filter through a 35 μm strainer cap into 5 mL tubes (Falcon) to avoid clogging the flow cytometer.
9. Analyze transduction efficiency using a BD Accuri C6 Plus flow cytometer (BD Biosciences) with appropriate gating and fluorescence settings. For EGFP detection, use FITC channel (excitation: 488 nm; emission filter: 533/30 nm or 670LP, as appropriate).
10. Analyze the data using FlowJo software (version 10.8.1). The percentage of EGFP-positive cells, detected via the FITC channel as a result of successful infection by each virus, can be plotted and normalized to AAV2-WT infectivity to directly compare the relative improvement in infectivity of the AAV–antibody conjugates. The AAV–VP1–Ab conjugate is expected to transduce SK-BR-3 more efficiently than AAV2-WT (Fig. 4a).

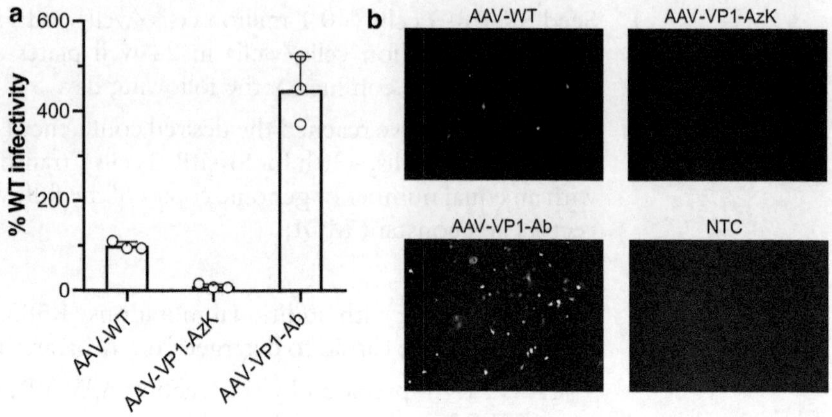

Fig. 4 (**a**) Representative infectivity measurement of AAV2 WT or AzK mutant virus without (AAV-VP1-AzK) or with conjugation with AntiHer2-Antibody (AAV-VP1-Ab), measured by the percentage of positive GFP cells by flow cytometry, upon infecting SK-BR-3 (HER2+) cells at a constant MOI (250), normalized to the percentage of WT AAV2 infectivity. (**b**) Fluorescence images associated with panel (**a**). NTC: Nontreated cells

4 Notes

1. Low-passage and healthy HEK293T cells are required for high-titer AAV production. Typically, one 15 cm dish can yield 1–3 × 10^{12} gc using our transfection protocol when using low-passage HEK293T cells.
2. Low-passage Expi293 cells are required for high antibody production. The addition of antibiotics reduces antibody titers in our experience. This protocol uses 0.25× antibiotic-antimycotic out of an abundance of caution, while still maintaining high antibody expression. It is possible to completely omit antibiotics in the growth medium to achieve higher antibody titers.
3. SK-BR-3 cells adhere strongly to the flask, so washing the cells with PBS is recommended to facilitate detachment. We recommend washing the cells with prewarmed 1× PBS (without Mg^{2+} or Ca^{2+}).
4. (i) pHelper encodes the necessary adenoviral helper genes (E2A, E4, VA). (ii) Each of the various pRep-Cap constructs encodes AAV2 Rep and Cap genes with an amber stop codon (TAG) introduced at the site of interest in the selected capsid protein. (iii) pIDTsmart-4xMbPylRS-4xPyOtR-ITR-GFP encodes an EGFP reporter gene flanked by AAV2 ITRs, an MbPylRS, and four copies of the optimized *M. mazei* pyrrolysine tRNA cassette for incorporating ncAA to the predefined TAG codon. Plasmids in (ii) and (iii) are available upon

request from the Chatterjee lab. Maps and sequences have been published [13–15].

5. In our experience, tRNA is typically the limiting factor for the ncAA incorporation. This plasmid system includes four copies of the optimized *M. mazei* pyrrolysyl tRNA, which provides enhanced ncAA incorporation efficiency [19]. Using optimized tRNA has been shown to greatly improve the production yield of recombinant proteins and AAVs. Other optimized tRNAs are also available for incorporating AzK to the AAV capsid proteins with high efficiency [20].
6. In our experience, PBS supplemented with higher salt concentration helps reduce AAV aggregation, minimize nonspecific binding, and improve viral particle recovery.
7. (i) pcDNA-AntiHer2-Ab-TAG encodes the trastuzumab light chain (LC) and heavy chain (HC) with a TAG codon at the site of interest. pIDTsmart-MbPylRS-4xPyOtR—encodes the *Methanosarcina barkeri* pyrrolysyl-tRNA synthetase (MbPylRS) and four copies of the optimized *Methanosarcina mazei* pyrrolysyl tRNA cassette. The plasmids in (i) and (ii) are available upon request from the Chatterjee lab. Maps and sequences of these plasmids have been published [15, 21].
8. The site of ncAA incorporation into AAV capsids is critical for virus production, transduction, and retargeting efficiency. The optimal ncAA incorporation is typically determined experimentally. We recommend starting with surface-exposed sites such as Q263, T454, T456, N587, and R588, which have worked well in our hands [15].
9. It is critical to perform transfection 16–24 h after splitting the cells. Delayed transfection significantly reduces AAV production. Low seeding density can result in poor cell growth, while high density can reduce transfection efficiency.
10. The plasmid ratio of pHelper: pRep-Cap: pGOI can significantly influence the ratio of full/empty capsid formation and overall genomic titer of AAV by transient transfection. For example, a previous study has indicated an optimal plasmid weight ratio of 1:3.52:0.50 (pHelper: pRep-Cap: pGOI) for AAV2 by transient transfection of suspension HEK293T cells, yielding enhanced genome titers by over twofold and reduced empty capsid formation by approximately 25% compared to the conventional 1:1:1 ratio [22].
11. HEK293T cells do not adhere strongly to tissue culture dishes. Therefore, adding the transfection mixture gently and evenly is important. Uneven addition or vigorous mixing can detach the cells.

12. We recommend including a control of transfected cells without ncAA supplementation. Virus production and transduction should depend on the presence of ncAA in the growth medium. For example, for the pIDTsmart-RC2-TAG construct, no virus should be packaged. For pIDTsmart-RC2-Δ VP1-CMV-VP1-TAG construct, virus should be produced at titers similar to wild-type AAV2 but exhibit minimal transduction due to the absence of the essential VP1 protein.
13. This step is time-sensitive, as AAV capsids are unstable under low-pH conditions. Applying gentle pressure to increase the flow rate of the elution fractions is recommended. Prolonged elution times can significantly reduce virus infectivity.
14. The resin can be regenerated and reused up to ten times. To regenerate the column:
    (i) Wash with 3 × 1 mL of elution buffer.
    (ii) Wash with 2 × 1 mL of column storage buffer, leaving 1 mL on top of the resin.
    (iii) Tightly cap both ends of the column and store at 4°C.
15. Avoid high-speed centrifugation to reduce shear force. During centrifugation, the virus tends to accumulate at the bottom near the membrane of the centrifugal filtration device, causing aggregation. Resuspending the virus in regular intervals (e.g., every 4 min) can ameliorate this issue. Avoiding over-concentration is important, as AAV tends to aggregate under these conditions.
16. Pluronic F68 reduces shear force during centrifugation, and higher NaCl concentrations help reduce AAV aggregation.
17. Our construct uses a single plasmid encoding both the antibody LC and HC genes as a starting point. However, optimizing the LC:HC ratio (e.g., 2:1 or 3:1) using separate plasmids has been shown to increase antibody titers.
18. The site of ncAA incorporation is important to get a good antibody titer. We typically target a surface-exposed residue in the Fab region.
19. Valproic acid functions as a histone deacetylase inhibitor, thereby increasing overall antibody expression.
20. It is recommended to prelabel AAV (as the limiting coupling partner) with tetrazine, due to its higher chemical stability relative to TCO.
21. The buffer can be supplemented with 1% DMSO (final concentration) to improve the solubility of the synthetic linkers. Avoid excessive DMSO, as it may denature AAV capsids and proteins of interest.

22. Prelabeled tetrazine handles on AAV capsids may cause some degree of nonspecific adsorption to filter membranes. The following membrane pretreatment is recommended:
    (i) Incubate the membrane of an Amicon centrifugal filter device (100 kDa, 4 mL) with 1 mL of centrifugal filter coating buffer (described in Sect. 2.8) for 10 min at room temperature. Remove the buffer.
    (ii) Fill the device with 5 mL of AAV buffer (described in Sect. 2.4). Centrifuge at 5000 × *g* for 4 min at room temperature. Discard the flow-through.
23. Due to the small amount of recombinant protein used for AAV capsid conjugation (low micromolar concentrations), excess protein is not removed following the AAV-protein conjugation. However, if needed, it can be removed by size exclusion chromatography.
24. MOIs should be empirically optimized for each cell type. Using a high MOI can lead to saturation effects and skew results, while a low MOI may result in insufficient infection. Suggested MOIs: 20–50 for HEK293T cells and 125–250 for SK-BR-3 cells.

Acknowledgment

This work was supported by NIH (R01GM124319 to AC).

References

1. Wang J-H, Gessler DJ, Zhan W et al (2024) Adeno-associated virus as a delivery vector for gene therapy of human diseases. Sig Transduct Target Ther 9:1–33
2. Ling Q, Herstine JA, Bradbury A et al (2023) AAV-based in vivo gene therapy for neurological disorders. Nat Rev Drug Discov 22:789–806
3. Liu D, Li T, Liu L et al (2024) Adeno-associated virus therapies: pioneering solutions for human genetic diseases. Growth Factor Rev 80:109–120
4. Li C, Samulski RJ (2020) Engineering adeno-associated virus vectors for gene therapy. Nat Rev Genet 21:255–272
5. Zwi-Dantsis L, Mohamed S, Massaro G et al (2025) Adeno-associated virus vectors: principles, practices, and prospects in gene therapy. Viruses 17:239
6. Kuzmin DA, Shutova MV, Johnston NR et al (2021) The clinical landscape for AAV gene therapies. Nat Rev Drug Discov 20:173–174
7. Pham Q, Glicksman J, Chatterjee A (2024) Chemical approaches to probe and engineer AAV vectors. Nanoscale 16:13820–13833
8. Becker J, Fakhiri J, Grimm D (2022) Fantastic AAV gene therapy vectors and how to find them—random diversification, rational design and machine learning. Pathogens 11:756
9. Deverman BE, Pravdo PL, Simpson BP et al (2016) Cre-dependent selection yields AAV variants for widespread gene transfer to the adult brain. Nat Biotechnol 34:204–209
10. Zhu D, Brookes DH, Busia A et al (2024) Optimal trade-off control in machine learning–based library design, with application to adeno-associated virus (AAV) for gene therapy. Sci Adv 10:eadj3786
11. Mével M, Bouzelha M, Leray A et al (2020) Chemical modification of the adeno-associated

virus capsid to improve gene delivery. Chem Sci 11:1122–1131
12. Leray A, Lalys P-A, Varin J et al (2024) Novel chemical tyrosine functionalization of adeno-associated virus improves gene transfer efficiency in liver and retina. Biomed Pharmacother 171:116148
13. Kelemen RE, Mukherjee R, Cao X et al (2016) A precise chemical strategy to alter the receptor specificity of the adeno-associated virus. Angew Chem Int Ed Engl 55:10645–10649
14. Erickson SB, Pham Q, Cao X et al (2024) Precise manipulation of the site and stoichiometry of capsid modification enables optimization of functional adeno-associated virus conjugates. Bioconjugate Chem 35:64–71
15. Pham Q, Glicksman J, Shahraeini S et al (2024) A facile chemical strategy to synthesize precise AAV-protein conjugates for targeted gene delivery. Mol Ther Oncol 33:201040. https://www.biorxiv.org/content/10.1101/2024.07.20.604406v1
16. Lang K, Chin JW (2014) Cellular incorporation of unnatural amino acids and bioorthogonal labeling of proteins. Chem Rev 114:4764–4806
17. Oliveira BL, Guo Z, Bernardes GJL (2017) Inverse electron demand Diels–Alder reactions in chemical biology. Chem Soc Rev 46:4895–4950
18. Erickson SB, Mukherjee R, Kelemen RE et al (2017) Precise photoremovable perturbation of a virus–host interaction. Angew Chem Int Ed Engl 56:4234–4237
19. Jewel D, Kelemen RE, Huang RL et al (2024) Enhanced directed evolution in mammalian cells yields a hyperefficient pyrrolysyl tRNA for noncanonical amino acid mutagenesis. Angew Chem Int Ed Engl 63:e202316428
20. Huang RL, Jewel D, Kelemen RE et al (2024) Directed evolution of a bacterial leucyl tRNA in mammalian cells for enhanced noncanonical amino acid mutagenesis. ACS Synth Biol 13:2141–2149. https://www.biorxiv.org/content/10.1101/2024.02.Address.581038v1
21. Loynd C, Singha Roy SJ, Ovalle VJ et al (2023) Electrochemical labelling of hydroxyindoles with chemoselectivity for site-specific protein bioconjugation. Nat Chem 16:389–397
22. Park S, Shin S, Lee H et al (2024) Enhancing the production of adeno-associated virus (AAV)2 and AAV9 with high full capsid ratio in HEK293 cells through design-of-experiment optimization of triple plasmid ratio. Biotechnol J 19:2300667

Chapter 2

Bioorthogonal Surface Modification of Measles and Vaccinia Viruses Using Metabolic Azide Reporters

Shan Hu, Thongpon Meethong, Monchupa Kingsak and Qian Wang

Abstract

Enveloped viruses, such as measles virus (MV) and vaccinia virus (VACV), acquire their lipid bilayer envelopes from host cell membranes during viral assembly. These membranes can be chemically modified to enable selective labeling and visualization. This chapter details a metabolic labeling strategy in which Vero cells, a prototypical host cell line, are supplemented with peracetylated azido sugars (Ac_4GalNAz, Ac_4ManNAz) and azido-choline analogs (AECho). These bioorthogonal metabolic reporters are incorporated into glycan and phospholipid via biosynthetic pathways, resulting in azide-functionalized glycoconjugates and phosphatidylcholine (PC) derivatives within both host and viral membranes. The surface-exposed azido groups on progeny virions can be further conjugated with fluorescent probes or other entities via strain-promoted azide-alkyne cycloaddition (SPAAC) or copper-catalyzed azide-alkyne cycloaddition (CuAAC) reactions. Fluorescent labeling enables direct visualization of viral particles by fluorescence microscopy. This chapter also provides detailed protocols for virus propagation, purification, and labeling, offering a versatile platform for live-cell imaging, viral trafficking studies, and the development of targeted delivery systems.

Keywords Enveloped viruses, Viral membrane, Metabolic labeling, Click chemistry, Bioorthogonal reaction, Azido-sugars, Azido-choline

1 Introduction

Enveloped viruses contain the viral genome (either RNA or DNA) and core proteins (called capsid) wrapped within lipid membranes, acquired from the host cell during virus assembly and budding [1]. Enveloped viral surface modification has significant potential to study viral infection mechanisms in living cells and to tune the virus–host interface, enabling targeted cancer therapy, oncolytic therapy, and gene therapy [2–6].

Current strategies for modifying viral surfaces include genetic [7], chemical [8], or metabolic approaches [3, 9–11], each with distinct mechanisms and ideal use cases. Over the past decade, the metabolic incorporation of unnatural substrates—tolerated by specific

Alexander Deiters (ed.), *Bioconjugation Methods*, Methods in Molecular Biology, vol. 3009,
https://doi.org/10.1007/978-1-0716-5112-4_2,

biosynthetic pathways—has been utilized to label cellular surfaces. Given the abundance of phosphatidylcholine (PC) and glycans in mammalian cell membranes, cells can incorporate unnatural sugars and choline derivatives through natural metabolic pathways (Fig. 1). Notable examples include *N*-azidoacetylgalactosamine (Ac_4GalNAz), *N*-azidoacetylmannosamine (Ac_4ManNAz), propargylcholine (PCho), and azidoethyl-choline (AECho), all of which introduce a bioorthogonal azide group.

Building on the success of metabolic labeling in cell surface studies, this approach has been extended to enveloped viruses. By leveraging the host cell's intrinsic biosynthetic machinery during viral assembly, unnatural sugars and lipid analogs—such as azido sugars and choline derivatives—can be incorporated into the viral envelope. The resulting surface-exposed azido groups on progeny virions can then be selectively targeted and visualized by specific chemical reactions called "click reactions," such as strain-promoted azide-alkyne cycloaddition (SPAAC) and copper-catalyzed azide-alkyne cycloaddition (CuAAC) (Fig. 2) [12, 13]. This enables efficient, non-invasive labeling of viruses for live-cell imaging, tracking, and targeted delivery applications.

In this chapter, we detail protocols for surface labeling of measles virus (MV) and vaccinia virus (VACV) through metabolic incorporation of azido-sugars (Ac_4ManNAz and Ac_4GalNAz) and AECho into Vero cells. The incorporation efficiency of AECho is quantified via lipid extraction, and successful labeling is validated using SPAAC and CuAAC reactions with corresponding fluorescent reagents, followed by fluorescence microscopy. Viral propagation and purification methods are described, and viral surface modification is confirmed through analogous labeling and imaging techniques.

2 Materials

Prepare and store all reagents at room temperature unless stated otherwise. All reagents, media, and equipment that come into contact with living cells must be sterile unless specified. Filter sterilization (0.2 μm) is preferred for all culture media. Always follow waste disposal regulations carefully when discarding materials.

2.1 General Supplies and Instruments

1. Sterile 10-cm Petri dish, 12-well plates, and 96-well plates.
2. CO_2 incubator with 5% CO_2 and 95% air at 37°C.
3. Cell scraper (Sterile, SL1250).
4. Refrigerator of 4°C and -80 °C.
5. Dounce homogenizer (Sigma-Aldrich).
6. Glass serological pipettes.
7. Sterile 50-mL centrifuge tubes.

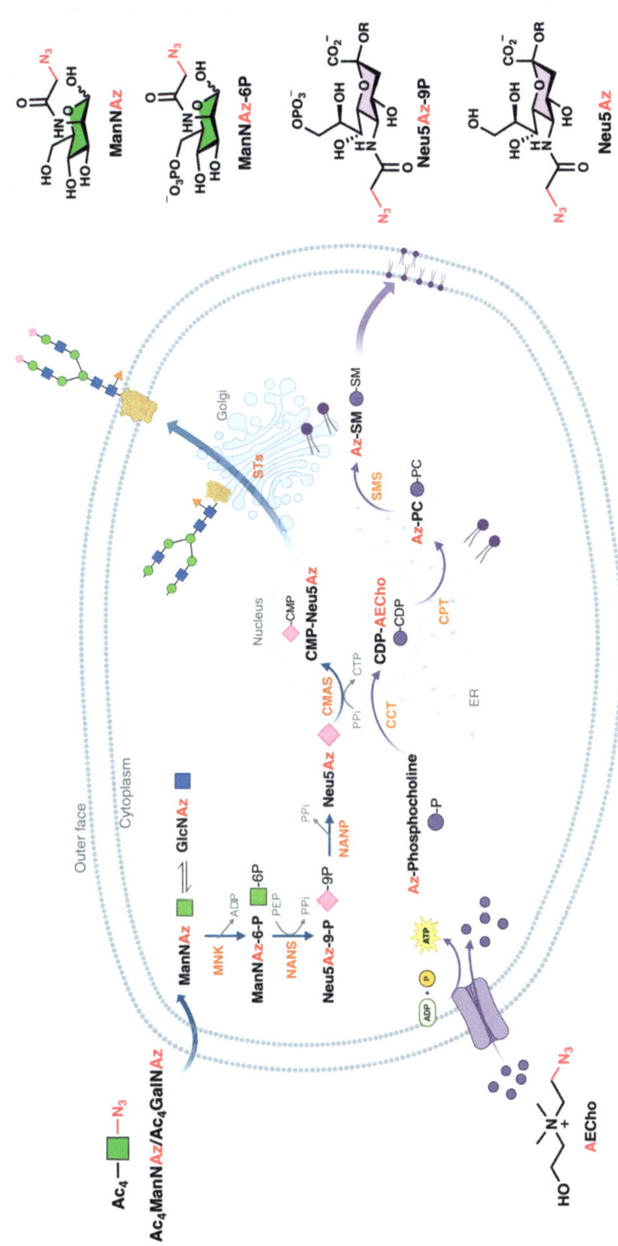

Fig. 1 Schematic illustration of the metabolic incorporation of unnatural choline derivatives and azido-functionalized sugars via the biosynthetic pathway. Figure created with BioRender.com

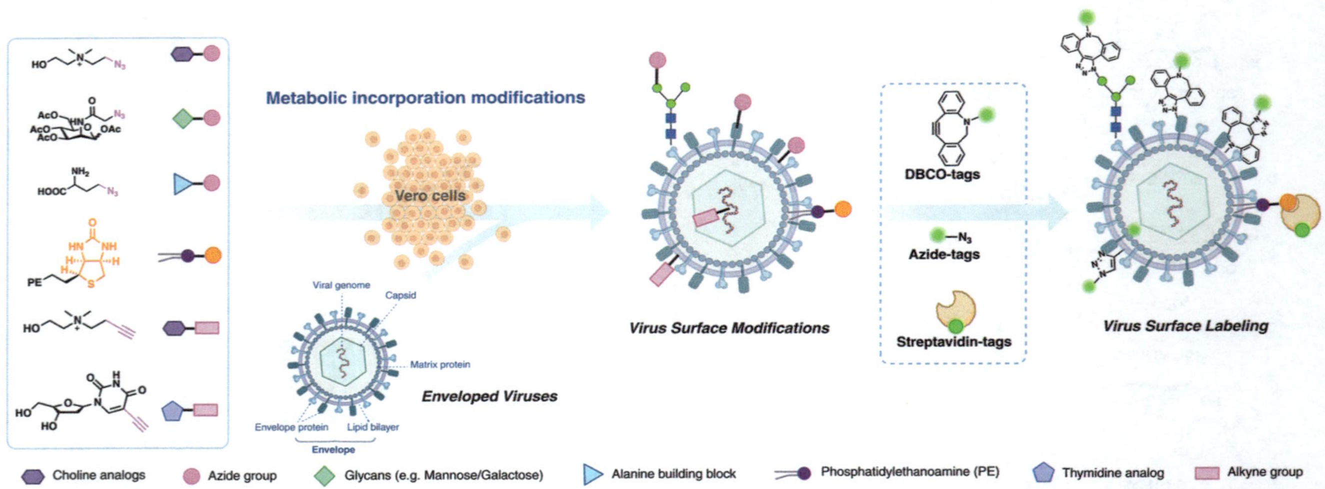

Fig. 2 Schematic illustration of metabolic incorporation of bioorthogonal molecules into enveloped viruses by biosynthetic pathway. Figure created with BioRender.com

8. 26-mL Ultracentrifuge tube bottles.
9. The 12-well cover glass slide.
10. The filter paper.
11. Carbon-coated copper grid (Sigma-Aldrich, 931519).
12. Transmission electron microscopy (TEM, JEOLJEM-1011).
13. Olympus IX81 fluorescence microscope (Shinjuku-ku, Tokyo, Japan).
14. HPLC column: Xbridge ® C18 3.5 μm (2.1*100 mm).
15. Q Exactive HF-X Orbitrap MS instrument (Thermo Fisher Scientific).
16. Sorvall RC5C Plus Refrigerated Centrifuge.
17. Beckman Coulter Optima L-100K Ultracentrifuge Floor Centrifuge (50.2 Ti rotor).
18. Microplate Reader (Molecular Devices SpectraMax 190).
19. Laminar flow cabinet.

2.2 Biological Materials and Chemicals

1. Vero cells line (U.S. Centers for Disease Control and Prevention).
2. Dulbecco's modified Eagle's medium (DMEM).
3. Fetal bovine serum (FBS).
4. Penicillin-streptomycin (10,000 U/mL).
5. Dulbecco's Phosphate-Buffered Saline (DPBS, 1×) without calcium and magnesium.
6. AECho (MedChemExpress).
7. Ac_4ManNAz and Ac_4GalNAz (Vector Laboratories).
8. Paraformaldehyde (PFA).
9. DBCO-Fluor 488 (Click Chemistry Tools (CCT), Scottsdale, AZ).
10. Biotin-PEG_4-DBCO (Vector Laboratories).
11. Rhodamine (TRITC) Streptavidin (Jackson ImmunoResearch Laboratories, INC.).
12. Bovine serum albumin (BSA).
13. Sodium azide (NaN_3).
14. Chloroform ($CHCl_3$).
15. Methanol (MeOH).
16. Sodium chloride (NaCl).
17. Acetonitrile (ACN).
18. Ammonium formate.
19. Isopropyl alcohol (IPA).
20. Trifluoroacetic (TFA).

21. Di-sodium hydrogen phosphate dihydrate.
22. Citric acid.
23. Tris(hydroxymethyl)aminomethane (Tris) hydrochloride.
24. Sucrose.
25. Triton X-100.
26. Propidium iodide.
27. 1% Alcian blue solution.
28. 2% Uranyl acetate.

2.3 Reagent Preparation

1. 10% FBS DMEM (complete medium): Mix 445 mL of DMEM with 50 mL of FBS and 5 mL of 100 U/mL penicillin-streptomycin in the Laminar flow cabinet.
2. 2% FBS DMEM: Mix 485 mL of DMEM with 10 mL of FBS and 5 mL of 100 U/mL penicillin-streptomycin in the Laminar flow cabinet.
3. 2% FBS DPBS: Mix 98 mL of DPBS with 2 mL of FBS in the Laminar flow cabinet.
4. AECho stock solution: Dissolve 100 mg of AECho with 628.1 μL of DPBS to prepare a 1 M solution.
5. AECho working solution: Dilute 50 μL of 1 M AECho stock solution into 10 mL of 10% FBS DMEM to prepare a 5 mM working solution.
6. $Ac_4ManNAz$ stock solution: Dissolve 5 mg of $Ac_4ManNAz$ in 1 mL of DPBS to prepare an 11.6 mM $Ac_4ManNAz$ stock solution.
7. $Ac_4ManNAz$ working solution: Prepare 100, 200, 300, and 400 μM solutions by diluting 43.1, 86.2, 129.3, and 172.4 μL of 11.6 mM $Ac_4ManNAz$ stock, respectively, into 10% FBS DMEM to a final volume of 5 mL.
8. $Ac_4GalNAz$ stock solution: Dissolve 5 mg of $Ac_4GalNAz$ in 1 mL of DPBS to prepare an 11.6 mM $Ac_4ManNAz$ stock solution.
9. $Ac_4GalNAz$ working solution: Prepare 100, 200, 300, and 400 μM solutions by diluting 43.1, 86.2, 129.3, and 172.4 μL of 11.6 mM $Ac_4GalNAz$ stock, respectively, into 10% FBS DMEM to a final volume of 5 mL.
10. PFA preparation: Dissolve 2, 3, or 4 g of PFA powder in 80 mL of DPBS at 60–65°C with stirring (add a few drops of 1 N NaOH if needed), then cool to room temperature, adjust the volume to 100 mL with DPBS, and filter sterilize to obtain 2%, 3%, or 4% PFA solution, respectively.
11. DBCO-Fluor 488 stock solution: Dissolve 1 mg of DBCO-Fluor 488 in 100 μL of DMSO to prepare a 10 mM stock solution, aliquot, and store at −20°C for long-term use.

12. DBCO-Fluor 488 working solution: Dilute 5 μL of DBCO-Fluor 488 stock solution into 4995 μL of DPBS to prepare a 10 μM working solution.
13. Biotin-PEG_4-DBCO stock solution: Dissolve 5 mg of Biotin-PEG_4-DBCO into 80 μL of DMSO to prepare an 83.3 mM stock solution.
14. Biotin-PEG_4-DBCO working solution: Dilute 14.4 μL of Biotin-PEG_4-DBCO stock solution in 12 mL of DPBS to prepare a 100 μM working solution.
15. Rhodamine streptavidin stock solution: Dissolve 1 mg of rhodamine-conjugated streptavidin in 2 mL of DPBS containing 30 mg of BSA and 1 μg of NaN_3 to prepare a 500 μg/mL solution.
16. Rhodamine streptavidin working solution: Dilute 120 μL of rhodamine streptavidin stock solution in 6 mL of DPBS to prepare a 10 μg/mL working solution.
17. Citrate phosphate buffer (CP buffer) stock solution: Mix 41.6 mL of 0.2 M disodium hydrogen phosphate dihydrate and 8.4 mL of 0.1 M citric acid to obtain 50 mL of 0.1 M CP buffer at pH 7.2.
18. CP buffer working solution: Dilute 20 mL of CP stock solution into 80 mL of Milli-Q water to prepare a 20 mM CP buffer working solution.
19. Tris buffer preparation: Dissolve 121.1 mg of Tris hydrochloride into 900 mL of Milli-Q water and adjust the pH to 8.0 by adding 1 M NaOH dropwise. Then, bring the final volume up to 1 L with Milli-Q water to prepare a 1 mM Tris buffer.
20. 60% and 20% Sucrose solution: Dissolve 60 g or 20 g of sucrose in 100 g of 1 mM Tris buffer or 20 mM CP buffer to prepare 60% and 20% (w/w) sucrose solutions, respectively.
21. 0.5% Triton X-100: Dissolve 0.5 mL of Triton X-100 in 100 mL of DPBS to prepare 0.5% Triton X-100 (v/v) solution.
22. 10 mg/mL propidium iodide (PI) solution: Dissolve 10 mg of propidium iodide in 1 mL of DPBS to prepare a 10 mg/mL PI solution.
23. 0.5% uranyl acetate (UA): Dilute 250 μL of 2% uranyl acetate into 750 μL of Milli-Q water to prepare a 0.5% UA solution.

3 Methods

This protocol outlines a step-by-step procedure for surface labeling of enveloped viruses through the metabolic incorporation of unnatural sugars and choline analogs. MV and VACV serve as model viruses, while the Vero cell line is used as the permissive host system. Unless otherwise specified, all procedures should be performed at room temperature.

3.1 Metabolic Incorporation and Live-Cell Imaging of Azido-Sugars in Vero Cells

This section outlines the protocol for the metabolic incorporation of azido-sugars (e.g., $Ac_4ManNAz$ and $Ac_4GalNAz$) into Vero cells, followed by labeling with specific reagents for cell surface visualization.

1. Plate 2 × 10^5 cells Vero cells at /mL density in complete medium in a 12-well plate (1 mL/well) and allowed to attach overnight.
2. Discard the old medium and wash with 1 mL of DPBS once.
3. When cells reach 60%–80% confluency (*see* **Note 1**), add 1 mL of prewarmed complete medium containing $Ac_4ManNAz$ or $Ac_4GalNAz$ at final concentrations of 100, 200, 300, and 400 μ M, respectively (*see* **Notes 2 and 3**).
4. Co-culture for 24 h in the incubator at 37°C.
5. After metabolic incorporation, wash the azide-modified cells with 1 mL of DPBS once and fix them with 500 μL of 4% (w/v) PFA for 15 min at room temperature.
6. Remove the excess amount of PFA with 1 mL of DPBS washes, three times.
7. Incubate with 1 mL of DBCO-Fluor 488 working solution for 1 h at room temperature.
8. Wash the labeled cells thoroughly with 1 mL of DPBS twice to remove the unbound dye and resuspend in 500 μL of DPBS for fluorescence imaging using an Olympus IX81 fluorescence microscope equipped with a spinning disk scanning unit confocal module (Fig. 3) [11].

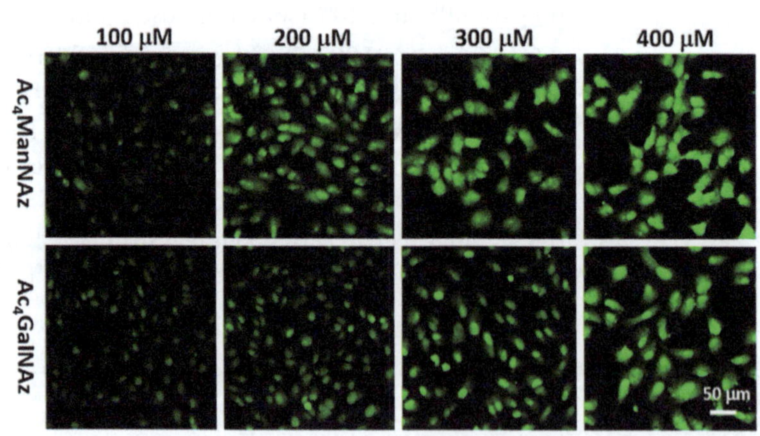

Fig. 3 Fluorescence labeling of the metabolically incorporated $Ac_4ManNAz$ and $Ac_4GalNAz$ in the Vero cells. Reprinted with permission from Bioconjugate Chem 26(9):1868–1872. Copyright © 2015 American Chemical Society

3.2 Metabolic Incorporation and Live-Cell Imaging of AECho in Vero Cells

This section outlines the protocol for the metabolic incorporation of AECho into Vero cells, followed by labeling with specific reagents for cell surface visualization.

1. Plate 2 × 10^5 cells Vero cells at /mL density in complete medium (10% FBS DMEM) in a 12-well plate (1 mL/well) and allowed to attach overnight.
2. Discard the old medium and wash with DPBS once.
3. When cells reach 60%–80% confluency (*see* **Note 1**), add 1 mL of pre-warmed complete medium containing AECho at final concentrations of 5 mM (*see* **Notes 2 and 3**).
4. Co-culture for 24 h in the incubator at 37°C.
5. After metabolic incorporation, wash the azide-modified cells with 1 mL of DPBS once and fix them with 500 μL of 4% (w/v) PFA for 15 min at room temperature.
6. Wash the excess amount of PFA with 1 mL of DPBS three times.
7. Incubate with 1 mL of Biotin-PEG_4-DBCO working solution for 1 h at room temperature.
8. Wash the labeled cells thoroughly with 1 mL of 2% FBS DPBS (*see* **Note 4**) twice to remove the unbound reagent.
9. Next, incubate the biotinylated cells with 500 μL of rhodamine streptavidin working solution for 30 min at room temperature in the dark.
10. After labeling, wash cells twice with 1 mL of 2% FBS DPBS (*see* **Note 4**) and resuspend in 500 μL of 2% FBS DPBS for fluorescence imaging using a fluorescence microscope equipped with a disk scanning unit confocal module (Fig. 4).

3.3 Lipid Extraction from Vero Cells

Following incorporation, PC and sphingomyelin (SM) are extracted from Vero cells and analyzed by LC-MS/MS (Figs. 5 and 6). The results of lipid incorporation efficiency are presented in Fig. 7 [13].

1. Plate 3 × 10^6 Vero cells in 10 mL of complete medium in a 10-cm cell culture dish and allow attachment for 24 h in a CO_2 incubator.
2. When cells reach 60%–80% confluency (*see* **Note 1**), discard the old medium and wash with 1 mL of DPBS once.
3. Add 10 mL of pre-warmed complete medium containing AECho at final concentrations of 5 mM (*see* **Notes 2 and 3**).
4. Co-culture for 24 h in the incubator at 37°C.
5. Discard the old medium and wash adhered cells with 1 mL of DPBS.

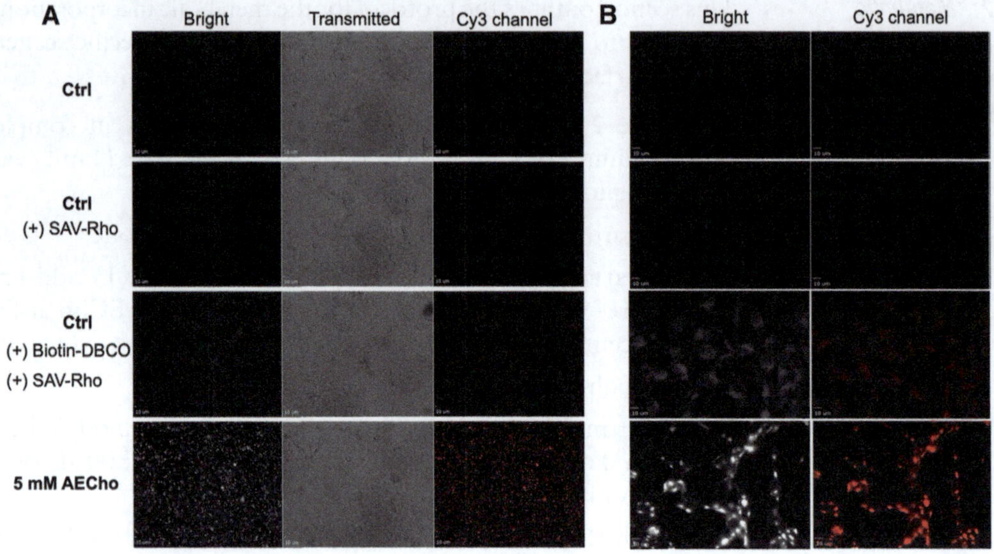

Fig. 4 Fluorescence imaging of Vero cells metabolically labelled with 5 mM AECho, visualized under bright-field, transmitted light, and Cy3 fluorescence channels. (**A**) Overview image acquired at 10× magnification. (**B**) Higher magnification image at 40×; scale bar: 10 μm

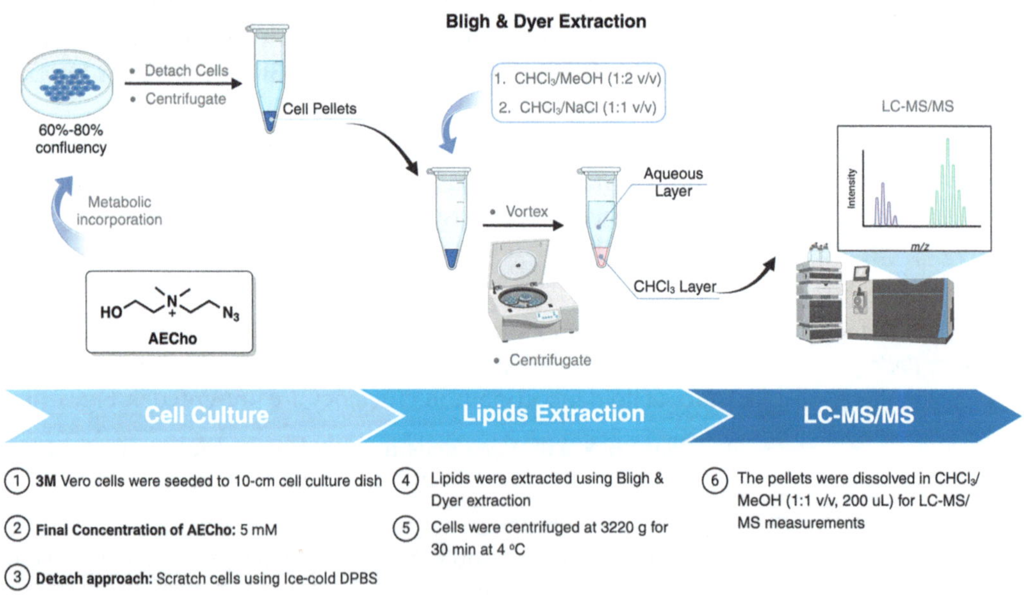

Fig. 5 Schematic overview of the lipid extraction protocol from metabolically incorporated cells. Figure created with BioRender.com

6. Add 1 mL of ice-cold DPBS (*see* **Note 5**), then scrape the cells into ice-cold DPBS. Adjust the final volume to 5 mL with additional ice-cold DPBS if needed in a 15 mL centrifuge tube, and centrifuge at 223 g for 5 min at room temperature.

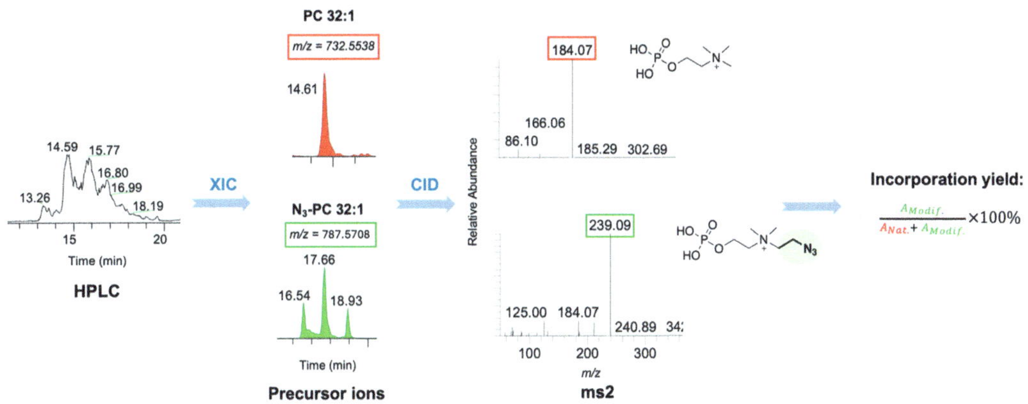

Fig. 6 Scheme for calculating the metabolic incorporation efficiency of AECho in Vero cells. LC-MS/MS spectra were acquired using a Q Exactive HF-X Orbitrap mass spectrometer. XIC: extracted ion chromatograms; CID: collision-induced dissociation

Head Group	Molecular Species	Formula	Species	CS [z]	Polarity	Precursor ions [m/z]	Fragments [m/z]	Incorporation yield
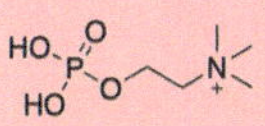	32:1 PC	$C_{40}H_{78}NO_8P$	+ H	1	Pos.	732.55378	184.07	-
	34:1 PC	$C_{42}H_{82}NO_8P$	+ H	1	Pos.	760.58508	184.07	-
	36:1 PC	$C_{44}H_{86}NO_8P$	+ H	1	Pos.	788.61638	184.07	-
	32:1 SM	$C_{37}H_{75}N_2O_6P$	+ H	1	Pos.	675.54355	184.07	-
	34:1 SM	$C_{39}H_{79}N_2O_6P$	+ H	1	Pos.	703.57485	184.07	-
	36:1 SM	$C_{41}H_{83}N_2O_6P$	+ H	1	Pos.	731.60615	184.07	-
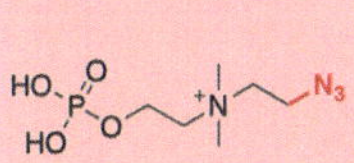	32:1 PC_N3	$C_{41}H_{79}N_4O_8P$	+ H	1	Pos.	787.57083	239.09	0.39%
	34:1 PC_N3	$C_{43}H_{83}N_4O_8P$	+ H	1	Pos.	815.60213	239.09	0.32%
	36:1 PC_N3	$C_{45}H_{87}N_4O_8P$	+ H	1	Pos.	843.63343	239.09	0.39%
	32:1 SM_N3	$C_{38}H_{76}N_5O_6P$	+ H	1	Pos.	730.56060	239.09	0
	34:1 SM_N3	$C_{40}H_{80}N_5O_6P$	+ H	1	Pos.	758.59190	239.09	0
	36:1 SM_N3	$C_{42}H_{84}N_5O_6P$	+ H	1	Pos.	786.62320	239.09	0

Fig. 7 Analysis of the metabolic incorporation of the native choline and AECho analogue in Vero cells. CS: charge states (*see* **Notes 8 and 9**)

7. The supernatant is then aspirated off, and the lipids are extracted from the resultant pellet using the modified Bligh & Dyer method [14].
8. Briefly, treat the cell pellet with 1 mL of a $CHCl_3$/MeOH mixture (1:2, v/v) and vortex for 1 min.
9. Then add 0.5 mL $CHCl_3$ and 0.5 mL of 1 M NaCl, followed by vortexing for an additional 1 min.
10. Phase separation will be achieved by centrifuging the mixture at 3220 × *g* at 4 °C for 30 min.

11. The lower lipid-enriched organic layer can be separated (*see* **Note 6**) and dried using a nitrogen stream at room temperature [13].
12. For all LC-MS/MS analyses, the dried lipids were dissolved in a mixture of $CHCl_3$ and MeOH (200 μL, 1:1 v/v solution) (*see* **Note 7**).
13. Use a Q Exactive HF-X Orbitrap MS instrument with an Xbridge® C18 column (3.5 μm, 2.1 × 100 mm). The mobile phase consisted of phase A: ACN/water (60/40) with 10 mM ammonium formate and 0.1% TFA, and phase B: IPA/water (90/10) with 10 mM ammonium formate and 0.1% TFA. The flow rate was set to 0.2 mL/min (*see* **Note 8**).
14. The metabolic incorporation efficiency of the AECho analog can be calculated using the MS2 relative intensity ratio between the native and azide-modified phosphine lipid head groups (Figs. 6 and 7) (*see* **Note 9**).

3.4 MV and N_3-MV Propagation and Purification in Vero Cells

Both viruses utilize the host's biosynthetic machinery for replication, involving sequential steps such as viral attachment, membrane fusion, uncoating, transcription and replication of the viral genome, protein synthesis, assembly, and nonlytic release of progeny virions. A hallmark of MV infection is the formation of multinucleated syncytia through virus-induced cell–cell fusion. This section outlines the protocol of MV (Fig. 8) propagation and purification in Vero cells.

1. Plate 3 × 10^6 Vero cells in 10 mL complete medium in a 10-cm cell culture dish and incubate at 37°C in the incubator.
2. When cells reach 60%–80% confluency (*see* **Note 1**), they are infected with MV at a multiplicity of infection (MOI) of 0.14 $TCID_{50}$ (see Sect. 3.6) per cell in 2% FBS DMEM. Maintain the infected cultures at 37°C until the syncytia, a cytopathic effect (CPE) of MV, can be observed in the dish (*see* **Note 10**).
3. After 80%–90% CPE, the infected culture media are removed and stored at −80°C for future viral propagation.
4. To purify MV (*see* **Note 11**), the infected media are collected into a 50-mL centrifuge tube, and the cell debris is removed by centrifugation at 3000 × *g* for 15 min at 4°C.
5. The MV in the clarified supernatant is purified using a sucrose cushion. Briefly, 4 mL of ice-cold 20% (w/w) sucrose in CP buffer is slowly underlaid beneath the infected supernatant in a 26-mL ultracentrifuge tube using a glass serological pipette, creating a two-layer system (*see* **Note 12**).
6. After centrifugation at 110,000 × *g* for 3 h at 4°C, the supernatant is quickly removed, and the virus pellet is observed at the bottom of the tube.

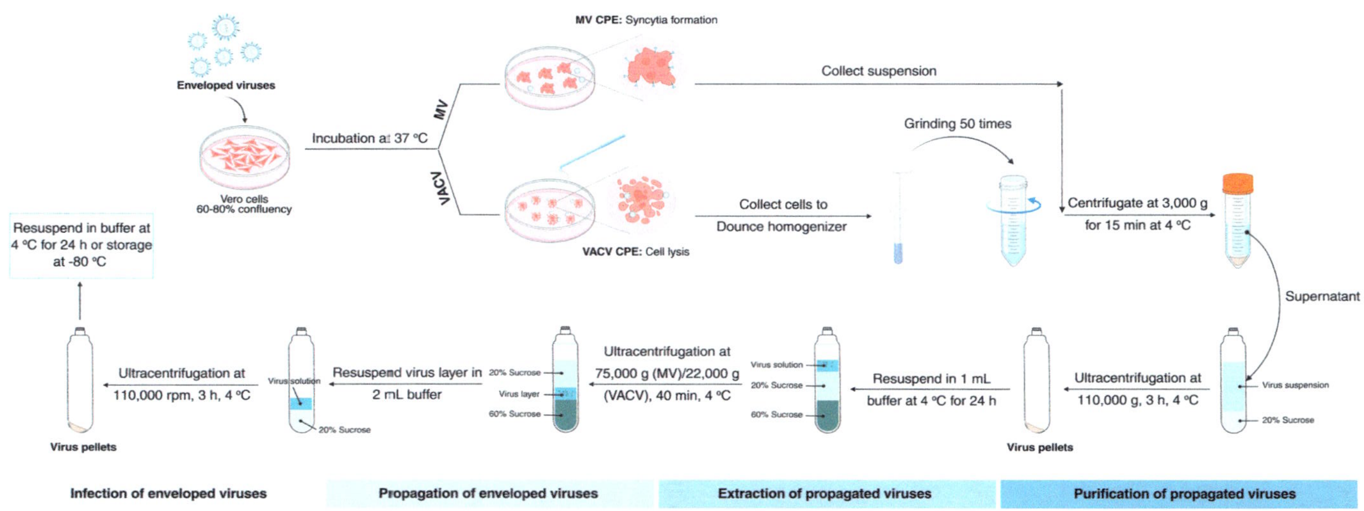

Fig. 8 Schematic overview of the protocol for MV, MV-N_3, VACV, and VACV-N_3 infection, propagation, and purification in Vero cells. Figure created with BioRender.com

7. Add 1 mL of 20 mM pH 7.2 CP buffer inside and align the tube horizontally in the 4°C refrigerator to resuspend MV into the buffer for 24 h (*see* **Note 13**) [15].
8. The purified MV can be kept at −80°C for long-term storage.
9. The MV preparation can be further purified via sucrose-gradient centrifugation. First, 4 mL of 20% (w/w) sucrose in CP buffer is layered into a 26-mL ultracentrifuge tube. Next, a concentrated 60% (w/w) sucrose solution in CP buffer is carefully overlaid beneath the 20% layer. The MV suspension is then gently overlaid on top of the gradient and centrifuged at 76,000 × *g*, 4°C for 3 h (optional, *see* **Note 14**).
10. Using a glass pipette, recover the interface between the 20% and 60% sucrose layers into a fresh 26-mL ultracentrifuge tube. Dilute the collected material with 2 mL of CP buffer, then repeat steps 6–8 (optional, *see* **Note 14**).
11. The N_3-MV can be propagated using the same protocol with an additional step. After culturing the Vero cells with 60%–80% confluency, the azido-substrates (choline or sugars) can be added and incubated for 24 h. Then, remove the cell culture media and wash the attached cells with 2 mL of DPBS, followed by the addition of MV.

3.5 VACV and N_3-VACV Propagation and Purification in Vero Cells

Both viruses also rely on the host's biosynthetic machinery for replication, involving sequential steps such as viral attachment, membrane fusion, uncoating, transcription and replication of the viral genome, protein synthesis, assembly of complex virion forms such as intracellular mature virions (IMVs) and extracellular enveloped virions (EEVs), and release of progeny virions through cell lysis (IMVs) or nonlytic exocytosis (EEVs). Unlike MV, VACV infection does not typically result in syncytia formation. This section outlines the protocol of VACV (Fig. 8) propagation and purification in Vero cells.

1. Firstly, plate Vero cells in a complete medium and incubate at 37°C in the incubator.
2. When cells reach 60%–80% confluency (*see* **Note 1**), they are infected with VACV at an MOI of 0.14 $TCID_{50}$ per cell in 2% FBS DMEM. Maintain the infected cultures at 37°C until the characteristic VACV-induced CPE appears (*see* **Note 15**).
3. After 80%–90% CPE, the infected culture media are removed and stored at −80°C for future viral propagation.
4. To purify VACV, the infected cells are scraped from the dish using a cell scraper and collected by rinsing with 2 mL of ice-cold DPBS twice.
5. The cells are subjected to homogenization using a Dounce homogenizer by grinding 50 times on ice (*see* **Note 11**).

6. Transfer the cell suspension to a 50-mL centrifuge tube, and remove the cell debris by centrifugation at 3000 × *g* for 15 min at 4°C.
7. The VACV in the clarified supernatant is purified using a sucrose cushion. Briefly, 4 mL of ice-cold 20% (w/w) sucrose in Tris buffer is slowly underlaid beneath the infected supernatant in a 26-mL ultracentrifuge tube using a glass serological pipette, creating a two-layer system (*see* **Note 12**).
8. After centrifugation at 110,000 × *g* for 3 h at 4°C, the supernatant is quickly removed, and the white virus pellet is observed at the bottom of the tube.
9. Add 1 mL of 1 mM pH 8.0 Tris buffer inside and align the tube horizontally in the 4°C refrigerator to resuspend VACV in the buffer for 24 h (*see* **Note 13**).
10. The purified VACV can be kept at −80°C for long-term storage [15].
11. The VACV preparation is further purified by sucrose-gradient centrifugation. First, 4 mL of 20% (w/v) sucrose in Tris buffer is layered into an ultracentrifuge tube. Next, 4 mL of 60% (w/v) sucrose in Tris buffer is gently underlaid beneath the 20% layer. The VACV suspension is then carefully overlaid on top and centrifuged at 22,000 × *g* and 4°C for 40 min (optional, *see* **Note 14**).
12. Using a glass pipette, recover the interface between the 20% and 60% sucrose layers into a fresh 26-mL ultracentrifuge tube. Dilute the collected material with 2 mL of Tris buffer, then repeat steps 6–8 (optional, *see* **Note 14**).
13. The N_3-VACV can be propagated using the same protocol with an additional step. After culturing the Vero cells with 60%–80% confluency, the azido-substrates (choline or sugars) can be added and incubated for 24 h. Then, remove the cell culture media and wash the attached cells with 2 mL of DPBS, followed by the addition of VACV.

3.6 Virus Titer Assays

The titers of control and modified enveloped viruses are quantified by a 50% tissue culture infective dose ($TCID_{50}$) assay.

1. Plate 2.5 × 10^4 cells/well Vero cells in 200 μL of 2% FBS DMEM in 96-well plates for 24 h at 37°C with a 5% CO_2 incubator.
2. The following day, remove the cell culture media and wash cells with 100 μL of DPBS three times.
3. Then, add 100 μL of the MV (native or N_3-modified, *see* **Note 16**), prepared as a 10-fold serial dilution in 2% FBS DMEM,

in quadruplicate to each well of the above 96-well plate and incubate in the incubator.

4. Count the wells showing virus-induced CPE and record at 12, 24, 48, 72, 84, and 96 h postinfection under a microscope.
5. Calculate the virus titer using the Reed–Muench method [16] explained in the following section and report in a unit of $TCID_{50}$/mL, which suggests metabolic labeling did not affect the propagation of MV in Vero cells (Fig. 9a). This can also be reported as a time-course study, which could show the infectivity efficiency of control and modified virus as shown in Fig. 9b [11].

3.7 Reed–Muench Method

1. Counting CPE in a 96-well plate: positive wells (showing CPE) and negative wells (no CPE) were counted (Fig. 10).
2. Cumulative sums: the cumulative sum of the positive wells can be calculated by adding the number of positive wells at the dilution to the number of positive wells at the dilution below. The cumulative sum of the negative wells can be calculated by adding the number of negative wells at the dilution to the number of negative wells at the dilution above. For example, the cumulative sum of the positive wells at 10^{-6} dilution is $0 + 0 + 2 = 2$. The cumulative sum of the negative wells at 10^{-7} dilution is $0 + 0 + 0 + 0 + 4 + 2 = 6$ (Fig. 11).
3. Calculating infection rates: The infection rate can be calculated using the formula below: The infection rate at 10^{-6} dilution is $\frac{2}{(2+2)} = 0.5$.

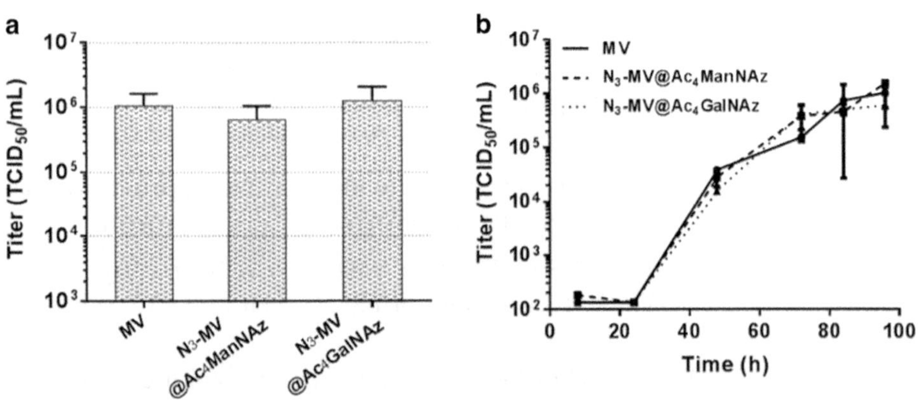

Fig. 9 (**a**) The virus titers of control MV and N_3-MV modified by Ac_4ManNAz and Ac_4GalNAz. (**b**) Time-course infectivity study of control MV and N_3-MV modified by Ac_4ManNAz and Ac_4GalNAz from 12, 24, 48, 72, 84, and 96 h. Reprinted with permission from Bioconjugate Chem 26(9):1868–1872. Copyright © 2015 American Chemical Society

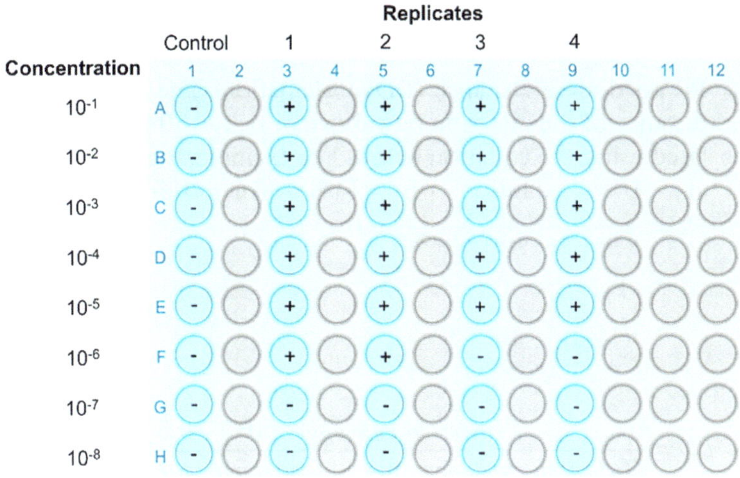

Positive wells	Negative wells
4	0
4	0
4	0
4	0
4	0
2	2
0	4
0	4

Fig. 10 Counting of positive (showing CPE) and negative (no CPE) wells on day 4 of virus infection

Concentration	Positive wells	Negative wells	Cumulative Positive wells	Cumulative Negative wells
10^{-1}	4	0	22	0
10^{-2}	4	0	18	0
10^{-3}	4	0	14	0
10^{-4}	4	0	10	0
10^{-5}	4	0	6	0
10^{-6}	2	2	2	2
10^{-7}	0	4	0	6
10^{-8}	0	4	0	10

Fig. 11 Cumulative sum of positive and negative wells under different concentrations

$$\textit{Infection rate} = \frac{\textit{Number of cumulative positive well}}{\textit{Number of cumalative positive wells} + \textit{Number of culmulative netagative wells}}$$

4. 50% Endpoint Determination: In this step, the dilution at a 50% infection rate will be used to calculate the $TCID_{50}$. However, if

there is no exact 50% infection rate, the first dilutions with the infection rate above 50% and below 50% will be used in step 5.

5. Proportionate Distance (PD) Calculation: The proportionate distance of the infection rate above or below 50% can be calculated using the equation below. For example, the infection rate at 10^{-6} dilution is 50% (or 0.5), therefore $PD = \frac{0.5 - 0.5}{0.5 - \textit{Infection rate below } 50\%} = 0$.

$$PD = \frac{(\textit{Infection rate above } 50\%) - 0.5}{(\textit{Infection rate above } 50\%) - (\textit{Infection rate below } 50\%)}$$

6. $TCID_{50}$ was calculated using the equation below. To determine the virus titer in $TCID_{50}$/mL, the inoculum volume must be considered.

$$\log(TCID_{50}) = \log(\textit{dilution with infection rate above or at } 50\%) + (-PD) \times \log(\textit{dilution factor})$$

For example:

$$\log(TCID_{50}) = \log(10^{-6}) + [0 \times \log(0.1)]$$

So,

$$TCID_{50} = 10^{-6}$$

If the inoculum volume is 0.1 mL, then (*see* **Note 17**):

$$\textit{Virus titer} = \frac{10^6}{0.1} = 10^7\, TCID_{50}/mL$$

3.8 Colocalization Efficiency of Azido-Labeled Viruses

To study how well chemical labeling works on the viral surface, two bioorthogonal methods—SPAAC and CuAAC—have been used [17]. These reactions allow azide-labeled viral components to bind with fluorescent probes for clear visualization. Fluorescence microscopy can be used to assess how well the labeled viral particles match with viral protein markers, giving insight into the accuracy and specificity of the labeling.

1. 1 mL of purified MV or N_3-MV (*see* **Note 16**) is dropped onto the cover glass slide and dried at 37°C for 1 h.
2. Wash the cover glass slide with DPBS to remove the excess purified MV or N_3-MV.
3. Fix the retained viruses with 1 mL of 3% (w/v) PFA for 30 min at room temperature and subsequently permeabilize with 1 mL of 0.5% (v/v) Triton X-100 in DPBS.
4. Add 1 mL of DBCO-Fluor 488 working solution and incubate at 37°C for 1 h to undergo the Click reaction.
5. Stain the nucleic acid of viruses with 0.5 mL of 10 μg/mL of PI solution for 15 min.
6. Wash out the excess fluorophores with 1 mL of DPBS.
7. The fluorescence images are captured by a Carl Zeiss LSM 700 confocal laser scanning microscope imaging system with an excitation wavelength of 488 nm and detected emission at 500–600 nm for DBCO-Fluor 488, and the excitation wavelength of 555 nm and detected emission at 560–700 nm were used for PI (Fig. 12) [11].

3.9 Transmission Electron Microscopy Imaging (TEM)

TEM is used to directly visualize the morphology of the labeled viruses and to confirm the structural integrity following chemical modification.

1. A 5 μL aliquot of native MV or N_3-MV (*see* **Note 16**) is applied to a carbon-coated copper grid pre-treated with 1% Alcian blue (*see* **Note 18**) for 5 min.

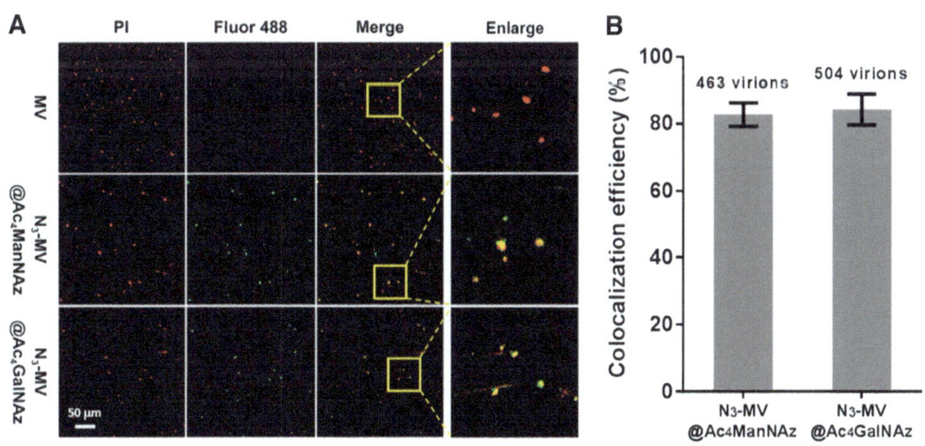

Fig. 12 (**A**) Fluorescence imaging of control virus and azide-labeled viruses stained with both DBCO-Fluor 488 (green) and propidium iodide (red), showing the accessible surface azido groups of MV particles derived from azido sugar-treated Vero cells. (**B**) Colocalization efficiency of azide-labeled viruses. Reprinted with permission from Bioconjugate Chem 29(9):1868–1872. Copyright © 2015 American Chemical Society

2. After a 10-min adsorption period, remove the excess virus using filter paper.
3. Then, fix the grid with 50 μL of 2% PFA for 5 min, followed by staining with 10 μL of 0.5% uranyl acetate (*see* **Note 19**) for 30 s.
4. Prepared samples are subsequently examined using TEM.

4 Notes

1. An optimal confluency range of 60%–80% ensures that cells are actively proliferating and maintain sufficient metabolic activity for efficient azido-analog incorporation. At lower confluency levels, cells may be stressed or insufficiently established, whereas at higher confluency levels, cells may approach growth arrest due to contact inhibition, reducing metabolic incorporation efficiency.
2. The concentration of azido-analogs should be adjusted according to cell types and experimental objectives.
3. A control group without the addition of unnatural substrates should always be included for comparison.
4. 2% FBS in DPBS is used to block the nonspecific binding with streptavidin dye during the labeling.
5. Ice-cold DPBS decreases protein degradation during the experiments.
6. When transferring the bottom $CHCl_3$ layer, carefully use a glass pipette to avoid disturbing the top layer.
7. For all control experiments, the procedure described above was followed without administration of any choline derivatives.
8. PC 32:1, PC 34:1, PC 36:1, SM 32:1, SM 34:1, and SM 36:1 are the major lipids involved in the metabolic incorporation of unnatural choline analogs.
9. Lipidomic analysis revealed that azido-lipids were the predominant metabolites derived from the azido-choline analog in live cells, with Vero cells showing the highest incorporation efficiency, reaching up to 0.39% (Fig. 10).
10. MV CPE includes syncytia formation, which could be easily visible under light microscopy as large clumps of nuclei in a single cell body. And usually, CPE can be observed 48–72 h post-infection in Vero cells. Harvesting at 80%–90% CPE ensures high viral titer while preserving virus integrity.
11. When infected cells were ground using the Dounce homogenizer, air bubbles should not be generated during this process.

12. Inject the sucrose solution gently and slowly using a pipette gun, ensuring a distinct sucrose layer forms at the bottom of the ultracentrifuge tube. Carefully place the tube into the centrifuge rotor to maintain layer integrity.
13. After ultracentrifugation, observe the white virus pellet at the bottom of the tube. Quickly discard the supernatant to prevent pellet resuspension. Mark the pellet's location by circling it on the tube with a marker. Then, align the tube horizontally to allow the buffer solution to gently cover the pellet area.
14. Purification by sucrose gradient is optional if only highly purified viruses are needed. The centrifugation speed differs for MV and VACV because of their size, shape, density, and complexity. VACV needs higher g-forces due to its heavier and more layered structure, whereas MV is less dense and easier to pellet or band at lower speeds.
15. VACV CPE includes cell rounding, detachment, granularity, and lysis, but not syncytia formation. And usually, CPE can be observed 48–72 h postinfection in Vero cells. Harvesting at 80%–90% CPE ensures high viral titer while preserving virus integrity.
16. MV could be replaced with other types of enveloped viruses, such as VACV.
17. The reciprocal of $TCID_{50}$ refers to the virus titer in terms of infectious dose per unit volume.
18. Alcian blue is a positively charged dye that creates a hydrophilic, charged surface. Alcian blue enhances the adherence and distribution of negatively charged particles (such as MV and VACV).
19. Uranyl acetate (UA) provides high electron density and superior image contrast, imparting a fine-grain appearance to the image due to uranium's high atomic weight (238).

Acknowledgments

We thank Dr. William E. Cotham at the Mass Spectrometry Center, University of South Carolina, for his assistance with the LC-MS/MS measurements. This work was partially supported by the NSF CBET-2330999 award. QW would also acknowledge the financial support by the National Science Foundation EPSCoR Program under NSF Award # OIA-2242812. TM would like to acknowledge the partial financial supported by The Royal Thai Government Scholarships in Science and Technology, Ministry of Higher Education, Science, Research and Innovation (MHESI), on behalf of Srinakharinwirot University.

References

1. Pettersson RF (1991) Protein localization and virus assembly at intracellular membranes. Curr Top Microbiol Immunol 170:67–106
2. Kingsak M, Meethong T, Jongkhumkrong J, Cai L, Wang Q (2023) Therapeutic potential of oncolytic viruses in the era of precision oncology. Biomater Transl 4:67–84
3. Banerjee PS, Ostapchuk P, Hearing P, Carrico I (2010) Chemoselective attachment of small molecule effector functionality to human adenoviruses facilitates gene delivery to cancer cells. J Am Chem Soc 132:13615–13617
4. Waehler R, Russell SJ, Curiel DT (2007) Engineering targeted viral vectors for gene therapy. Nat Rev Genet 8:573–587
5. Hedley SJ, Chen J, Mountz JD, Li J, Curiel DT, Korokhov N, Kovesdi I (2006) Targeted and shielded adenovectors for cancer therapy. Cancer Immunol Immunother 55:1412–1419
6. Matthews QL, Yang P, Wu Q, Belousova N, Rivera AA, Stoff-Khalili MA, Waehler R, Hsu H-C, Li Z, Li J, Mountz JD, Wu H, Curiel DT (2008) Optimization of capsid-incorporated antigens for a novel adenovirus vaccine approach. Virol J 5:98
7. Singh M, Cattaneo R, Billeter MA (1999) A recombinant measles virus expressing hepatitis B virus surface antigen induces humoral immune responses in genetically modified mice. J Virol 73:4823–4828
8. Bruckman MA, Kaur G, Lee LA, Xie F, Sepulveda J, Breitenkamp R, Zhang X, Joralemon M, Russell TP, Emrick T, Wang Q (2008) Surface modification of tobacco mosaic virus with "click" chemistry. ChemBioChem 9:519–523
9. Huang B-H, Lin Y, Zhang Z-L, Zhuan F, Liu A-A, Xie M, Tian Z-Q, Zhang Z, Wang H, Pang D-W (2012) Surface labelling of enveloped viruses assisted by host cells. ACS Chem Biol 7:683–688
10. Huang L-L, Lu G-H, Hao J, Wang H, Yin D-L, Xie H-Y (2013) Enveloped virus labelling via both intrinsic biosynthesis and metabolic incorporation of phospholipids in host cells. Anal Chem 85:5263–5270
11. Zhao X, Cai L, Adogla EA, Guan H, Lin Y, Wang Q (2015) Labelling of enveloped virus via metabolic incorporation of azido sugars. Bioconjugate Chem 26:1868–1872
12. Hang HC, Yu C, Kato DL, Bertozzi CR (2003) A metabolic labelling approach toward proteomic analysis of mucin-type O-linked glycosylation. Proc Natl Acad Sci 100:14846–14851
13. Jao CY, Roth M, Welti R, Salic A (2009) Metabolic labelling and direct imaging of choline phospholipids in vivo. Proc Natl Acad Sci USA 106:15332–15337
14. Stith BJ, Hall J, Ayres P, Waggoner L, Moore JD, Shaw WA (2000) Quantification of major classes of Xenopus phospholipids by high performance liquid chromatography with evaporative light scattering detection. J Lipid Res 41:1448–1454
15. Bellini WJ, Trudgett A, McFarlin DE (1979) Purification of measles virus with preservation of infectivity and antigenicity. J Gen Virol 43:633–639
16. Reed LJ, Muench H (1938) A simple method of estimating fifty per cent endpoints. Am J Epidemiol 27:493–497
17. Jewett JC, Bertozzi CR (2010) Cu-free click cycloaddition reactions in chemical biology. Chem Soc Rev 39:1272–1279

Chapter 3

Synthesis of Linkage-Specific Polyubiquitin Probes Using Unnatural Amino Acids

Matthew T. Buhl and T. Ashton Cropp

Abstract

Activity-based probes based on polyubiquitin chains are powerful tools for studying ubiquitin-conjugating pathways and the substrate specificity of deubiquitinase (DUB) enzymes. These probes can be synthesized through a genetically encoded unnatural amino acid (UAA), N^{ε}-L-thiaprolyl-L-lysine (ThzK). This chapter details the chemical synthesis of ThzK and preparation of diUb probes, leveraging ThzK to enable the production of site-specifically installed isopeptide bonds. The resulting probes facilitate mechanistic studies of ubiquitination and deubiquitination specificity and have applications in understanding diseases associated with ubiquitination dysfunction.

Keywords Ubiquitin, Activity-based probes, DUBs, Deubiquitinase, Unnatural amino acid

1 Introduction

Ubiquitin (Ub) is a 76-amino acid globular protein that serves as a post-translational modification on lysine residues of target proteins. After initial Ub conjugation, subsequent polyubiquitin chains can be assembled via isopeptide linkages involving the C-terminal glycine of a distal Ub and any of seven lysine residues [1]. The type of linkage can determine the functional outcome of ubiquitination, influencing cellular processes like proteasomal degradation, signaling, and DNA repair [2]. Deubiquitinases (DUBs) play a critical role by reversing this process by cleaving ubiquitin chains, yet the substrate specificities of the array of DUBs are not completely known [3, 4]. Deubiquitinases and their polyubiquitin substrates are involved in homeostatic regulation, cellular signaling, and proteolysis of degraded, damaged, or misfolded proteins.

Traditional methods for synthesizing ubiquitin chains with designed linkages include enzymatic ligation and/or solid-phase peptide synthesis [5–7]. Recent advancements using genetic code

Alexander Deiters (ed.), *Bioconjugation Methods*, Methods in Molecular Biology, vol. 3009,
https://doi.org/10.1007/978-1-0716-5112-4_3,

expansion offer new opportunities to control linkage specificity. Genetic incorporation of N^{ε}-L-thiaprolyl-L-lysine (ThzK) into ubiquitin enables the synthesis of diUb probes with precise linkages, such as K48, which are critical for studying DUB interactions [8, 9]. ThzK is useful because of its ability to be deprotected at the thiazolidine ring to produce CysK, which has a 1,2-aminothiol that can be used in native chemical ligation with a Ub C-terminal thioester. This leaves a cysteine scar, which can be converted to a dehydroalanine residue [9]. Using these activity-based diUb probes, the dehydroalanine warhead is able to covalently and irreversibly react with an active site cysteine present in the deubiquitinase enzyme(s) when serving as a substrate. These probes are particularly valuable for elucidating mechanisms of ubiquitin signaling, natural substrate specificity, and the role of DUBs and their substrates in disease. In this protocol, we describe the chemical synthesis of ThzK (Fig. 1), and methods for producing polyUb probes (Fig. 2). These protocols could be adapted for the preparation of other Ub-containing conjugates of any target or linkage combination.

2 Materials

2.1 Competent Cells and Plasmids

1. *E. coli* strains capable of unnatural amino acid incorporation, such as BL21 (DE3) cells for T7-promoter-based expression.
2. pPylT_UbK48TAG (tet resistance, arabinose inducible expression of Ub with stop codon in place of K48, C-terminal 6× -histidine tag).
3. pBK_ThzKRS (amp resistance, expresses PylS variant specific for ThzK) [8].
4. pTXB1_synUb1-75 (am resistance, T7 promoter, expresses codon optimized gene encoding wild-type Ub, amino acids 1-75, as a fusion protein to intein-chitin binding domain).

2.2 Media

LB or 2xYTmedia supplemented with ThzK-OMe, IPTG and/or arabinose, and appropriate antibiotics.

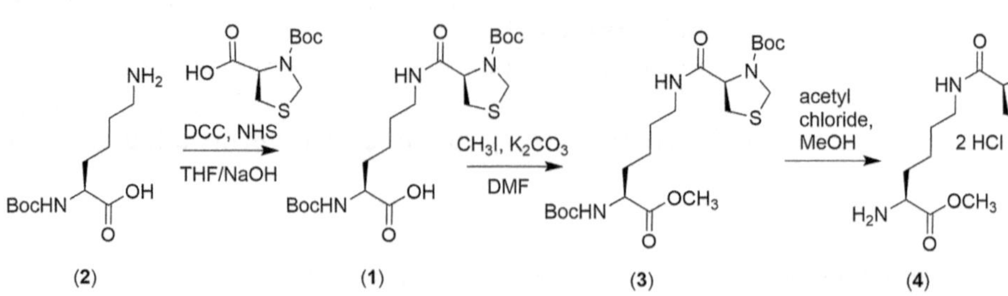

Fig. 1 Chemical synthesis of ThzKOMe (**4**) for use in protein synthesis. **4** is metabolized to active ThzK in vivo

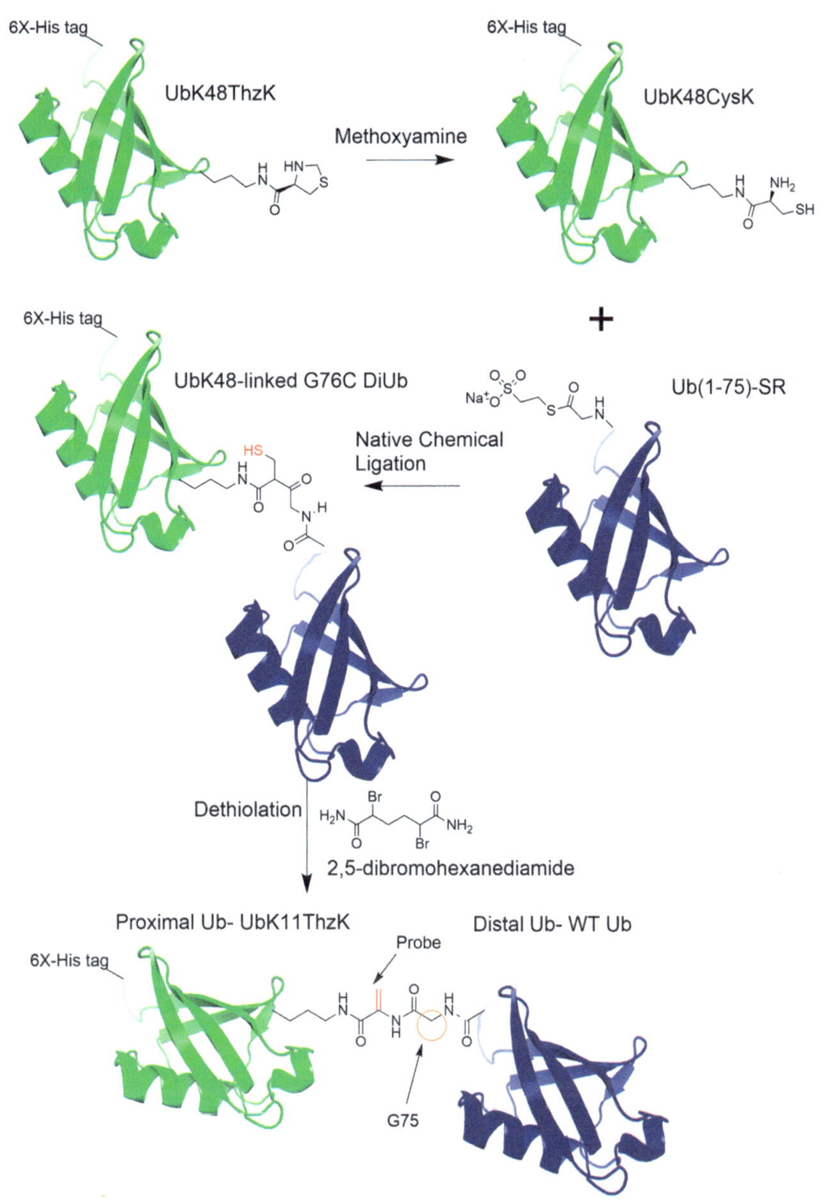

Fig. 2 Production of diUB-based probes is facilitated by genetic encoding of ThzK. The proximal Ub (green) carries a his tag for purification and immobilization. The Cys residue derived from the remnants of ThzK becomes the equivalent of Gly76 of the distal Ub (blue)

1. **Lysogeny Broth (LB)**: 10 g/L tryptone, 5 g/L yeast extract, 10 g/L sodium chloride, pH 7.0.
2. **2xYT media:** 16 g/L tryptone, 10 g/L yeast extract, 5 g/L sodium chloride, pH 7.0.

2.3 Buffers and Solutions

1. Ligation buffer: 50 mM Tris-HCl, 150 mM sodium chloride, pH 7.5.
2. Nickel purification lysis buffer: 20 mM imidazole, 300 mM sodium chloride, 50 mM sodium phosphate or Tris, pH 8.0.
3. Nickel purification wash buffer: 50 mM imidazole, 300 mM sodium chloride, 50 mM sodium phosphate or Tris, pH 8.0.
4. Nickel purification elution buffer: 300 mM imidazole, 300 mM sodium chloride, 50 mM sodium phosphate or Tris, pH 8.0.
5. UbThzK dialysis buffer: 25 mM Tris, 50 mM sodium chloride, 1 mM DTT, pH 8.0.
6. SDS-PAGE running buffer: 50 mM Tris, 200 mM glycine, 0.1% SDS, pH 8.3.
7. Stacking gel buffer: For 100 mL, 6.06 g Tris base, 4 mL 10% SDS, pH 6.8.
8. Resolving gel buffer: For 100 mL, 18.17 g Tris base, 4 mL 10% SDS, pH 8.8.
9. Protein storage buffer: 50 mM HEPES, 100 mM sodium chloride, 10% glycerol, pH 7.
10. Native chemical ligation buffer: 6 M guanidinium HCl, 200 mM sodium phosphate, 100 mM mercaptophenylacetic acid, 60 mM TCEP, pH 7.1.
11. Dehydroalanine (Dha) conversion buffer: 50 mM Tris-HCl, 150 mM sodium chloride, pH 7.5 + 2,5-dibromohexane diacetamide (DBHDA).
12. IMPACT column/lysis buffer: 20 mM HEPES, 500 mM sodium chloride, 0.5% Triton-x100, 1 μM TCEP, and 20 μM PMSF, pH 8.0.
13. IMPACT cleavage buffer: Column buffer + 10 mM TCEP + 60 mM MESNa, pH 8.0.
14. ThzK conversion buffer: 6 M guanidinium chloride, 200 mM sodium phosphate, pH 7.2.

2.4 Chemicals

1. ThzK-OMe: unnatural amino acid for protein expression.
2. O-methylhydroxylamine.
3. Sodium 2-mercaptoethanesulfonate (MESNa).
4. 2,5-Dibromohexane diacetamide.
5. Chitin resin.
6. Nickel-NTA resin.
7. Pierce Snakeskin 3 kDa MW cutoff dialysis tubing.

2.5 SDS-PAGE Gels and Associated Reagents

3 Methods

3.1 Chemical Synthesis of ThzKOMe (Fig. 1) (see Note 1)

The synthesis strategy employed to generate L-(thioproyl)-L-lysine methyl ester starts from α-Boc-thioproline and Nα-Boc-L-lysine. First, the α-Boc-thioproline is activated at the carboxy group by creating a better leaving group with NHS in the presence of DCC by reacting at room temperature. The product of that reaction is immediately used, because of its reactivity, by dissolving it in THF and reacting it with Nα-Boc-L-lysine, which has been dissolved in 1 M sodium NaOH solution. This reaction is performed overnight at room temperature. Following this step, the product is methylated by dissolving it in DMF then using methyl iodide in the presence of sodium carbonate for 24 h. Finally, two methods of deprotection to remove the labile Boc group, without removing the methyl ester, are described. This includes forming HCl in situ in a solution of methanol and acetyl chloride or by using HCl in anhydrous dioxane. The products are characterized using NMR spectroscopy.

1. Nα-Boc-L-(α-Boc-thiaprolyl)-L-lysine(**1**): Dissolve 5.01 g of α -Boc-thioproline and 2.47 g of N-hydroxysuccinimide (NHS) in 10 mL of acetonitrile. Slowly add 4.41 g of N-N′-dicyclohexylcarbodiimide (DCC) and allow it to react for 4 h at room temperature. Filter the product and remove solvent via rotary evaporation to provide activated thioproline (6.08 g, 86%) and use without further purification or characterization.
2. Separately dissolve 8.04 g of Nα-Boc-L-lysine (**2**) in 20 mL of 1 M NaOH solution and dissolve 6.0 g of the NHS-activated α -Boc-L-thioproline from the previous step in 20 mL of THF. Slowly add the α-Boc-L-thioproline solution to the α-Boc-L-lysine solution and allow it to stir overnight at room temperature. Remove the THF by rotary evaporation, then acidify the aqueous layer with 1 M HCl until the pH reaches 4.5. Extract four times with 20 mL of ethyl acetate, pool the ethyl acetate extractions, dry over anhydrous $MgSO_4$, and filter. Remove the ethyl acetate under rotary evaporation to yield a yellow to orange oil (6.3 g,70%) Nα-Boc-L-(α-Boc-thiaprolyl)-L-lysine(**1**) ^{1}H NMR, (400 MHz, $CDCl_3$) δ 12.5ppm (s, 1 H), 7.2 ppm (s, 1 H), 6.8 ppm (s, 1 H), 4.5 ppm (m, 1 H), 4.1 ppm (q, J = 5.3 Hz, 2 H), 3.0 ppm (m, 2 H), 2.0 ppm (s, 2 H), 1.7 ppm (m, 2 H), 1.4ppm (s, 9 H), 1.2 ppm (m, 2 H).

3. Dissolve α-Boc-L-(thiaprolyl)-L-lysine methyl ester (**3**): 6.0 g of α-Boc-L-(thiaprolyl)-L-lysine from the previous step in 45 mL of dimethylformamide, to which add 7.89 g of potassium carbonate and 1.2 mL of neat methyl iodide. Stir the mixture for 24 h at room temperature and filter off the precipitate. Add 10 mL of ethyl acetate and 10 mL of water to quench the reaction, then extract with 4× with 10 mL ethyl acetate. Remove the ethyl acetate fraction via separatory funnel, wash three times with 20 mL of water, twice with 20 mL of brine, and separate the organic layer followed by removing the solvent by rotary evaporation, yielding a thick yellow oil, α-Boc-L-(thiaprolyl)-L-lysine methyl ester (6.1 g, 95%) (**3**). ^{1}H NMR, (400 MHz, $CDCl_3$) δ: 7.2 ppm (s, 1 H), 6.6 ppm (s, 1 H), 5.2 ppm (d,J = 3.9 Hz, 1 H), 4.6 ppm (d,J = 4.7 Hz, 2 H), 4.2 ppm (s, 1 H), 4.1 ppm (q, 5.3 Hz, 2 H), 3.2 ppm (m, 4 H) 3.0 ppm (s, 2 H), 2.0 ppm (s, 3 H), 1.7 ppm (m, 1 H), 1.5 ppm (m, 1 H), 1.4 ppm (s, 9 H), 1.3 ppm (m, 2 H).
4. ThzKOMe (**4**) can be deprotected by one of two methods:

   i. Deprotection of α-Boc-L-(thioprolyl)-L-lysine methyl ester using acetyl chloride and methanol: Dissolve 0.46 g (1 eq) of α-Boc-L-(thiaprolyl)-L-lysine methyl ester in 15 mL of methanol. On ice, slowly add an excess of 0.25 mL (30 eq) of acetyl chloride, and allow to stir for 2–4 h, while allowing the ice to melt and the reaction to slowly warm to room temperature. Using rotary evaporation, remove the acetyl chloride and methanol, yielding an orange to pink solid (0.22 g, 82%). If the solid is oily/sticky, rinse the flask with anhydrous diethyl ether while swirling, pour off the liquid, and allow the residual diethyl ether to evaporate in the hood yielding an orange to pink solid, L-(thiaprolyl)-L-lysine methyl ester hydrochloride salt. Store in a desiccator at room temperature or dissolve to a concentration of 200 mM ThzKOMe in 200 mM sodium hydroxide solution (**4**).

   ii. Method 2: Deprotection of α-Boc-L-(thiaprolyl)-L-lysine using HCl in dioxane: Dissolve 3.5 g of α-Boc-L-(thiaprolyl)-L-lysine in 5 mL of anhydrous dioxane. Then, add 2 mL of anhydrous 4 N HCl dissolved in dioxane to give a concentration of ~1 N, and allow to stir at room temperature for 1–2 h. Remove HCl and dioxane with rotary evaporation to yield an orange to pink solid (2.8 g, 92%). If the solid is sticky or oily, rinse the flask with anhydrous diethyl ether while swirling, pour off the excess diethyl ether, and allow the residual diethyl ether to evaporate in the hood yielding an orange to pink solid, L-(thiaprolyl)-L-lysine methyl ester hydrochloride salt. It can be stored in a desiccator at room temperature or dissolved to a concentration of 200 mM ThzKOMe (**4**). ^{1}H NMR, (400 MHz, D_2O) δ: 4.7 ppm (t, J = 4.7 Hz, 1 H); 4.5 ppm (q, J = 9.8 Hz, 2 H); 4.2 ppm (t, J = 4.5 Hz, 1 H); 3.9 ppm (s, 3 H); 3.6 ppm (m, 1 H); 3.3 ppm (m, 3 H); 2.0 ppm (m, 2 H); 1.6 ppm (m, 2 H); 1.5 ppm (m, 2 H).

3.2 Expression and Purification of Ubiquitin Containing Site-Specific ThzK (Fig. 2)

1. Transform two plasmids into competent *E. coli* cells that encode the desired ubiquitin mutant. For example, pPylT_UbK48TAG (tetracycline resistant) and a plasmid with an evolved pyrrolysyl-tRNA synthetase (Pyl-RS) and pyrrolysyl-tRNA pair specific for ThzK, such as pBkThzKRS (ampicillin resistant), are transformed into DH-10β cells (*see* **Note 2**).
2. For arabinose inducible plasmid (pPylT_UbK48TAG): Inoculate a 1 L culture with 5 mL of overnight starter culture, include 1 mL of 50 mg/mL tetracycline and 100 mg/mL ampicillin, or relevant antibiotics, grow to an OD_{600} of 0.6, then add 2 mM ThzKOMe (*see* **Note 3**) and induce protein expression with 0.2% arabinose at room temperature overnight. ThzKOMe is hydrolyzed in vivo, and has been shown to result in improved protein yields [10].
3. IPTG-inducible plasmids (Ub~SR): Inoculate 1 L of culture with 5 mL of overnight starter culture, include 1 mL of 50 mg/mL ampicillin or relevant antibiotic, grow to an OD600 of ~0.6, then induce with 1 mL of 1 M of IPTG and allow to express 4–6 h.
4. Harvest cells by centrifugation at 12,000 × *g* for 20 min, decant supernatant, resuspend in cold PBS, centrifuge again, and pour off supernatant. Resuspend the cells in nickel purification lysis buffer and 100 μg/mL lysozyme and incubate on ice for 30 min. Lyse using sonication. Sonication settings: 40% power in 30-s pulses with 30 s of rest in between. Perform sonication on ice.
5. Purify ubiquitin containing ThzK using nickel-affinity chromatography. Elute using imidazole concentration of 300 mM, or utilize a gradient (20–500 mM) if using an automated column. Using either a column or resin in conical tube, allow clarified lysate, the supernatant left after centrifuging the lyses cells, to bind to the resin for at least 1 h at 4°C. Remove flowthrough by centrifugation or allowing column to drain (save 2 μL aliquots of lysate, flowthrough, and wash for SDS-PAGE analysis). Wash resin with 40 column volumes of wash buffer, and then elute with 500 μL of elution buffer per liter of culture ten times. Analyze purity of the elution by SDS-PAGE. UbK48ThzK is 9.5 kDa.

6. Immediately dialyze protein into UbThzK dialysis buffer using snakeskin dialysis tubing, and lyophilize protein to dryness.

3.3 Expression and Purification of Ub~SR 1-75

1. Transform *E. coli* cells, BL21 (DE3), with pTXB1_synUb1-75. Select for incoming plasmid with 100 µg/mL ampicillin.
2. Inoculate 1 L of culture with 5 mL of overnight starter culture. Grow to an OD600 of ~0.6, then induce with 1 mL of 1 M IPTG solution and allow to express 4–6 h.
3. Harvest cells by centrifugation for 10 min at 12,000 x *g*. Once cells have been pelleted, resuspended the pellet in 20 mL of IMPACT column buffer/lysis buffer and transfer to 50 mL falcon tube.
4. Centrifuge for 30 min at 12,000 x *g* to pellet cell debris.
5. Equilibrate chitin column with five column volumes of column buffer for 20 min. Use 15 mL of settled chitin resin (30 mL of 50% slurry in ethanol) per liter of culture.
6. Load clarified lysate, the supernatant, to the chitin column slowly at a rate of 1–2 mL per minute and allow to drain. Collect flowthrough for analysis by SDS-PAGE.
7. Wash column with 20 column volumes of IMPACT column buffer/lysis buffer. The resin should be white by the end.
8. Collect a 10 µL aliquot of the chitin resin itself for later analysis of binding of the fusion protein by SDS-PAGE. Resin can be boiled in SDS loading dye and run directly on a gel. The Ub-intein-CBD fusion protein is 36 kDa.
9. Add 3 column volumes of IMPACT cleavage buffer (column buffer supplemented with 60 mM of MESNa) (*see* **Note 4**), allow to drain without fully drying out the resin, and then incubate at 4°C for 16–40 h to allow intein-mediated cleavage of the Ub thioester from the column.
10. Drain and collect the protein in 1 mL fractions. Analyze by SDS-PAGE. Ub1-75SR is 8.6 kDa.
11. The chitin resin can be analyzed post-cleavage to determine extent of cleavage. The cleaved intein-CBD remnant is 28 kDa.
12. Immediately dialyze the eluted protein into storage buffer using 3.5 kDa MWCO snakeskin dialysis tubing.

3.4 Conversion of ThzK to CysK

1. Dissolve lyophilized UbK48ThzK in ThzK conversion buffer to a final concentration of 1 mg protein/mL of buffer. Then add O-methylhydroxyamine to a final concentration of 400 mM. Incubate at 37°C for >4 h.
2. Converted UbK48CysK is purified by reversed-phase HPLC and lyophilized to dryness.

3. Confirm the conversion using mass spectrometry (ESI-MS), which will result in a mass loss of 36 Da.

3.5 Synthesis of CysK-Linked DiUb

1. Dissolve the entirety of purified UbK48CysK to a final concentration of 750 μM in native chemical ligation buffer.
2. Separately, dissolve a twofold molar ratio, relative to purified UbK48CysK, of purified Ub1-75SR to a final concentration of 1500 μM in native chemical ligation buffer
3. Mix the two proteins in a 1:1 volume ratio (resulting in a twofold molar excess of Ub1-75SR) and incubate at room temperature for 2–4 h.
4. Dialyze the product into 1× PBS buffer supplemented with 1 mM DTT.
5. Purify the ligated diUb by nickel-affinity chromatography (which will bind the proximal, His-tagged Ub) as described above.
6. Protein sample was desalted using a PD-10 minitrap column (GE Healthcare), previously equilibrated with 10 mM PBS.
7. Analyze the ligated product by SDS-PAGE. The ligated diUb is 18.1 kDa.

3.6 Functionalization of diUb Probes

1. The purified diUb is added to a solution of PBS to a final concentration of 0.5 mg/mL.
2. 1 mg of 2,5-dibromohexane diacetamide (DBHDA) is added to the solution of diUb, and shaken at 37°C for 4 h.
3. Upon cooling to room temperature, the DBHDA precipitates, and it is pelleted and removed by centrifugation at 12,000 × *g*. The supernatant is removed to yield the Ub probe.
4. The functionalized probe can be frozen and stored at −80°C and used for future pull-down studies by immobilization using the attached 6×-histidine tag, or through other methods (*see* **Note 5**).

4 Notes

1. Conditions are provided for a moderate-scale chemical synthesis, but the reagents can be scaled larger or smaller for the desired amount.
2. One can either transform both plasmids simultaneously, selecting for both, or sequentially transform, by doing a single transformation, and then making competent cells, and transforming with a second plasmid. Sequential transformations are best.

3. Ensure media contains at least 2 mM ThzK-OMe to achieve high incorporation efficiency. ThzK-OMe is hygroscopic so material stored for long periods may absorb moisture. Monitor cell growth as excessive ThzK may affect viability.
4. Use freshly prepared MESNa for efficient cleavage of ubiquitin-intein fusion proteins. Ensure the pH of the buffer is correct *after* the addition of MESNa, as it can alter the pH.
5. Store Dha-functionalized probes at −80°C to maintain reactivity.

Acknowledgments

Matthew T. Buhl and T. Ashton Cropp are supported by NSF grant award # 2109051.

References

1. Komander D, Rape M (2012) The ubiquitin code. Annu Rev Biochem 81:203–229
2. Oh E, Akopian D, Rape M (2018) Principles of ubiquitin- dependent signaling. Annu Rev Cell Dev Biol 34:137–162
3. Amerik AY, Hochstrasser M (2004) Mechanism and function of deubiquitinating enzymes. Biochim Biophys Acta Mol Cell Res 1695:189–207
4. Clague MJ, Urbé S, Komander D (2019) Breaking the chains: deubiquitylating enzyme specificity begets function. Nat Rev Mol Cell Biol 20:338–352
5. Pickart CM, Raasi S (2005) Controlled synthesis of polyubiquitin chains. Methods Enzym 399:21–36
6. Castaneda C, Liu J, Chaturvedi A, Nowicka U, Cropp TA, Fushman D (2011) Nonenzymatic assembly of natural polyubiquitin chains of any linkage composition and isotopic labeling scheme. J Am Chem Soc 133:17855–17868
7. Kumar KS, Bavikar SN, Spasser L, Moyal T, Ohayon S, Brik A (2011) Total chemical synthesis of a 304 amino acid K48-linked tetraubiquitin protein. Angew Chem Int Ed Engl 50:6137–6141
8. Nguyen DP, Elliott T, Holt M, Muir TW, Chin JW (2011) Genetically encoded 1,2-aminothiols facilitate rapid and site-specific protein labeling via a bio-orthogonal cyanobenzothiazole condensation. J Am Chem Soc 133:11418–11421
9. Zhou H, Carpenter T, Fu X, Jin B, Ody B, Hassan MS, Jacobs SE, Cheung J, Nicholson EM, Turlington M, Zhao B, Lorenz S, Cropp TA, Yin J (2022) Linkage-specific synthesis of di-ubiquitin probes enabled by the incorporation of unnatural amino acid ThzK. ChemBioChem 23:e2022001
10. Zhou H, Cheung JW, Carpenter T, Jones SK, Luong NH, Tran NC, Jacobs SE, Galbada Liyanage SA, Cropp TA, Yin J (2020) Enhancing the incorporation of lysine derivatives into proteins with methylester forms of unnatural amino acids. Bioorganic Med Chem Lett 30:126876

Chapter 4

Synthesis of Ub-ACA, a Fluorogenic DUB Substrate via ACPL Method

Saibal Chanda and Wenshe Ray Liu

Abstract

This chapter details the step-by-step synthesis of ubiquitin-7-amino-4-coumarinyl-acetic acid (Ub-ACA) using the activated cysteine-directed protein ligation (ACPL) technique. Ub-ACA is a fluorogenic substrate for deubiquitinases (DUBs). It can be readily synthesized via ACPL to a high quantity by exchanging a *C*-terminal cysteine residue in a recombinantly expressed Ub-G76C mutant with Gly-ACA, a process mediated simply by 2-nitro-5-thiocyanatobenzoic acid (NTCB). This chapter provides protocols for protein expression, thiol activation, fluorophore conjugation, and product purification, yielding Ub-ACA with high purity suitable for enzymatic assays.

Keywords Ubiquitin, Deubiquitinase, 7-amino-4-coumarinyl-acetic acid, Fluorogenic substrate, Ub-ACA, Activated cysteine-directed protein ligation, ACPL, DUB

1 Introduction

Ubiquitin is a small, evolutionarily conserved protein present in nearly all eukaryotic organisms [1]. It regulates diverse cellular processes by covalently attaching to target proteins [2], thereby modulating their activity, localization, or stability. This posttranslational modification, is known as ubiquitination [3]. The posttranslational modification of proteins with ubiquitin plays a pivotal role in regulating proteostasis, signal transduction, cell cycle progression, and immune responses [4, 5]. As such this is a target for therapeutic intervention [6–8]. This ubiquitination modification is dynamically controlled by a cascade of enzymes that includes ubiquitin-activating (E1), conjugating (E2), and ligating (E3) enzymes, along with deubiquitinases (DUBs) that reverse ubiquitination. Many of these enzymes, particularly within the DUB superfamilies (USP, UCH, OTU, and MJD) [9], utilize an active-site cysteine for catalysis. The ability to monitor DUB activity with molecular precision is essential for understanding their biological roles and for developing therapeutic inhibitors.

Alexander Deiters (ed.), *Bioconjugation Methods*, Methods in Molecular Biology, vol. 3009,
https://doi.org/10.1007/978-1-0716-5112-4_4,

Fluorogenic substrates, such as ubiquitin-7-amino-4-methylcoumarin (Ub-AMC), have historically enabled real-time DUB activity assays by releasing a fluorescent AMC moiety upon enzymatic cleavage [10]. The synthesis of Ub-AMC is typically achieved by reacting Gly-AMC with a recombinantly expressed $Ub^{1\text{-}75}$-intein that naturally forms a thioester between $Ub^{1\text{-}75}$ and its intein fusion partner. This process is called expressed protein ligation. There are two limitations related to this existing approach. The first is a complication related to handling the purification of $Ub^{1\text{-}75}$-intein that undergoes hydrolysis at the thioester site, and the second is the very low solubility of Gly-AMC, which limits the yield of the overall synthesis. To overcome these limitations, we replaced the expressed protein ligation process with the activated cysteine-based protein ligation (ACPL, shown in Fig. 1) technique [11, 12] that involves only a small molecule-triggered exchange of a *C*-terminal cysteine followed by a 6×His tag with an amine-containing compound and therefore is easy to handle and replaced Gly-AMC with Glycine-7-amino-4-coumarinyl-acetic acid (Gly-ACA as shown in Fig. 2) that retains similar excitation/emission characteristics of AMC while offering superior water solubility, chemical stability, and signal-to-noise ratio in enzymatic assays [13].

In this chapter, we describe a detailed and reproducible protocol for the synthesis of Ub-ACA via the ACPL approach (Fig. 2). We include the design and expression of the Ub-G76C mutant, its activation with 2-nitro-5-thiocyanatobenzoic acid (NTCB) at the introduced cysteine residue, and subsequent ligation with Gly-ACA

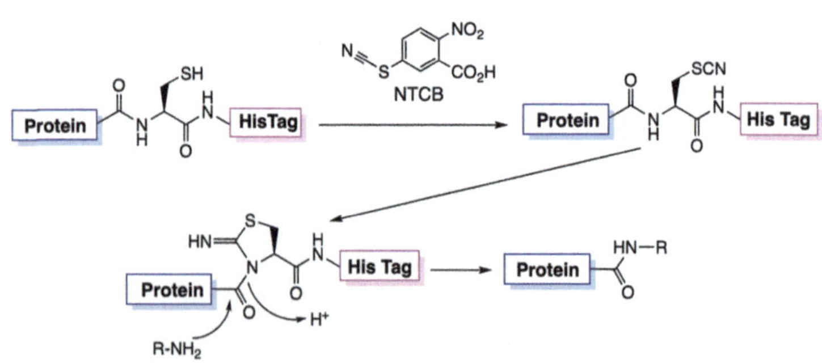

Fig. 1 The ACPL reaction mechanism

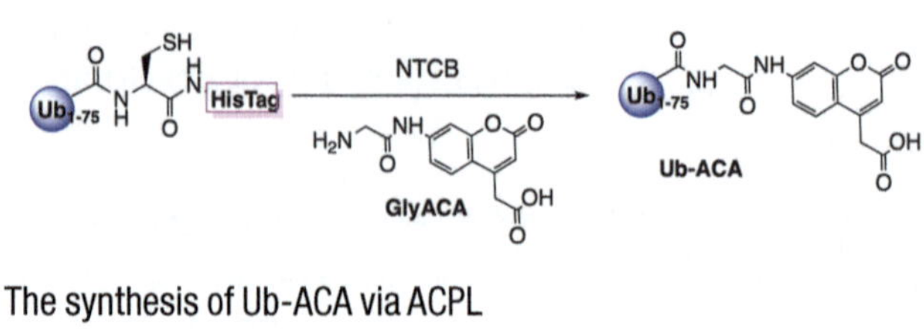

Fig. 2 The synthesis of Ub-ACA via ACPL

under mild alkaline conditions. The resulting Ub-ACA probe is suitable for quantitative fluorogenic DUB assays in recombinant systems, cell lysates, and tissue extracts. This methodology not only enables the preparation of Ub-ACA but is also generalizable to other ubiquitin and ubiquitin-like (Ubl) proteins modified with amine-containing nucleophiles, facilitating broader applications in enzymology and chemical biology.

2 Materials

2.1 Reagents

Prepare all solutions using either ultrapure water (prepared by purifying deionized water, to attain a sensitivity of 18 MΩ-cm at 25°C) or analytical grade reagents as specified. Store all reagents at designated temperatures. Diligently follow all waste disposal regulations when disposing waste materials.

1. pET duet vector harboring Ub-G76C-6×His expression plasmid in BL21 (DE3) *E. coli* cells.
2. 2xYT medium: Ready-to-use. Use 1 L medium for expressing the protein.
3. Ampicillin: Prepare 100 μg/mL aliquots in water, store at −20°C.
4. Isopropyl Thio-β-Galactoside (IPTG): Dissolve 2.38 g of IPTG in 10 mL of Milli-Q water to prepare 1 M stock. Store at −20°C.
5. Disodium hydrogen Phosphate (Na_2HPO_4): Prepare freshly before use. Store at room temperature.
6. Imidazole: Prepare freshly before use. Store at room temperature.
7. Sodium Chloride (NaCl): Prepare freshly before use. Store at room temperature.
8. Lysis Buffer: Dissolve 7.1 g Na_2HPO_4, 29.22 g NaCl, 340.4 mg imidazole, 276 mg TCEP into 800 mL of Milli-Q water, adjust the pH till 7.8, make up the volume till 1 L, filter and store at room temperature (Final concentration: 50 mM Na_2HPO_4, 500 mM NaCl, 5 mM imidazole, 1 mM TCEP).
9. Wash Buffer: Dissolve 7.1 g Na_2HPO_4, 29.22 g NaCl, 2.04 g imidazole, 276 mg TCEP into 800 mL of Milli-Q water, adjust the pH till 7.8, make up the volume till 1 L, filter and store at room temperature (Final concentration: 50 mM Na_2HPO_4, 500 mM NaCl, 30 mM imidazole, 1 mM TCEP).
10. Elution Buffer: Dissolve 7.1 g Na_2HPO_4, 14.6 g NaCl, 20.4 g imidazole, 276 mg TCEP into 800 mL of Milli-Q water, adjust

the pH till 7.8, make up the volume till 1 L, filter and store at room temperature (Final concentration: 50 mM Na_2HPO_4, 250 mM NaCl, 300 mM imidazole, 1 mM TCEP).

11. NTCB: Prepare 500 mM stock solution of NTCB in DMSO. Store at −20°C.
12. Gly-ACA: Weigh directly before use at required concentration.
13. Phosphate Buffer Saline (PBS, 20×): 202.8 mM Na_2HPO_4, 35.2 mM KH_2PO_4, 54 mM KCl, 2.74 M NaCl, pH 7.4: Store at 4°C. Dissolve 50 mL of 20× PBS to 950 mL of Milli-Q water to prepare 1× PBS solution.
14. TCEP: Prepare 500 mM stock solution of TCEP in Milli-Q water. Store at −20°C.
15. 50 mM ammonium bicarbonate buffer: Dissolve 4 g of ammonium bicarbonate into 1 L of Milli-Q water. Filter and store at 4°C.
16. ACA Assay Buffer: Dissolve 605.7 mg of Tris, 584.5 mg NaCl, 14.6 mg EDTA, 30.8 mg DTT, 10 mg BSA (Final concentration: 50 mM Tris, 100 mM NaCl, 0.5 mM EDTA, 2 mM DTT, 0.1% BSA (v/v)) in 70 mL of Milli-Q water. Adjust the pH to 7.6, make up the volume till 100 mL, filter and store at 4°C.
17. Hi-trap and Hi-prep26/10 desalting columns: use for buffer exchange purpose, store at 4°C.
18. Deubiquitinases: UCHL1, UCHL5, and USP25. All three DUBs were acquired commercially.
19. 3KDa MWCO Amicon ultracentrifugation filter.
20. Pierce 660 protein assay kit (Thermo Scientific, Cat: 22662): use for measuring protein concentration at 660 nm.
21. Microsyringe filters.
22. Tris: Prepare freshly before use. Store at room temperature.
23. Sodium Chloride (NaCl): Prepare freshly before use. Store at room temperature.
24. Disodium hydrogen Phosphate (Na_2HPO_4): Prepare freshly before use. Store at room temperature.
25. Imidazole: Prepare freshly before use. Store at room temperature.
26. Dithiothreitol (DTT): Prepare freshly before use. Store at −20°C.
27. Ethylenediaminetetraacetate (EDTA): Prepare freshly before use. Store at room temperature.
28. Bovine Serum Albumin (BSA): Prepare freshly before use. Store at 4°C.

29. Ni-NTA Agarose: Store at 4°C.
30. Tetrahydrofuran (THF): Store at room temperature.
31. Methanol: Store at room temperature.
32. Dichloromethane (DCM) anhydrous: Store at room temperature.
33. Trifluoroacetic acid (TFA): Store at room temperature.
34. Hydrogen balloon.
35. Celite pad: Used for filtration purpose.
36. Palladium (Pd) catalysts.

3 Methods

The ACPL method is a chemo selective strategy for *C*-terminal protein modification that enables the installation of diverse nucleophiles—such as small molecules, fluorophores, or peptides—onto a recombinant protein bearing a *C*-terminal cysteine residue. Unlike traditional expressed protein ligation [14], which requires intein fusion constructs or enzymatic catalysis, ACPL is a purely synthetic methodology, does not involve enzymes, and can be carried out under mild aqueous conditions. This technique features a nitrile donor molecule, e.g., NTCB, to activate a protein cysteine as shown in Fig. 1 to form a 1-acyl-2-iminothiazolidine intermediate that subsequently undergoes a nucleophilic acyl substitution reaction with a small molecule amine. The overall reaction leads to replacing cysteine with the amine at the *C*-terminus of the target protein.

3.1 Expression and Purification of Ub-G76C-6×His

1. An overnight culture of *E. coli* BL21(DE3) cells harboring a pETDuet expression vector coding Ub-G76C-6×His was inoculated (1:100 dilution) into a 2xYT medium containing 100 μg/mL ampicillin.
2. Cells were allowed to grow at 37°C until OD600 reached 0.6–0.9 and 1 mL IPTG (From 1 M stock, Final concentration: 1 mM) was added to induce protein expression.
3. The cells were allowed to grow in a shaker at 37°C for 8 h or 18°C for 16 h and collected by centrifugation (10,000 × *g*, 30 min, 4°C).
4. The cell pellets were resuspended in a 100 mL lysis buffer and lysed by sonication.
5. The cell lysates were clarified by centrifugation (20,000 × *g*, 30 min, 4°C) and the supernatant was decanted.
6. 6 M HCl solution was then added dropwise into the supernatant to adjust pH to 2 to drive the formation of white

precipitate inside the solution. The precipitate was pelleted by centrifugation (20,000× *g*, 30 min, 4°C) and removed.

7. The pH of the supernatant was then adjusted back to 7.8 by adding 6 M NaOH. This supernatant was loaded onto 4 mL high-affinity Ni^{2+}-NTA resins (pre-equilibrated with 50 mL lysis buffer) and incubated with shaking for 1 h at room temperature and then washed with 10 mL of wash buffer five times to remove any unbound/unwanted proteins.
8. The bound Ub-G76C-6×His was eluted by 7 mL of elution buffer and then desalted into a 50 mM ammonium bicarbonate (ABC) buffer using HiPrep 26/10 desalting column.
9. All proteins were concentrated using 3 kDa MWCO Amicon ultracentrifugation filter devices and concentrations were measured using a Pierce 660 protein assay kit.
10. Eventually, all proteins were aliquoted at 500 μM concentration (100 μL each), lyophilized, and stored at −80°C.

3.2 Synthesis of Gly-ACA

1. Using an established synthesis method [15], benzyl 2-(7-(2-((*tert*-butoxycarbonyl)amino)acetamido)-2-oxo-2H-chromen-4-yl)acetate (1) was synthesized first.
2. 2-(7-(2-((*tert*-butoxycarbonyl)amino)acetamido)-2-oxo-2H-chromen-4-yl)acetic acid (2): To a stirred solution of 1 (2.0 g, 3.43 mmol) in THF: Methanol (1:1, 20 mL) was added 10% Pd/C (160 mg). The reaction flask was sealed, and the air inside was removed under the Schlenk line. A hydrogen balloon was then attached to the reaction flask and the reaction mixture was stirred at room temperature for 2 h. After the completion of deprotection, the reaction mixture was filtered with a pad of Celite. The filtrate was concentrated in vacuo and the residue was used for the next step without further purification.
3. 2-(7-(2-aminoacetamido)-2-oxo-2H-chromen-4-yl)acetic acid or Gly-ACA (3): To a stirred solution of 2 (1.50 g, 3.98 mmol) in anhydrous DCM (10 mL) at 0°C was added TFA (3 mL). The reaction mixture was stirred at room temperature for 3 h. After completion of the reaction, remove the solvent in vacuo and wash with hexane to afford Gly-ACA (Fig. 3) as a white solid.

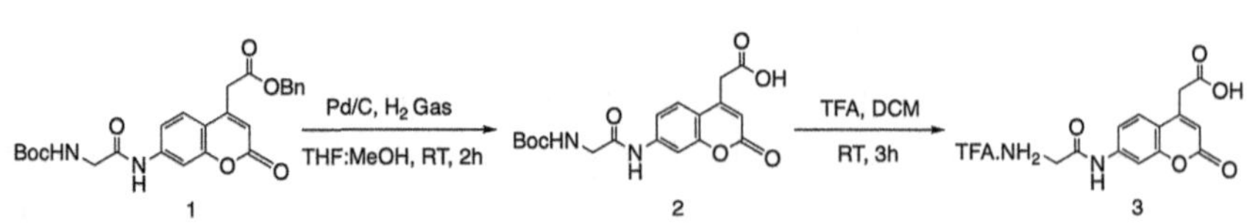

Fig. 3 The synthesis of Gly-ACA

3.3 Ub-ACA Synthesis Using ACPL

1. Dissolve 68.75 mg of Gly-ACA (500 mM) in 250 μL of 1× PBS buffer, pH 7.4 (*see* **Note 1**). Dropwise add 6 M NaOH to adjust the pH to 9.5 (*see* **Note 2**) and mix very well.
2. At this stage, add 100 μL of 500 μM of Ub-G76C-6×His, 5 μL of NTCB (from 500 mM stock), and 1 μL of TCEP (from 500 mM stock) sequentially and mix very well (*see* **Note 3**). Make up the volume to 500 μL using PBS buffer (final concentrations: 500 μM Ub-G76C-6×His, 500 mM Gly-ACA, 5 mM NTCB, 1 mM TCEP). Incubate the reaction mixture at 37°C overnight.
3. Equilibrate a hi-trap desalting column in 50 mM ammonium bicarbonate buffer. Inject the incubated reaction mixture into the column and buffer exchange using FPLC (*see* **Note 4**). Collect and combine the fractions containing the purified, ligated product.
4. Equilibrate a 2 mL nickel NTA agarose resin slurry in 20 mL of 50 mM ammonium bicarbonate buffer. Add 100 μL of the resin slurry into the combined fractions and incubate at 4°C for 30 min to capture any unreacted Ub-G76C-6×His. Store the remaining resin at 4°C and reuse later.
5. Centrifuge the sample at 16,000 × *g* for 5 min to precipitate the resin and collect the supernatant containing Ub-ACA (*see* **Note 5**), filter through micro syringe filters.
6. Concentrate using 3 KDa MWCO Amicon ultracentrifugation filter (centrifuge at 3000 × *g* for 20 min), collect the top fraction containing the probe.
7. Measure the concentration by pierce 660 (absorbance at 660 nm) assay reagents (follow manufacturer's protocol) and store aliquots at 50 μM concentration at −80°C.

3.4 Applications of Ub-ACA

Ub-ACA is a versatile fluorogenic probe designed for activity assays of deubiquitinases (DUBs). Upon cleavage of the ACA moiety, it emits a fluorescent signal (Ex: 380 nm, Em: 460 nm), enabling sensitive, real-time monitoring of enzymatic activity.

In cell or tissue lysates, Ub-ACA enables detection of endogenous DUB activity with minimal background. It can reveal tissue-specific activity patterns and is useful for evaluating pharmacological inhibition or genetic perturbation. Ub-ACA is well-suited for high-throughput screening (HTS) platforms due to its homogeneous assay format, rapid kinetics, and broad linear range, supporting inhibitor discovery and screening campaigns.

1. In a 96-well plate, add 1 μL (from a 5 μM stock prepared in PBS, pH 7.4) of three cysteine-based DUBs such as UCHL1, UCHL5 and USP25 to a final concentration of 50 nM in 100 μL

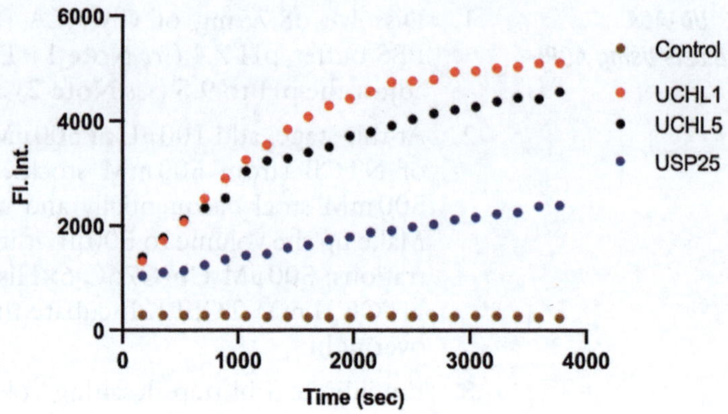

Fig. 4 DUB-catalyzed ACA release from Ub-ACA Indicated by fluorescence increase

of ACA assay buffer (*see* **Note 6**). Incubate the mixture at 30°C.

2. Add 800 nL of Ub-ACA (from a 50 μM stock, final concentration 400 nM) and take fluorescence measurements (Ex: 380 nm, Em: 460 nm) for 1.5 h at every 2 min interval (Fig. 4).

4 Notes

1. Set a designated volume for the reaction (e.g., 500 μL). Do not add excess buffer.
2. Adjust the pH slowly. Avoid adding a higher volume of 6 M NaOH, otherwise, sudden increase in pH may hamper the chemical ligation process.
3. After adding TCEP and NTCB, it is possible that the pH may decrease. Adjust the pH back if that happens.
4. It is recommended to do the buffer exchange process at lower temperature such as 4°C. This prevents the protein/probe from degradation and maintains its stability.
5. While collecting the supernatant, do not disturb the nickel resin; use a needle syringe to aspirate out the supernatant.
6. Always use BSA in fluorescence assay buffer. By preventing non-specific interactions, BSA helps to lower background fluorescence and improve signal-to-noise ratios in real-time fluorescence assays.
7. Generate a curve by plotting the fluorescence intensity at Y-axis and run time at X-axis.

Acknowledgments

This work was supported in part by the Welch Foundation (Grant A-1715 to W.R.L.), National Institutes of Health (Grant R35GM145351), Cancer Prevention and Research Institute of Texas (Grant RP230345), and Harry E. Bovay, Jr. Foundation.

References

1. Swatek KN, Komander D (2016) Ubiquitin modifications. Cell Res 26(4):399–422. https://doi.org/10.1038/cr.2016.39
2. Trempe JF (2011) Reading the ubiquitin postal code. Curr Opin Struct Biol 21(6):792–801. https://doi.org/10.1016/j.sbi.2011.09.009
3. Komander D, Rape M (2012) The ubiquitin code. Annu Rev Biochem 81:203–229. https://doi.org/10.1146/annurev-biochem-060310-170328
4. Ciechanover A, Schwartz AL (1998) The ubiquitin-proteasome pathway: the complexity and myriad functions of proteins death. Proc Natl Acad Sci U S A 95(6):2727–2730. https://doi.org/10.1073/pnas.95.6.2727
5. Ciechanover A (1998) The ubiquitin-proteasome pathway: on protein death and cell life. EMBO J 17(24):7151–7160. https://doi.org/10.1093/emboj/17.24.7151
6. Schwartz AL, Ciechanover A (1999) The ubiquitin-proteasome pathway and pathogenesis of human diseases. Annu Rev Med 50:57–74. https://doi.org/10.1146/annurev.med.50.1.57
7. Hegde AN, Upadhya SC (2007) The ubiquitin-proteasome pathway in health and disease of the nervous system. Trends Neurosci 30(11):587–595. https://doi.org/10.1016/j.tins.2007.08.005
8. Wang J, Maldonado MA (2006) The ubiquitin-proteasome system and its role in inflammatory and autoimmune diseases. Cell Mol Immunol 3(4):255–261
9. Estavoyer B et al (2022) Mechanisms orchestrating the enzymatic activity and cellular functions of deubiquitinases. J Biol Chem 298(8):102198. https://doi.org/10.1016/j.jbc.2022.102198
10. Nicholson B et al (2008) Characterization of ubiquitin and ubiquitin-like-protein isopeptidase activities. Protein Sci 17(6):1035–1043. https://doi.org/10.1110/ps.083450408
11. Yu G, Qiao Y, Blankenship LR, Liu WR (2022) Protein synthesis via activated cysteine-directed protein ligation. Methods Mol Biol 2530:159–167. https://doi.org/10.1007/978-1-0716-2489-0_11
12. Qiao Y et al (2020) Expressed protein ligation without intein. J Am Chem Soc 142(15):7047–7054. https://doi.org/10.1021/jacs.0c00252
13. Chanda S et al (2025) Ubiquitin azapeptide esters as next-generation activity-based probes for cysteine enzymes in the ubiquitin signal pathway. J Am Chem Soc 147:17817–17828. https://doi.org/10.1021/jacs.5c01732
14. Muir TW, Sondhi D, Cole PA (1998) Expressed protein ligation: a general method for protein engineering. Proc Natl Acad Sci U S A 95(12):6705–6710. https://doi.org/10.1073/pnas.95.12.6705
15. Heiner S, Heiner D, Kuhn A, Kunz H (2006) Hydrophilic photolabelling of glycopeptides from murine liver-intestine (LI) cadherin recognition domain. Bioorg Med Chem 14(18):6149–6164

Check for updates

Chapter 5

Native Chemical Ligation to Install Chemical Protein Modifications

Eldon R. Hard, Afraah Javed and Matthew R. Pratt

Abstract

Native chemical ligation is utilized for the generation of homogenous samples of full-length proteins with site-specific chemical modifications. It is commonly used for the site-specific installation of post-translational modifications, in particular those that to-date cannot be readily incorporated by other means. Here, we describe the assembly of recombinantly expressed and/or chemically synthesized protein fragments via this chemoselective bioconjugation reaction. Derivative ligation methods are also discussed to enable the greatest amount of flexibility for the generation of a desired protein.

Keywords Native chemical ligation (NCL), Expressed protein ligation (EPL), Protein semisynthesis, Post-translational modifications (PTMs), Expressed protein selenoester ligation (EPSL), Diselenide-selenoester ligation (DSL), Maleimide protection/deprotection

1 Introduction

Native chemical ligation (NCL) is a chemoselective technique that involves the coupling of a C-terminal thioester to an N-terminal cysteine or thiol-containing residue under mild aqueous conditions with reportedly decent yields. This reaction generates a thioester-linked intermediate that rearranges spontaneously to form a peptide containing a native peptide bond to a cysteine residue through an S-N acyl shift (Fig. 1). This method was first introduced in 1994 by Kent and colleagues [1–3]. NCL has a plethora of applications in the field of chemical biology, as it is used to synthesize full-length native proteins with site-specifically modified residues, as well as in chemoselective labeling [4], or liposomal drug delivery studies [5]. Proteins that do not contain a native cysteine residue, which is necessary for the ligation, can take advantage of the metal-based or nonmetal-based desulfurization or deselenization methods to convert cysteine or selenocysteine to alanine residues, respectively [6–8]. Other thiol-containing noncanonical amino acids (such as

Alexander Deiters (ed.), *Bioconjugation Methods*, Methods in Molecular Biology, vol. 3009,
https://doi.org/10.1007/978-1-0716-5112-4_5,

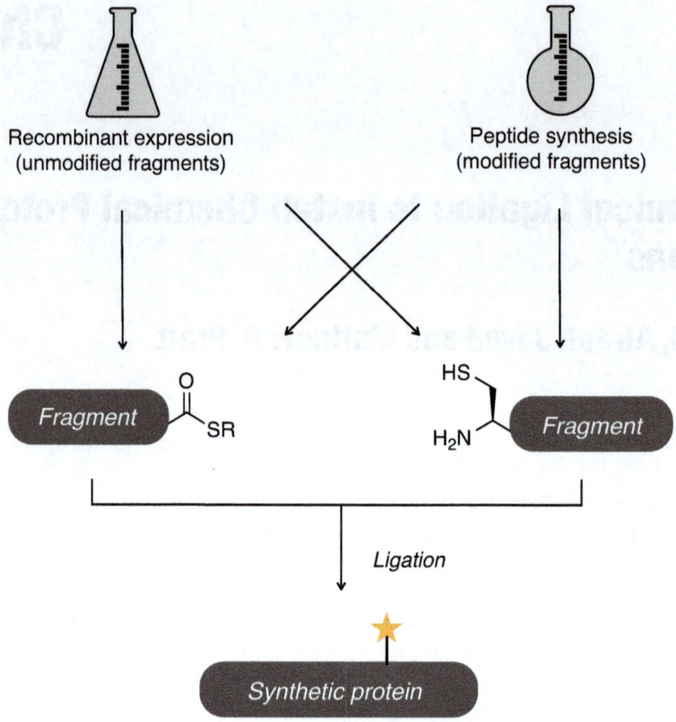

Fig. 1 Native Chemical Ligation for protein synthesis. (Left) Recombinant and synthetic protein fragments can be selectively joined through the reaction of a C-terminal thioester and an N-terminal cysteine, resulting in a native amide bond. Modifications can be incorporated into the synthetic fragment(s) using solid-phase peptide synthesis. (Right) Preparation of complete NCL buffer from crude ligation buffer and 10× stock buffers

penicillamine and mercaptoproline) can also be used to increase synthetic access to NCL.

Expressed Protein Ligation (EPL), an extension of NCL, is generally employed to generate large site-specifically modified proteins, as it allows for the use of recombinant fragments that overcome the length limitations of peptide synthesis. Traditional NCL can already combine recombinant protein-fragments containing an N-terminal cysteine with chemically synthesized peptides with a C-terminal thioester. EPL takes advantage of the intein-extein protein splicing mechanisms of *E. coli* cells to selectively express protein fragments with C-terminal thioesters, extending the use of synthetic and recombinant segments at either the N- or C-terminus, or even the middle of protein targets [9–12]. Using this technique, we can produce stoichiometric amounts of native proteins with site-specifically modified residues, such as the addition of post-translational modification (PTMs) such as O-GlcNAcylation or phosphorylation. Both NCL and EPL require the presence of a thioester for the ligation reaction to take place.

When designing ligation sites, careful consideration should be given to the C-terminal residue and following N-terminal residue at the ligation junction. Ligation junction residues that are sterically hindered in nature, hydrophobic, or bulky, may need longer reaction times and will be generally more difficult to ligate [13, 14]. C-terminal residues such as valine, leucine, isoleucine, proline, and threonine are known to take additional time to ligate [14–16]. Asparagine and glutamine should also be avoided as a C-terminal thioester as they can undergo cyclization to form the corresponding imide. Glutamate and aspartate should not be used as they similarly cyclize to form C-terminal anhydrides that generate a mixture of isopeptide and standard bonds [17]. Additionally, large, post-translationally-modified residues should also not be positioned immediately at the ligation site unless absolutely necessary.

2 Materials

Prepare all solutions using ultrapure (UP) water (18 MΩ) H_2O at 25°C and at least proteomics grade reagents. Prepare and store all reagents at the manufacturer's recommended temperature unless otherwise specified. It is recommended to sparge/degas each buffer before use with N_2 gas to reduce the risk of oxidation of reagents, cysteines, or methionines. Follow all waste disposal regulations and procedures when disposing of chemical wastes.

2.1 Native Chemical Ligation/Expressed Protein Ligation

1. Crude Ligation Buffer: 6 M guanidine hydrochloride (GnHCl), 200 mM sodium phosphate (phosphate) (*see* **Note 1**), pH >7.2. Add about 2 mL of water to a 15 mL conical tube. Weigh out 2.86 g of GnHCl and add to the tube. Measure out 141.99 mg of sodium phosphate dibasic (or 139.99 mg of sodium phosphate monobasic) and vortex until dissolved. Adjust pH >7.2 with sodium hydroxide. Make up to 5 mL with water. Store for up to 48 h at 25°C.
2. 10× Stock TCEP Buffer: 6 M GnHCl, 200 mM phosphate, 250 mM Tris (2-carboxyethyl) phosphine hydrochloride (TCEP HCl), pH 7.0. Aliquot 1 mL of crude ligation buffer into a microcentrifuge tube and add 71.66 mg of TCEP-HCl, vortexing until completely dissolved. Use sodium hydroxide to pH to 7.0. Store at −20°C for up to 12 months.
3. 10× Stock MPAA Buffer: 6 M GnHCl, 200 mM sodium phosphate, 250 mM 4-mercaptophenyl acetic acid (MPAA), pH 7.0. Aliquot 1 mL of crude ligation buffer into a microcentrifuge tube and add 42.05 mg of MPAA (*see* **Note 2**), vortexing until completely dissolved. Use sodium hydroxide to adjust pH to 7.0. Store at −20°C for up to 12 months.
4. Complete Ligation Buffer: 6 M GnHCl, 200 mM phosphate, 25 mM TCEP-HCl, 25 mM MPAA, pH 7.2. Aliquot 800 μL of crude ligation buffer into a microcentrifuge tube. Add 100 μL of 10× Stock TCEP Buffer and 100 μL of 10× MPAA stock

solution to the microcentrifuge tube. Use sodium hydroxide or hydrochloric acid to adjust the pH to 7.2. Prepare this buffer fresh before every use.

5. Lyophilized thioester containing N-terminal protein fragment.
6. Lyophilized cysteine/thiol-containing C-terminal protein fragment.
7. Thermomixer or other agitation instrument.
8. Analytical C4 RP-HPLC column.

2.2 Acetylacetone (acac) Activation and Ligation

1. Activation Buffer: 6 M GnHCl, 200 mM MPAA, pH 3.0. Add 2 mL of water to a 15 mL conical tube. Weigh out 2.86 g of GnHCl and dissolve in the water. Fill up to 5 mL with water. pH the buffer to 3.0 using saturated sodium hydroxide dropwise. Take 500 μL of solution and add 17 mg of MPAA and vortex/bath sonicate until no apparent solids remain. Prepare this buffer fresh before use.
2. 1:20 acac dilution: In a separate microcentrifuge tube, add 19 μL of 6 M activation buffer, and 1 μL of neat acetylacetone. The molarity of neat acetylacetone is 9.737 M, and the molarity of the 1:20 dilution is 0.487 M. Prepare this solution immediately before use.
3. 2× Ligation Buffer: 6 M GnHCl, 400 mM Sodium Phosphate, 50 mM TCEP HCl, pH 7.5. Add 2 mL of water to a 15 mL conical tube. Weigh out 2.86 g of GnHCl and dissolve in the water. Weigh out 71.66 mg of TCEP HCl and 283.98 mg of sodium phosphate dibasic. Fill up to 5 mL with water. pH the buffer to 7.5 using saturated sodium hydroxide dropwise. Prepare this buffer fresh before use.
4. Lyophilized hydrazide containing protein fragment.
5. Lyophilized cysteine/thiol-containing protein fragment.
6. Thermomixer or other agitation instrument.

2.3 Expressed Protein Selenoester Ligation

1. Selenoesterification Reaction Buffer: 6 M GnHCl, 200 mM HEPES, 200 mM TCEP, 50 mM diphenyldiselenide (DPDS), pH 2.0 and add 2 mL of water to a 15 mL conical tube. Weigh out 2.86 g of GnHCl and dissolve in the water. Weigh out 238.3 mg of HEPES and 286.65 mg of TCEP HCl and dissolve in the solution via bath sonication. Add 54.59 mg of DPDS and fully dissolve. Buffer should be made fresh before use.
2. 1:20 acac Dilution: In a separate microcentrifuge tube, add 19 μL of 6 M selenoesterification reaction buffer, and 1 μL of neat acetylacetone. The molarity of neat acetylacetone is 9.737 M, and the molarity of the 1:20 dilution is 0.487 M. Prepare this solution immediately before use.

3. Diselenide-Selenoester Ligation Buffer: 6 M GnHCl, 200 mM HEPES, 200 mM TCEP HCl, pH = 5.0. Buffer is stable at room temperature for 48 h. Add 2 mL of water to a 15 mL conical tube. Weigh out 2.86 g of GnHCl and dissolve in the water. Weigh out 238.3 mg of HEPES and 286.65 mg of TCEP HCl and dissolve in the solution via bath sonication. Add saturated sodium hydroxide in 10 μL to pH = 5.0. Buffer is stable at 4°C for 48 h.
4. Deselenization Buffer: 6 M GnHCl, 200 mM HEPES, 60 mM TCEP HCl, 60 mM glutathione (GSH). Add 2 mL of water to a 15 mL conical tube. Weigh out 2.86 g of GnHCl and dissolve in the water. Weigh out 238.3 mg of HEPES and dissolve. Measure out 86.00 mg of TCEP HCl and fully dissolve. Degas buffer with N_2 gas, then add 92.20 mg of reduced GSH. Buffer should be made fresh immediately before use.
5. Lyophilized hydrazide containing protein fragment.
6. Lyophilized diselenide containing protein fragment.
7. Thermomixer or other agitation instrument.

2.4 N-methylmaleimide Protection/Deprotection

1. Maleimide Protection Buffer: 6 M GnHCl, 200 mM sodium phosphate, 2 mM TCEP (from 10× stock), pH 7.2. Add about 2 mL of water to a 15 mL conical tube. Weigh out 2.86 g of GnHCl and dissolve in the water. Weigh out 141.99 mg of sodium phosphate dibasic, add to the solution and vortex/bath sonicate until dissolved. Add 8 μL of 10× TCEP stock solution and vortex. Adjust pH with sodium hydroxide to 7.2. This buffer is stable for roughly 48 h at 4°C.
2. Maleimide Deprotection Buffer: 6 M GnHCl, 200 mM sodium phosphate, pH 7.2. Add about 2 mL of water to a 15 mL conical tube. Weigh out 2.86 g of GnHCl and add to the tube. Measure out 141.99 mg of sodium phosphate dibasic and vortex until dissolved. Adjust pH >7.2 with sodium hydroxide. Make up to 5 mL with water. This buffer is stable for roughly 48 h at 25°C.
3. Stock $PdCl_2$ Solution: To a microcentrifuge tube, add 594 μL of maleimide deprotection buffer. Add 10.5 mg of $PdCl_2$ and agitate at 37°C for 10 min to fully dissolve. Prepare this buffer immediately before use.
4. Stock $MgCl_2$ Solution: To a microcentrifuge tube, add 300 μL of maleimide deprotection buffer. Dissolve 30 mg of $MgCl_2$ and vortex. Prepare this solution immediately before use.
5. Quenching Stock Solution: 1.6 M dithiothreitol (DTT), 6 M GnHCl, 200 mM sodium phosphate. To a microcentrifuge tube, add 25 mg of DTT and dissolve it in 100 μL of maleimide deprotection buffer. Prepare this solution immediately before use.
6. Thermomixer or other agitation instrument.

2.5 HPLC

1. HPLC Buffer A: 0.1% TFA in H_2O. Combine 999 mL ultrapure water with 1 mL trifluoroacetic acid. Store at room temperature.
2. HPLC Buffer B: 0.1% TFA and 90% CH_3CN in H_2O. Combine 98 mL ultrapure water with 900 mL acetonitrile and 1 mL trifluoroacetic acid.
3. C4 analytical column: Jupiter 5 um C4 300 A, LC column 150 × 4× 4.6 mm from Phenomenex or equivalent column
4. Agilent 1200+ series or equivalent HPLC system

3 Methods

3.1 Native Chemical Ligation

Carry out all procedures at room temperature unless otherwise specified. All pH measurements were made using pH-indicator paper pH 1.0–5.0, or 6.4–8.0. If using a protein hydrazide, skip to section 3.2. If using a protein selenoester or protecting a native cysteine for desulfurization, skip to Sects. 3.3 or 3.4, respectively.

1. Prepare 5 mL of crude ligation buffer.
2. Create your 10× TCEP stock buffer and 10× MPAA stock buffer.
3. Transfer 800 μL of crude ligation buffer into a new microcentrifuge tube. Add 100 μL of the 10× TCEP stock buffer to a final concentration of 25 mM TCEP. Add 100 μL of the 10× MPAA stock buffer to a final concentration of 25 mM MPAA.
4. Dissolve the **cysteine/thiol-containing** protein fragment in the desired volume of complete ligation buffer for a final concentration of 5 mM of the post-translationally modified fragment (*see* **Note 3**). This is the last chance to check the pH and adjust it before adding thioester. Ensure that the pH is between 7.0 and 7.2 to avoid hydrolysis of the thioester (*see* **Note 4**).
5. Transfer your cysteine/thiol-containing protein solution via wetting the pipette (*see* **Note 5**) into the thioester containing fragment. Mix well and ensure quantitative transfer by spinning down both tubes and retransfer any solution leftover in the original cysteine/thiol-containing fragment tube.
6. Adjust reaction pH to 7.2 a final time, if necessary (*see* **Note 6**).
7. To monitor reaction via RP-HPLC, transfer 4 μL of the reaction to a fresh microcentrifuge tube, and add 16 μL TCEP stock solution. Let sit for 5 min. This sample is a snapshot of the "time zero" (T0) of the reaction. Inject all 20 μL onto an equilibrated C4 analytical column and run at 1 mL/min over

60 min with a gradient of 0%–70% HPLC Buffer B. Save the HPLC trace for future reference.

8. Parafilm to secure the lid of the microcentrifuge reaction tube.
9. Set to rotate/mix at room temperature (25°C) overnight (*see* **Notes 7, 8, and 9**).
10. The next day, monitor reaction progress via RP-HPLC: transfer 4 μL of the reaction mixture to a fresh microcentrifuge tube, and add 16 μL of 10× TCEP stock buffer. Let sit for 5 min. Inject all 20 μL onto equilibrated C4 analytical column and run at 1 mL/min over 60 min with a gradient of 0%–70% HPLC Buffer B.
11. During the HPLC run, manually collect samples from each peak (at 210 nm) for mass spectrometry analysis and identification. In comparison to the T0 trace, expect that new peaks appear, and that peaks present in the T0 trace diminish or disappear completely (Fig. 2A) (*see* **Note 10**).
12. Once the desired product appears, it can be isolated using semipreparative or preparative RP-HPLC and characterized via mass spectrometry (Fig. 2B), SDS-PAGE, and analytical RP-HPLC (Fig. 2C) for correct molecular weight and at least 95% purity by area under the curve of all peaks on the 210 nm signal on the HPLC chromatogram.

3.2 Acetylacetone Activation and Ligation (see Note 11) (Fig. 3)

1. Prepare 5 mL of 6 M GnHCl.
2. To a 1 mL aliquot of GnHCl, add 33.6 mg (200 mM) MPAA, and bath sonicate until no apparent solids remain (10+ min).
3. Dissolve your hydrazide fragment in the activation buffer to a final concentration of 9 mM and pH to 3.0–3.5.
4. To initiate the reaction, add 2.5 eq of 1:20 acac dilution for every 1 eq of hydrazide fragment.
5. Take an activation T0 h sample (2 μL [about 20–30 μg of protein] in 18 μL TCEP stock solution) to monitor reaction progress and stir/agitate for 3 h at room temperature. The gradient should be a 0%–70% buffer B over 60 min.
6. Parafilm to secure the lid of the microcentrifuge reaction tube.
7. Take an activation T3 h sample (2 μL of sample in 18 μL TCEP stock solution) to monitor thioester formation. The starting material peak should be diminished, and new peaks should appear.
8. To the remaining 4 mL of 6 M GnHCl solution, add 227.136 mg of dibasic phosphate (400 mM) and 57.33 mg (50 mM) TCEP and pH to 7.2 to make the 2× ligation buffer.
9. Dissolve your cysteine fragment in the fresh 2× ligation buffer and combine it with your newly formed thioester fragment via

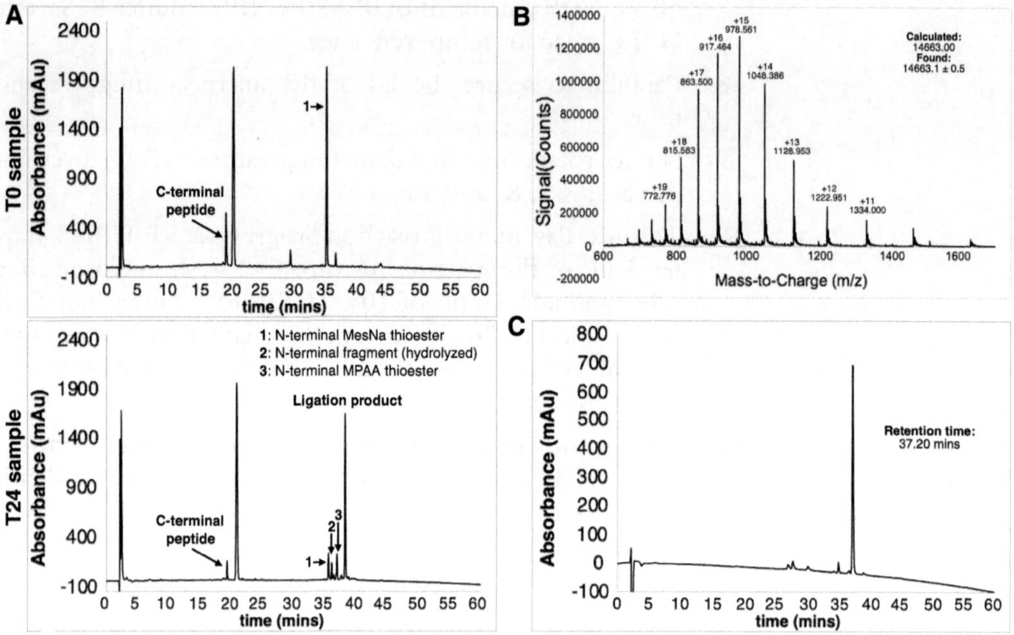

Fig. 2 Example HPLC traces monitoring ligation reaction, with labeled protein fragment peaks. (**A**) Starting components of ligation reaction include C-terminal peptide with N-terminal cysteine and N-terminal MesNa thioester fragment. After 24 h, components of ligation reaction include less C-terminal peptide and N-terminal MesNa thioester fragment, along with new peaks of byproducts and the desired ligation product. (**B**) Mass spectra of final product after RP-HPLC purification. (**C**) RP-HPLC analytical of final product after purification

wetting the pipette. The final concentration of your PTM protein fragment should be 4–5 mM.

10. Adjust pH of the solution to 7–7.2 and take ligation T0 h sample (4 μL of sample in 16 μL TCEP stock solution) and parafilm to secure the lid of the microcentrifuge reaction tube.
11. Stir/agitate the solution overnight. There should be the presence of a new peak corresponding to the product fragment in the HPLC chromatogram.
12. After incubation overnight, take ligation T16 h sample (4 μL of sample in 16 μL TCEP stock solution) for HPLC analysis. The peaks corresponding to the cysteine fragment and hydrazide fragment should decrease while new product peak formation should be present. The reaction time can be extended another 24 h or heated to 37°C if little-to-no reaction progress occurs.
13. The product should be isolated via RP-HPLC and characterized via mass spectrometry.

3.3 Expressed Protein Selenoester Ligation (see Note 12)

1. Prepare 5 mL of selenoesterification reaction buffer. Bath sonicate until DPDS fully dissolves and no visible solid remains. Sparge buffer for at least 15 min.

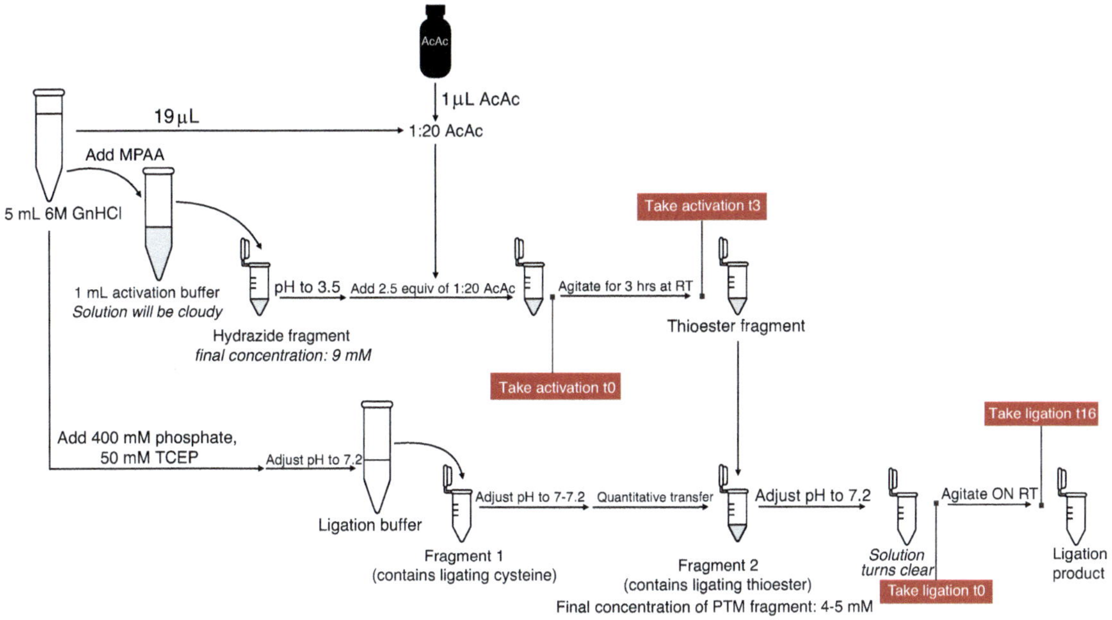

Fig. 3 Schematic of AcAc activation and Ligation. Stepwise schematic of each step of the acetylacetone activation and ligation reaction. Steps where timepoints should be taken have been highlighted in red. Cloudy solutions are highlighted in grey, while clear solutions are clear

2. Dissolve protein hydrazide in selenoesterification reaction buffer to a final concentration between 250 μM and 1 mM and pH to 2.0.
3. Initiate the reaction by adding 5 eq of 1:20 acac dilution with respect to your protein hydrazide fragment. Take a T0 h time point (8 μL [20–30 ng of protein] of reaction, 16 μL DTT stock solution) for RP-HPLC analysis.
4. Parafilm to secure the lid of the microcentrifuge reaction tube.
5. Agitate the reaction at room temperature for 3 h, after which take a T3 time point (4 μL of reaction, 16 μL DTT Stock solution) for RP-HPLC analysis.
6. Confirm the presence of the selenoesterified product via mass spectrometry and the presence of a new peak on the HPLC chromatogram. If the product is present and in good conversion, increase the pH to 5.0.
7. Dissolve the selenopeptide in the reaction, to a final ratio of 2:1 of selenoester to selenopeptide containing the PTM.
8. Take a T0 time point (4 μL of reaction, 16 μL DTT Stock solution) for RP-HPLC analysis. The selenopeptide peak should appear on the chromatogram.
9. Parafilm to secure the lid of the microcentrifuge reaction tube.
10. Agitate the reaction at room temperature for 1 h.

11. Take T1 time point (4 μL of reaction, 16 μL DTT Stock solution) for RP-HPLC analysis. If the reaction appears completed, the product can be isolated. If the reaction is not completed, it can be left to incubate longer at room temperature or heating at 37°C.
12. To isolate the product, either purify via RP-HPLC or extract the DPDS using 10× volume diethyl ether.
13. If the product was RP-HPLC purified and lyophilized, in a separate 15 mL conical tube, add 60 mM TCEP HCl and 60 mM reduced GSH to the deselenization buffer, and sparge the deselenization buffer with N_2. After sparging, pH the buffer to 6.5–7.0.
14. If the product was not isolated, after extractions add 60 mM TCEP HCl and 60 mM reduced GSH and pH to 6.5–7.0 using saturated sodium hydroxide 4 μL at a time. The final concentration of protein should be around 250 μM.
15. Dissolve the product to a final concentration of 250 μM, and parafilm to secure the lid of the microcentrifuge reaction tube.
16. Agitate the reaction for 16 h at 25°C. Take a T0 time point for RP-HPLC analysis (4 μL reaction in 16 μL 10× TCEP stock).
17. If multiple peaks appear after 16 h, the product may be degrading due to leftover ether in the aqueous phase. If this happens, purify the full reaction via HPLC and lyophilize.
18. After 16 h, analyze the reaction via RP-HPLC and mass spectrometry for the presence of the product.
19. The product should be isolated via RP-HPLC and characterized via mass spectrometry.

3.4 Maleimide Protection/Deprotection (see Note 13) (Fig. 4)

3.4.1 Maleimide Protection

1. Prepare 5 mL of maleimide protection buffer.
2. Dissolve your protein to a final concentration of 2 mM.
3. If using a protein thioester, the reaction should be done at −20°C for 30 min to avoid hydrolysis. If using a protein hydrazide, the reaction can proceed at room temperature for 30 min.
4. Dissolve 3 eq of maleimide to initiate the reaction, and pH to 7.2. Take a T0 time point (4 μL of reaction, 16 μL of 10× TCEP stock solution).

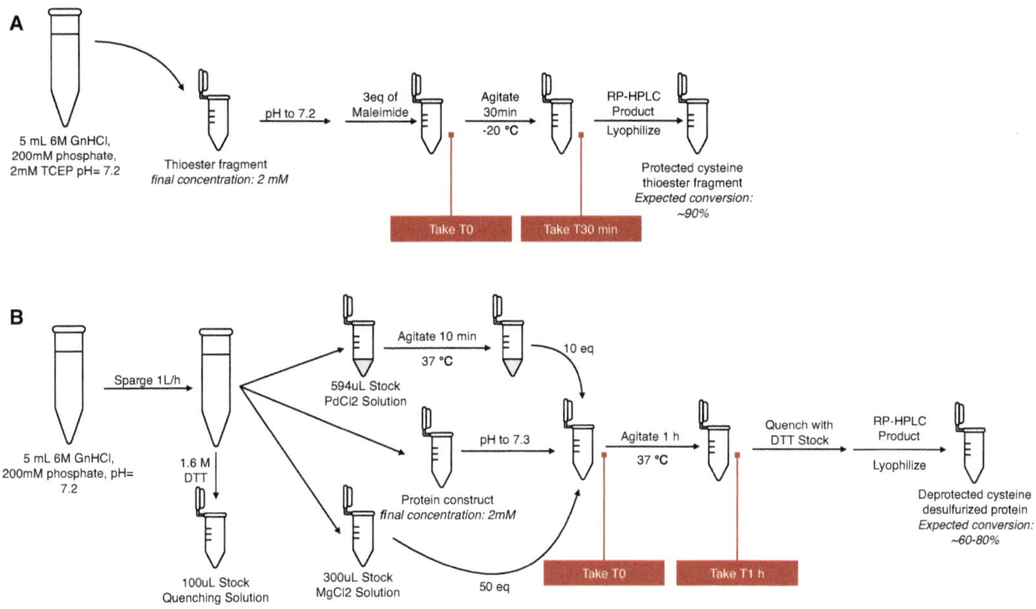

Fig. 4 Schematic of Maleimide Protection and Deprotection. (**A**) Stepwise schematic of each step of the maleimide protection. (**B**) Stepwise schematic of each step of the maleimide deprotection. Steps where time-points should be taken have been highlighted in red. Cloudy solutions are in gray, while colorless solutions are clear

5. Parafilm to secure the lid of the microcentrifuge reaction tube and agitate the reaction for 30 min.
6. After 30 min, take a T30 min sample to check the progress of the reaction. If the protection is complete, the protein fragment can be RP-HPLC purified, or buffer exchanged thrice into crude ligation buffer to remove the excess maleimide.

3.4.2 Maleimide Deprotection

1. Prepare 5 mL of maleimide deprotection buffer.
2. Degas/Sparge crude ligation buffer with N_2 to remove oxygen at a rate of 1 h per 1 L.
3. Prepare $PdCl_2$ and $MgCl_2$ stock solutions in crude ligation buffer.
4. Dissolve your protein in the desired amount of crude ligation buffer and pH to 7.3.
5. Add 10 eq of $PdCl_2$ (74.2 μL of stock) and 50 eq of $MgCl_2$ (35.2 μL of stock) to the degassed protein solution to initiate the reaction.
6. Take a T0 timepoint and use a 0%–70% Buffer B gradient over 1 h. Parafilm to secure the lid of the microcentrifuge reaction tube.
7. Allow the reaction to run for 1 h at 37°C with agitation.

8. After 1 h, run a time point to observe a shift in the HPLC chromatogram and mass spectrum. While the gradient is running, keep the reaction at room temperature to stop the reaction.
9. If the reaction is finished/has achieved over 75%–80% completion, quench with 10 μL of DTT stock solution and vortex. If the reaction has not achieved this level of completion, incubate again for another 30 min, and run another time point. Letting the reaction proceed longer than 2 h may result in oxidation and/or degradation of your protein.
10. Product should be isolated the same day via RP-HPLC and characterized by mass spectrometry and analytical RP-HPLC.

4 Notes

1. Sodium Phosphate Monobasic (NaH_2PO_4) with Guanidine HCl will create an acidic solution around pH 2. Sodium Phosphate Dibasic (Na_2HPO_4) with Guanidine HCl will create a basic solution around pH 7.8–8. Dibasic phosphate also dissolves much more slowly and will need to bath sonicate/vortex until all particles are dissolved.
2. MPAA will not dissolve in crude ligation buffer under acidic conditions. As the pH increases to 7.0, it will slowly dissolve.
3. The posttranslationally modified (PTM) fragment should be the limiting reagent. Ideally, the ratio of the PTM fragment to the other protein fragment should be 1:2, respectively.
4. Protein fragments purified via HPLC contain TFA/FA salts and will lower the pH of the buffer when dissolved. The pH is recommended to be checked after dissolving the first protein fragment.
5. Since many proteins adhere to plastic, protein Lo-bind tubes are recommended as well as "wetting the pipette" tip before transferring protein solutions to aid in quantitative transfer. This technique involves slowly pipetting up and down thrice with the desired volume of solution before transferring the solution to a new vessel.
6. pH 7.2 is the ideal pH for NCL, however lower pHs can be used depending on the protein fragments. For fragments prone to hydrolysis, a pH as low as 6.5 can be used, however, the reaction speed is also slowed [17].
7. Depending on the protein fragments of interest, the ligation may be done overnight at 37°C [17].

8. We've found that bulky residues may provide difficulties in NCL, such as penicillamine, an amino acid similar to valine with an additional thiol on the beta carbon. The reaction may be allowed to go for up to 48 h, while taking timepoints every 8 h to ensure reaction progression. Coupling directly to the post-translationally modified residue is not recommended, as it may be too sterically hindered to react with the corresponding reactive group (thioester or cysteine).
9. Protein thioesters are often prone to hydrolysis, which may be observed experimentally during ligation reactions. Careful adjustment of the pH to 6.8–7.0 during thiolysis and purification of the protein thioester will maximize the amount of productive thioester going into a ligation reaction. Careful adjustment of the pH to 7.0–7.2 when setting up the ligation reaction can reduce the amount of unproductive hydrolyzed byproduct.
10. **Non-Cysteine Ligations** Although most proteins are fairly tolerable of cysteine mutations, there will be cases in which a cysteine residue is not ligation accessible. In which case, there are thiol-containing amino acids can be used for the ligations followed by desulfurization [18, 19]. Thiol-containing amino acids can be found for Val [16], Asn [20], Asp [21, 22], Arg [23], Gln [24], Glu [25], Ile [26], Leu [27], Lys [28], Phe [29, 30], Pro [31, 32], Thr [33]. In an example case where the desired ligation site involves an N-terminal proline, a γ -thioproline can be used. (2S,4R)-Fmoc-Mpt(Trt)-OH is incorporated into peptides using traditional solid phase peptide synthesis following Pfp ester activation. Only this diastereomer successfully yields a ligation product, as it adopts an S→ N acyl transfer intermediate position that does not face steric hindrance. We recommend Pfp-activating your N-terminal unnatural amino acid if it is sterically hindered, or if you are ligating to a sterically hindered amino acid. After the peptide is purified, the ligation reaction is set up in the same manner as traditional native chemical ligation. Nonmetal based desulfurization readily converts the ligation product to the desired protein with the native amino acid.
11. **Ligations using Hydrazides**. Masked peptide thioesters can be generated as C-terminal hydrazides, allowing them to be inert until their conversion to the corresponding peptide or protein thioester [18]. In the azide oxidation activation method, the peptide hydrazide is first oxidized by sodium nitrite ($NaNO_2$) to form a peptide azide under low pH (3.0–4.0) aqueous conditions. Subsequent addition of a thiol compound, namely 4-mercaptophenylacetic acid (MPAA) [34, 35] at pH = 7.0, will then covert this intermediate into the

corresponding thioester. This system affords an orthogonal mechanism to peptidyl thioesters as the peptide hydrazide is inert, allowing for N-to-C terminal ligations. A newer activation method, discussed below, involves taking advantage of the Knorr pyrazole synthesis [36], which we term "acetylacetone activation and ligation." This chemoselective synthesis results in access to peptide acyl pyrazoles which are weakly active and can be exchanged with exogenous thioesters when under the conditions of NCL. Under acidic conditions, hydrazone imines form with high chemoselectivity and rapid kinetics, and in the presence of stoichiometric amounts of acetyl acetone, thermodynamics drive the conversion of peptide hydrazides to acyl pyrazoles without disruption of sensitive functional group [36]. This method can be effectively used during Fmoc-SPPS, and in a sequential one-pot reaction. We find these conditions to be generally robust, repeatable, and more mild conditions, therefore we recommend it for most C-terminal hydrazide applications. For a schematic of the reaction, see Fig. 3.

12. **Expressed Protein Selenoester Ligation**. In cases where the protein of interest has a natural cysteine(s) necessary for proper folding or function but a cysteine at the ligation site is undesired, expressed protein selenoester ligation (EPSL) can be used. This method developed by Kulkarni et al. can be used for one-pot semisynthesis of post-translationally modified proteins [37]. This method takes advantage of the rapid kinetics of ligations between N-terminal selenopeptides generated via Fmoc-SPPS [38] and aryl selenoesters generated from expressed intein fusions. Based on the previous work of diselenide-selenoster ligations (DSL) [38], EPSL also allows for chemoselective deselenization giving a corresponding alanine residue while retaining native cysteine residues. EPSL involves generating the expressed protein-intein fusion, followed by hydrazinolysis of the intein to the corresponding protein hydrazide [39]. Leveraging the acetylacetone activation and ligation chemistry [36], the acyl hydrazide can be converted into a selenoester, and used in DSL with a selenopeptide. After RP-HPLC or DPDS extraction, the resulting protein construct can be incubated with 60 mM TCEP HCl, and 60 mM GSH at room temperature for 16 h to afford the selectively deselenized product.

13. **Maleimide protection/deprotection**. N-methylmaleimide protecting groups can be used to site-specifically protect cysteine side chains on the protein during the desulfurization process [40]. These groups can be added and removed orthogonally to the NCL process. Effectively, this is an

alternative strategy to retain native cysteine residues while transforming the cysteine at the ligation site to alanine. Additionally, charged amino acids can be coupled onto carboxylic acid derivatives of N-methylmaleimide to aid in RP-HPLC separation. Of note, when a charged peptide maleimide is attached to a protein fragment, diastereomers may appear on the HPLC trace as two separate peaks with the same mass. The addition of the protecting group occurs under pH 7.2, with three equivalents of the maleimide per cysteine in a protein/peptide fragment [40]. The reaction is extremely fast and should be finished within 30 min. The product material will shift on the HPLC, usually to the right of the starting material. Removal of the maleimide occurs under a noncatalytic mechanism involving palladium chloride and magnesium chloride [41]. This reaction proceeds at 37°C for 1 h at pH 5.5 with agitation. Prior work done by Vamisetti et al. found that 10 eq of $PdCl_2$ and 50 eq of $MgCl_2$ gives the best reaction results [41]. If the reaction is proceeding slowly an additional 10 eq and 50 eq of $PdCl_2$ and $MgCl_2$, respectively, can be added to the reaction. The final concentration of your protein should be 2 mM. For a schematic of protection and deprotection, see Fig 4.

References

1. Agouridas V, El Mahdi O, Diemer V, Cargoët M, Monbaliu JCM, Melnyk O (2019) Native chemical ligation and extended methods: mechanisms, catalysis, scope, and limitations. Chem Rev 119(12):7328–7443. https://doi.org/10.1021/acs.chemrev.8b00712
2. Dawson PE, Muir TW, Clark-Lewis I, Kent SB (1994) Synthesis of proteins by native chemical ligation. Science 266(5186):776–779. https://doi.org/10.1126/science.7973629
3. Kent SBH (2009) Total chemical synthesis of proteins. Chem Soc Rev 38(2):338–351. https://doi.org/10.1039/b700141j
4. Yeo DSY, Srinivasan R, Chen GYJ, Yao SQ (2004) Expanded utility of the native chemical ligation reaction. Chem Eur J 10:4664–4672
5. Cho CJ et al (2025) Diacylation of peptides enables the construction of functional vesicles for drug-carrying liposomes. Angew Chem Int Ed 64(20):e202421932. https://doi.org/10.1002/anie.202421932
6. Jin K, Li X (2018) Advances in native chemical ligation–desulfurization: a powerful strategy for peptide and protein synthesis. Chem Eur J 24(66):17397–17404. https://doi.org/10.1002/chem.20180206
7. Metanis N, Keinan E, Dawson PE (2010) Traceless ligation of cysteine peptides using selective deselenization. Angew Chem Int Ed 49(39):7049–7053. https://doi.org/10.1002/anie.201001900
8. Wan Q, Danishefsky SJ (2007) Free-radical-based, specific desulfurization of cysteine: a powerful advance in the synthesis of polypeptides and glycopolypeptides. Angew Chem Int Ed 46(48):9248–9252. https://doi.org/10.1002/anie.200704195
9. Vila-Perelló M (ed) (2020) Expressed protein ligation: methods and protocols. Humana Press. https://doi.org/10.1007/978-1-0716-0434-2
10. Muir TW, Sondhi D, Cole PA (1998) Expressed protein ligation: a general method for protein engineering. Proc Natl Acad Sci U S A 95(12):6705–6710. https://doi.org/10.1073/pnas.95.12.6705
11. Muir TW (2003) Semisynthesis of proteins by expressed protein ligation. Annu Rev Biochem 72:249–289. https://doi.org/10.1146/annurev.biochem.72.121801.161900
12. Perler FB, Davis EO, Dean GE, Gimble FS, Jack WE, Neff N, Belfort M (1994) Protein splicing elements: Inteins and exteins – a definition of terms and recommended nomenclature. Nucl Acids Res 22(7):1125–1127. https://doi.org/10.1093/nar/22.7.1125

13. Tan Z, Shang S, Danishefsky SJ (2010) Insights into the finer issues of native chemical ligation: an approach to cascade ligations. Angew Chem Int Ed. 49:9500–9503
14. Hackeng TM, Griffin JH, Dawson PE (1999) Protein synthesis by native chemical ligation: expanded scope by using straightforward methodology. Proc Natl Acad Sci U S A 96:10068–10073
15. Chen J, Wan Q, Yuan Y, Zhu J, Danishefsky SJ (2008) Native chemical ligation at valine: a contribution to peptide and glycopeptide synthesis. Angew Chem Int Ed 47:8521–8524
16. Haase C, Rohde H, Seitz O (2008) Native chemical ligation at valine. Angew Chem Int Ed 47:6807–6810
17. Javed A et al (2024) O-GlcNAc modification of HSP27 alters its protein interactions and promotes refolding of proteins through the BAG3/HSP70 co-chaperone. Protein Sci 33:1–12. https://doi.org/10.1002/pro.5173
18. Pratt M, Muir T (2023) Protein Synthesis and Engineering. In Advanced Chemical Biology. Essay, Wiley-VCH, pp 167–192
19. Mitchell NJ, Sayers J, Kulkarni SS, Chem DC (2017) Accelerated protein synthesis vis one-pot ligation-deselenization chemistry. ChemBioChem 2:703–705
20. Sayers J, Thompson RE, Perry KJ et al (2015) Thiazolidine-protected β-thiol asparagine: applications in one-pot ligation-desulfurization chemistry. Org Lett 17(19):4902–4905
21. Villain M, Gaertner H, Botti P (2003) Native chemical ligation with aspartic and glutamic acids as C-terminal residues: scope and limitations. Eur J Org Chem 2003(17):3267–3272
22. Thompson RE, Chan B, Radom L et al (2013) Chemoselective peptide ligation-desulfurization aspartate. Angew Chem Int Ed 52(37):9723–9727
23. Malins LR, Cergol KM, Payne RJ (2013) Peptide ligation-desulfurization chemistry at arginine. ChemBioChem 14(5):559–563
24. Siman P, Karthikeyan SV, Brik A (2012) Native chemical ligation at glutamine. Org Lett 14(6):1520–1523
25. Cergol KM, Thompson RE, Malins LR et al (2014) One-pot peptide ligation-desulfurization at glutamate. Org Lett 16(1):290–293
26. Pasunooti KK, Yang R, Banerjee B et al (2016) 5-Methylisoxazole3-carboxamide-directed palladium-catalyzed γ-C(sp(3))-H acetoxylation and application to the synthesis of γ-mercapto amino acids for native chemical ligation. Org Lett 18(11):2696–2699
27. Wang X, Sanchez J, Stone MJ, Payne RJ (2017) Sulfation of the human cytomegalovirus protein UL22A enhances binding to the chemokine RANTES. Angew Chem 129(29):8610–8614
28. Merkx R, de Bruin G, Kruithof A et al (2013) Scalable synthesis of γ-thiolysine starting from lysine and a side by side comparison with δ-thiolysine in non-enzymatic ubiquitination. Chem Sci 4(12):4494–4495
29. Crich D, Banerjee A (2007) Native chemical ligation at phenylalanine. J Am Chem Soc 129(33):10064–10065
30. Malins LR, Payne RJ (2012) Synthesis and utility of β-selenol-phenylalanine for native chemical ligation-deselenization chemistry. ChemBioChem 2:703–705
31. Townsend SD, Tan Z, Dong S et al (2012) Advances in proline ligation. J Am Chem Soc 134(8):3912–3916
32. Shang S, Tan Z, Dong S, Danishefsky SJ (2011) An advance in proline ligation. J Am Chem Soc 133(28):10784–10786
33. Chen J, Wang P, Zhu J et al (2010) A program for ligation at threonine sites: application to the controlled total synthesis of glycopeptides. Tetrahedron 66(13):2277–2283
34. Fang GM, Li YM, Shen F, Huang YC, Li J, Bin, Lin Y, Liu L (2011) Protein chemical synthesis by ligation of peptide hydrazides. Angew Chem Int Ed 50(33):7645–7649. https://doi.org/10.1002/anie.201100996
35. Li H, Dong S (2017) Recent advances in the preparation of Fmoc-SPPS-based peptide thioester and its surrogates for NCL-type reactions. Sci China Chem 60(2):201–213. https://doi.org/10.1007/s11426-016-0381-1
36. Flood DT, Hintzen JCJ, Bird MJ, Cistrone PA, Chen JS, Dawson PE (2018) Leveraging the Knorr pyrazole synthesis for the facile generation of thioester surrogates for use in Native Chemical Ligation. Angew Chem 130(36):11808–11813. https://doi.org/10.1002/ange.201805191
37. Kulkarni SS et al (2022) Expressed protein selenoester ligation. Angew Chem Int Ed 61:1–6. https://doi.org/10.1002/anie.202200163
38. Harris KM, Flemer S, Hondal RJ (2007) Studies on deprotection of cysteine and selenocysteine side-chain protecting groups. J Pept Sci 13:81–93. https://doi.org/10.1002/psc.795
39. Kulkarni SS, Watson EE et al (2019) Diselenide–selenoester ligation for chemical protein synthesis. Nat Protoc 14:2229–2257. https://doi.org/10.1038/s41596-019-0180-4
40. Gamage K et al (2024) O-GlcNAc modification of α-synuclein can alter monomer dynamics to control aggregation kinetics. ACS Chem Neurosci 15:3044–3052. https://doi.org/10.1021/acschemneuro.4c00301
41. Vamisetti GB et al (2020) On-demand detachment of succinimides on cysteine to facilitate (semi)synthesis of challenging proteins. J Am Chem Soc 142:19558–19569. https://doi.org/10.1021/jacs.0c07663

Check for updates

Chapter 6

Efficient and Site-Specific Protein Bioconjugation Quantification in Mammalian Cells

Gabriella Belsito, Olivia Shade and Alexander Deiters

Abstract

Techniques for the quantification of protein bioconjugation in live cells have been limited. This new method allows for quantification to be measured in cellulo. Inverse electron-demand Diels-Alder (IEDDA) cycloaddition reactions are utilized to label biomolecules. IEDDA compatible unnatural amino acids (UAAs) and their chloroalkane-derivatized conjugates are labeled with the HaloTag enzyme, which is used to quantify the bioconjugation reaction through a gel shift of the conjugated protein of interest. This method is efficient and able to be adjusted to the specific protein of interest. It utilizes common instrumentation and techniques that are found in most labs. This tool's versatility is highlighted by the fact that a variety of UAA/bioconjugation pairs can be used at different incorporation sites in any protein of interest.

Keywords Protein bioconjugation, Genetic code expansion, Click chemistry, Tetrazine ligation

1 Introduction

Protein bioconjugation is a method that can be used with a variety of substrates for many applications, such as drug discovery and delivery [1, 2], protein imaging [3, 4], biochemical and biophysical assay development, and the study of protein function in native environments [5–7]. In order to observe real-time protein bioconjugation in live cells, fast, robust, and specific reactivity is necessary [8, 9]. Traditional methods for the quantification of bioconjugation reactions include using fluorescent labels for in-gel fluorescence or fluorescent microscopy readouts, and mass spectrometry for nonfluorescent ligands [10, 11]. Qualitative evaluation of protein bioconjugation has been done by using fluorescent labels and monitoring the reaction through imaging. Quantitative analysis techniques of protein bioconjugation in cells, however, are limited by the permeability and intracellular distribution of the labeling fluorophore and the

Alexander Deiters (ed.), *Bioconjugation Methods*, Methods in Molecular Biology, vol. 3009,
https://doi.org/10.1007/978-1-0716-5112-4_6,

use of fluorescent proteins as cellular targets [12–14]. Our method, described below, allows for site-specific quantification that can be utilized with any protein.

Unnatural amino acids (UAAs) can be used to expand the genetic code for site-specific installation of small reactive handles into proteins [15, 16]. This methodology involves the use of an aminoacyl-tRNA synthase and $tRNA_{CUA}$ pair, both orthogonal to the host organism. Through transient transfection, a gene of interest that has been modified with an amber stop codon is delivered to mammalian cells along with the necessary orthogonal machinery [15, 16]. UAA mutagenesis provides complete specificity regarding the protein of interest (POI) and the site of UAA incorporation [6].

Of the various biorthogonal labeling reactions [12–14, 17], the inverse electron demand Diels-Alder (IEDDA) cycloaddition between a tetrazine and a strained alkene are near ideal due to its fast kinetics, high conversion rates, excellent biorthogonality, and nontoxic reaction components [13]. Many IEDDA reactive UAAs have been encoded through genetic code expansion, such as *trans*-cyclooctenes (TCOK, sCOK) [18–20], bicyclononyne (BCNK) [20, 21], norbornene (NorK) [18, 22, 23], tetrazine-phenylalanine (TetF) [10, 24]. Additionally, cell lines have been generated for optimized incorporation efficiency and stable expression of UAA mutagenesis machinery [25].

Here, we utilize TetF for subsequent labeling by strained *trans*-cyclooctene-chloroalkane (sTCO-CA). Only short incubation times (minute scale) are needed for bioconjugation to occur, and toxicity of the synthetic molecules is negligible. IC_{50} values for TetF and sTCO-CA of 753 and 6.32 μM, respectively, after incubation of HEK293T cells for 72 h [26]. No toxicity was observed when the incubation with sTCO-CA was reduced to 4 h [26].

In the methodology described here, the bioorthogonal UAA on the POI surface reacts directly with its designed reactive partner bearing a chloroalkane handle. Following cell lysis, the lysate is incubated with recombinant HaloTag enzyme, which covalently binds to the chloroalkane handle, now bioconjugated to the UAA-containing protein. The specific reactivity and high cell permeability of the chloroalkane provides distinct advantages compared to that of fluorescent dyes commonly used for bioconjugation quantification in cellulo [12–14, 17, 27]. Finally, simple western blotting is used to quantify the extent of protein bioconjugation, as a higher molecular weight conjugate can only be formed if the initial protein bioconjugation is successful (Fig. 1).

This approach utilizes soluble, cell permeable small molecules to quantify bioconjugation at various regions on the same protein of interest (Fig. 2). As protein bioconjugation continues to evolve, it is important to accurately quantify the amount of bioconjugation that occurs at a specific location on a protein surface, as the resulting reactivity can affect downstream results or designs. We have observed

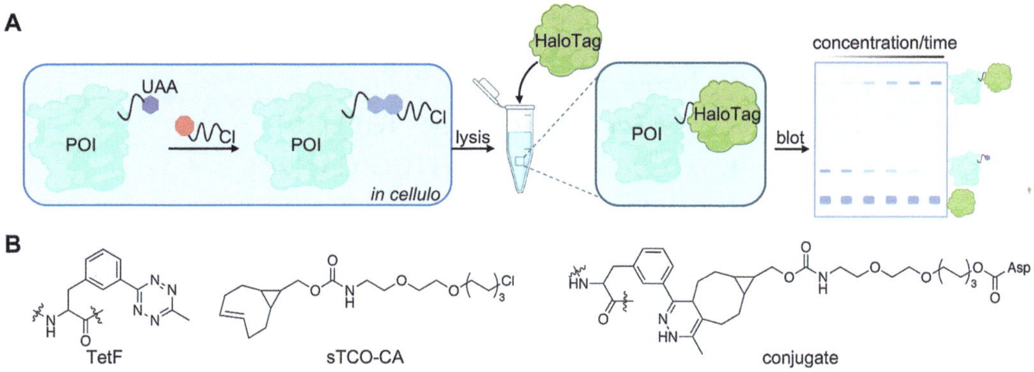

Fig. 1 (**A**) Cartoon representation of in cellulo protein bioconjugation followed by cell lysis and HaloTag labeling for analysis of bioconjugation through western blot. (**B**) Chemical structures of tetrazine phenylalanine (TetF), strained *trans*-cyclooctene chloroalkane (sTCO-CA), and the resultant conjugate between the two, shown after reaction with the HaloTag protein. Fig. 1A was adapted with permission from [26]. Copyright 2022 American Chemical Society

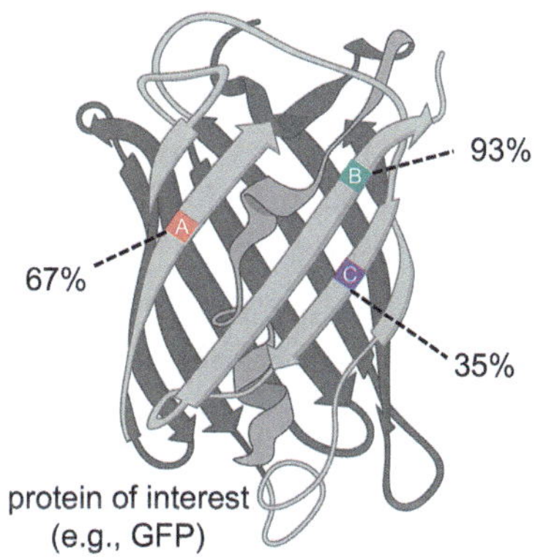

Fig. 2 Cartoon depicting variable conjugation efficiency dependent on the location of conjugation site on the protein of interest

different bioconjugation efficiencies among different regions on the same protein [26], thus highlighting the need for this quantification method.

In this chapter, we provide detailed instructions for the quantitative analysis of intercellular, protein-specific bioconjugation reactions using only basic equipment.

2 Materials

1. Dulbecco's Modified Eagle Medium (DMEM), 10% fetal bovine serum (Stored in fridge and warmed to 37°C before use).

2. Gibco™ Opti-MEM™ Reduced Serum Medium (OMEM, Fisher 22-600-050, stored in fridge and warmed to 37°C before use).
3. Tetrazine phenylalanine (TetF) [10].
4. sTCO-Chloroalkane (sTCO-CA) [28].
5. pAcBac-TetFRS [10].
6. pmCherry-EGFP-TAG and pERRα-HA-TAG mutants [26].
7. PBS: 0.256% (w/v) $Na_2HPO_4 \bullet 7\ H_2O$, 0.8% (w/v) sodium chloride, 0.2% (w/v) potassium chloride, 0.2% (w/v) KH_2PO_4, in Milli-Q water, stored in fridge.
8. TBS: 0.121% (w/v) Tris base, 0.876% (w/v) sodium chloride, in Milli-Q water.
9. 10 mg/mL Poly-D-Lysine (PDK).
10. Linear polyethyleneimine (LPEI).
11. RIPA buffer: 50 mM Tris-HCl, pH 8.0, 150 mM sodium chloride, 1% (v/v) NP-40, 0.5% (w/v) sodium deoxycholate, 0.1% (w/v) sodium dodecyl sulfate, stored in fridge.
12. HaloTag protein [26].
13. 4× SDS-PAGE sample loading buffer: 200 mM Tris-HCl, pH 6.5, 400 mM DTT, 8% (w/v) SDS, 6 mM bromophenol blue, 4 M glycerol.
14. Transfer buffer: 0.3% (w/v) Tris base, 1.44% (w/v) glycine, in Milli-Q water, pH 8.3.
15. TBST: 0.121% (w/v) Tris base, 0.876% (w/v) sodium chloride, 0.09% (v/v) Tween-20, in Milli-Q water.
16. Blocking buffer: 5% milk in TBST or 5% BSA in TBST, depending on the primary antibody.
17. Rabbit mAb anti-HA primary antibody (Cell Signaling Technologies 3724S).
18. Rabbit pAb anti-GAPDH primary antibody (ProteinTech 50-172-6351).
19. Anti-rabbit IgG, HRP-linked Antibody (Cell Signaling Technologies 7074S).
20. TBST: 100 mL 10× TBS, 1 g Tween-20, diluted to 1 L with Milli-Q water, stored in fridge.
21. 0.1 M Tris buffer: 12.1 g Tris base, 1000 mL Milli-Q water, pH to 9.35 with conc. hydrochloric acid.
22. Developing solution: 6 mg of luminol, 7 mg of *p*-iodophenol, 7.5 mL of 0.1 M Tris buffer, pH 9.35.
23. Enhancer solution: 1 μL of 30% hydrogen peroxide, 10 mL of 0.1 M Tris buffer, pH 9.35.

2.1 Equipment

1. Micropipettes.
2. Incubator.
3. Shaker.
4. Sterile laminar flow hood.
5. Electrophoresis power supply.
6. Electrophoresis chamber.
7. ChemiDoc Imaging System.
8. ImageJ Software.

3 Methods

3.1 Incorporation of UAAs into the POI

1. Create a solution of PolyD-lysine (PDK) by diluting 100 μL of a 10 mg/mL solution (100×) into 10 mL of Milli-Q water.
2. PDK treat a sterile, flat-bottom 48-well cell culture plate. Add 125 μL of the PDK solution to each well, ensuring the entire bottom of the well is covered. Incubate at room temperature for 10 min.
3. Remove the PDK solution and wash each well with 125 μL of Milli-Q water.
4. Remove the Milli-Q water and dry the plates in a biosafety hood for 1 h with the plate lids opened slightly.
5. Passage the mammalian cells. Under sterile conditions, first remove all of the media from the plate with a large pipette. Gently rinse the plate with 1 mL of trypsin. Remove the trypsin wash and discard it. Then, add another 1 mL of trypsin to the plate, agitating the cells and covering them thoroughly. Once the cells start to lift, use a p1000 pipette to gently lift all the cells from the bottom of the plate, being careful to avoid bubbles.
6. After rinsing all the cells from the plate, dilute the 1 mL of lifted cells into 9 mL of fresh complete medium.
7. Transfer 1 mL of diluted cells (step 6) to a new plate containing 9 mL of fresh complete medium. Gently swirl to mix the cells on the plate and place it into an incubator.
8. Allow the cells to grow at 37°C and 5% CO_2 to maintain the cell line for future experiments. Grow the cells to about 90% confluency before passaging again.
9. Seed HEK293T cells (50,000 cells/well) into the PDK-treated 48-well plate, using a final volume of 250 μL of cell solution per well. Incubate overnight.

10. When the cells reach 80%–90% confluency, replace the media in each well with 250 μL of fresh DMEM supplemented with 0.25 mM tetrazine phenylalanine (TetF) or other UAA of choice by creating a working solution containing 1% DMSO by diluting 12.5 μL of a stock solution (100 mM TetF in DMSO) into 5 mL of antibiotic-free DMEM.
11. To test bioconjugation via biological triplicates, exchange the media in 6 separate wells with 250 μL of the UAA-supplemented media.
12. In a sterile 1.7-mL microcentrifuge tube, prepare a transfection master mix solution. For the transfection of six wells, add 240 μL of Opti-MEM to the tube. Then, combine 1600 ng of the plasmid encoding the synthetase and tRNA (pAcBac-TetFRS for TetF, Fig. 3), and 800 ng of the plasmid encoding the protein of interest, pmCherry-EGFP-TAG mutants or pERRα-HA-TAG mutants. Finally, add 12 μL of a 1 mg/mL stock of linear polyethyleneimine (LPEI).
13. Mix the transfection mixture by pipetting up and down, and then incubate for 10 min at room temperature.

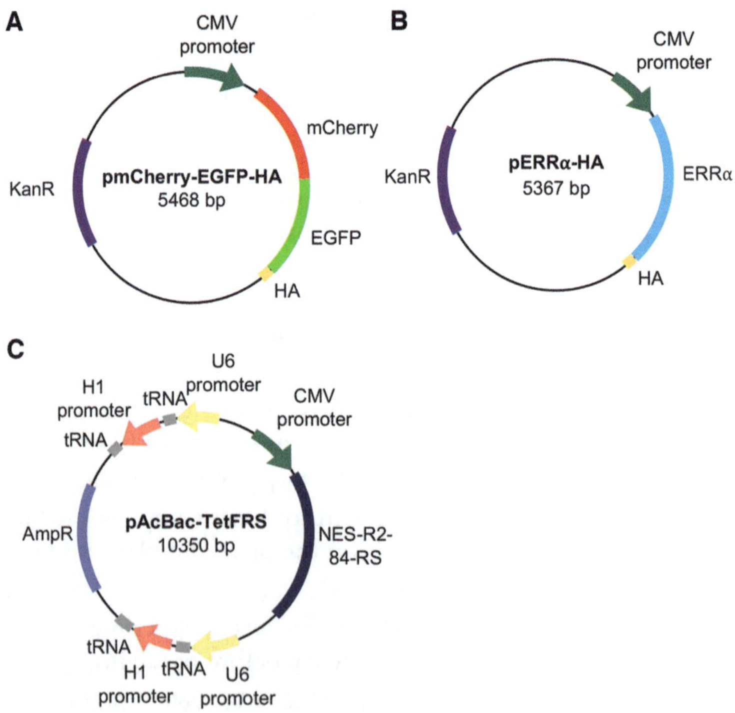

Fig. 3 Maps of the plasmids used in this approach. (**A**) pmCherry-EGFP-TAG-HA, (**B**) pERRα-TAG-HA, and (**C**) pAcBac-TetFRS. This figure was reproduced with permission from [26]. Copyright 2022 American Chemical Society

14. Add 40 μL of the transfection mixture dropwise to each well already prepared with UAA-containing, antibiotic-free medium.
15. Incubate cells for 24–48 h at 37°C with 5% CO_2.
16. For fluorescent proteins, confirm successful UAA incorporation via fluorescence microscopy. For nonfluorescent proteins, skip to conjugation protocol.

3.2 Conjugation

1. Remove UAA-containing medium and wash cells by applying 250 μL of fresh complete medium to each well. Immediately remove wash.
2. Apply another 250 μL of fresh complete medium to each well and incubate at 37°C for 30 min to ensure removal of remaining UAA.
3. Create a 10 μM working solution of sTCO-CA containing 0.1% DMSO by diluting 3 μL of a 10 mM stock solution in DMSO into 3 mL of DMEM. *See* **Note 1** for sTCO-CA storage tips.
4. Create a negative control solution by adding the same volume (3 μL) of DMSO to 3 mL of complete medium.
5. Remove the second wash from Step 2, then apply the medium supplemented with or without chloroalkane to the respective wells. Incubate the cells at 37°C for 1 h.
6. Remove the media and wash the cells with 250 μL of fresh complete medium. Incubate for 1 h at 37°C with 5% CO_2.

3.3 HaloTag Conjugation

1. Remove media following step 6 of the conjugation protocol and wash the cells once with 200 μL of ice-cold PBS to remove residual medium.
2. Place cells on ice.
3. Determine the amount of lysis buffer needed: (number of wells) (60 μL/well) = volume/μL. Increase the needed volume by 100 μL to ensure enough lysis buffer is made.
4. Prepare the amount of lysis buffer calculated in step 3 by supplementing ice-cold RIPA buffer with 100× protease inhibitor cocktail.
5. Add 60 μL of lysis buffer (step 4) to each well, then incubate the plate at 250 rpm shaking for 15–30 min.
6. Transfer lysates to 1.7 mL microcentrifuge tubes.
7. Pellet cell debris by centrifugation at 21,000 × *g* for 10 min at 4°C.
8. Combine 50 μL of the supernatant with 10 μL of 80 μM HaloTag and incubate at 37°C overnight. *See* **Note 2** for HaloTag storage tips.

3.4 Western Blot

1. Combine the 60 μL of lysate-HaloTag mixture with 20 μL of 4× SDS-PAGE sample loading buffer.
2. Heat samples at 95°C for 10 min.
3. Separate proteins by 10% (v/v) SDS-PAGE gel electrophoresis (80 V for 20 min, 150 V for 80 min) with an ice pack inside the tank next to the gel cassette. *See* **Note 3** for tips on how to optimally load the gel.
4. Perform a protein transfer to a 0.45 μm PVDF membrane at 80 V for 2 h using ice cold transfer buffer, keeping the transfer tank in an ice bath throughout.
5. Cut the membranes at the necessary molecular weight marker using a straight edge, such that the loading control and the protein of interest are separated.
6. Block the membranes for 1 h with 3 mL of blocking buffer at room temperature with rocking.
7. Probe the blots with rabbit mAb anti-HA (1:1000 dilution) and rabbit pAb anti-GAPDH (1:5000 dilution) primary antibodies in 3 mL of blocking buffer overnight with rocking at 4°C.
8. Wash each membrane three times by incubating the blots with 3 mL of ice-cold TBST with rocking, incubating 5 min for each wash.
9. Incubate membranes for 1 h with secondary antibody diluted in 3 mL of TBST.
10. Wash each membrane three times with 3 mL of ice-cold TBST with rocking, incubating 5 min for each wash.
11. Combine 5 mL of the developing solution with 5 mL of the enhancer solution (see Sect. 2 Steps 17 and 18 for recipes). Invert this solution three to four times, and then, pour it onto the membrane to be developed. Incubate in the developing solution for 5 min. *See* **Note 4** for tips on preventing the membrane from drying out before/during imaging.
12. Image the blots using a BioRad ChemiDoc system with automated exposure times. Use Cy5 and Cy3 filters to image the ladder, and Chemi filters to image the chemiluminescence of the antibody-labeled protein of interest.

3.5 Western Blot Analysis

1. Analyze the bands of the western blot using ImageJ software.
2. Use the rectangle tool to select the leftmost band in the blot that corresponds to nonconjugated protein. Use the "Rectangle" tool to draw a rectangle over the entirety of

the band, avoiding overlap with adjacent lanes. Once selected, click Analyze > Gels > Select First Lane, or key in Command + 1. *See* **Note 5** to determine which bands to use for quantifying conjugation.

3. Click inside the yellow rectangle and drag until a second equivalent yellow rectangle appears. Drag this rectangle to the next lane. Then, select Analyze > Gels > Select Second Lane, or key in Command + 2.
4. Repeat step 3 until every band is outlined with a rectangle.
5. Create a plot of the highlighted bands in a coordinate plane by going to Analyze > Gels > Plot Lanes, or key in Command + 3. This will allow values to be integrated.
6. Add a baseline to aid with quantification and subtract the gray background commonly found in Western blots. Use the "Straight Line" tool to draw a level line across the bottom of each peak.
7. Use the "Wand" tool and click inside each peak to measure the area under each curve.
8. Transfer the populated values of each lane to an Excel file for record keeping and further analysis.
9. Repeat the above outlined quantification for the loading control blot and transfer the resultant values to the same Excel file.
10. Normalize the amount of POI to its respective loading control by dividing the area of the experimental band by the area of the respective loading control band for each lane in Excel. This will adjust for any variations in pipetting, cell health, cell lysis, etc.
11. Normalize the experimental values to the negative control (DMSO-treated cells) for experiments done with multiple replicates by calculating the average of each negative control using the "Average" function in Excel. Then, divide each lane value by the average of the negative wells. This determines the amount of nonconjugated protein remaining after treatment with sTCO-CA.
12. To determine the amount of protein conjugated to sTCO-CA, subtract the average of the remaining protein (from step 23) from the average of the normalized negative control and multiply by 100%.
13. Determine the standard error from three biological replicates (Fig. 4).

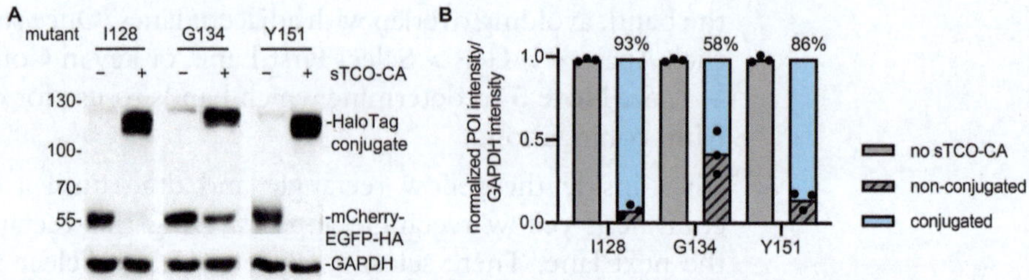

Fig. 4 Results from the experiment including (**A**) western blot and (**B**) bar graph depiction of the quantification of protein bioconjugation of sTCO-CA with EGFP protein surface mutations I128TetF, G134TetF, and Y151TetF. This figure was reproduced with permission from [26]. Copyright 2022 American Chemical Society

4 Notes

1. Over time, the *trans* isomer of sTCO-CA can isomerize back to its *cis* isomer. To prevent freeze-thaw cycles from affecting the total amount of *trans* isomer and to prolong the lifetime of the active compound, a 10× DMSO stock should be stored as single-use aliquots (10 μL each) at –80°C. To further monitor the isomerization from *trans* to *cis*, a stock of freshly made sTCO-CA can be prepared in DMSO-d_6 and frozen at the same time as the aliquoted compound for NMR analysis over time.
2. We observed that recombinant HaloTag reaches the end of its shelf-life after 2–3 months of storage at –80°C, after which we noticed an increase in the formation of a double band at the POI-HaloTag conjugate molecular weight. To reduce the likelihood of double banding, remake HaloTag every 2–3 months.
3. On either side of the loaded samples, add PageRuler Protein Ladder as a molecular weight marker and to assist in cutting the membrane after transfer. Use of a freshly prepared gel with the appropriate concentration of acrylamide (dependant on the molecular weight of the protein of interest), in combination with a stacking gel has led to the most consistent western blotting results. Further, be sure to select an acrylamide concentration such that the protein of interest and loading control proteins are sufficiently separated during electrophoresis.
4. After the developer solution has been applied to the membrane for the specified amount of time, pour off the solution into the requisite waste container. Then, using tweezers, carefully remove the developed membrane from its container and place it into a clear, colorless ziploc bag. This will prevent the membrane from drying out during imaging, in the event that the membrane needs to be stripped and reprobed.

5. The lower molecular weight band corresponding to the protein not conjugated by sTCO-CA is used for quantification. The higher molecular weight conjugate band suffers from poor transfer efficiency and can therefore lead to inaccurate quantification values.

Acknowledgments

This work was supported by the National Science Foundation (CHE-1904972). The plasmid pAcBac-TetFRS was provided by the GCE4All Biomedical Technology Development and Dissemination Center supported by the National Institute of General Medical Science (RM1GM144227). Some figure elements were created with BioRender.com.

References

1. Roy S, Cha JN, Goodwin AP (2020) Nongenetic bioconjugation strategies for modifying cell membranes and membrane proteins: a review. Bioconjug Chem 31(11):2465–2475. https://doi.org/10.1021/acs.bioconjchem.0c00529
2. Martinez DG, Huttelmaier S, Bertoldo JB (2020) Unveiling druggable pockets by site-specific protein modification: beyond antibody-drug conjugates. Front Chem 8:586942. https://doi.org/10.3389/fchem.2020.586942
3. Hira J, Uddin MJ, Haugland MM, Lentz CS (2020) From differential stains to next generation physiology: chemical probes to visualize bacterial cell structure and physiology. Molecules 25(21):4949. https://doi.org/10.3390/molecules25214949
4. Lieser RM, Yur D, Sullivan MO, Chen W (2020) Site-specific bioconjugation approaches for enhanced delivery of protein therapeutics and protein drug carriers. Bioconjug Chem 31(10):2272–2282. https://doi.org/10.1021/acs.bioconjchem.0c00456
5. Kalia J, Raines RT (2010) Advances in bioconjugation. Curr Org Chem 14:138–147
6. Spicer CD, Davis BG (2014) Selective chemical protein modification. Nat Commun 5:4740. https://doi.org/10.1038/ncomms5740
7. Liu J, Cui Z (2020) Fluorescent labeling of proteins of interest in live cells: beyond fluorescent proteins. Bioconjug Chem 31(6):1587–1595. https://doi.org/10.1021/acs.bioconjchem.0c00181
8. Moghaddam-Taaheri P, Karlsson AJ (2018) Protein labeling in live cells for immunological applications. Bioconjug Chem 29(3):680–685. https://doi.org/10.1021/acs.bioconjchem.7b00722
9. Reddy NC, Kumar M, Molla R, Rai V (2020) Chemical methods for modification of proteins. Org Biomol Chem 18(25):4669–4691. https://doi.org/10.1039/d0ob00857e
10. Jang HS et al (2020) Access to faster eukaryotic cell labeling with encoded tetrazine amino acids. J Am Chem Soc 142(16):7245–7249. https://doi.org/10.1021/jacs.9b11520
11. Bednar RM et al (2021) Genetic incorporation of two mutually orthogonal bioorthogonal amino acids that enable efficient protein dual-labeling in cells. ACS Chem Biol 16(11):2612–2622. https://doi.org/10.1021/acschembio.1c00649
12. Lang K, Chin JW (2014) Cellular incorporation of unnatural amino acids and bioorthogonal labeling of proteins. Chem Rev 114(9):4764–4806. https://doi.org/10.1021/cr400355w
13. Nikić-Spiegel I (2020) Expanding the genetic code for neuronal studies. ChemBioChem 21(22):3169–3179. https://doi.org/10.1002/cbic.202000300
14. Saleh AM et al (2019) Non-canonical amino acid labeling in proteomics and biotechnology. J Biol Eng 13:43. https://doi.org/10.1186/s13036-019-0166-3
15. Stephanopoulos N, Francis MB (2011) Choosing an effective protein bioconjugation strategy. Nat Chem Biol 7(12):876–884. https://doi.org/10.1038/nchembio.720
16. Reinkemeier CD et al (2021) Synthesis and evaluation of novel ring-strained noncanonical amino acids for residue-specific bioorthogonal reactions in living cells. Chemistry 27(19):6094–6099. https://doi.org/10.1002/chem.202100322

17. Oliveira BL, Guo Z, Bernardes GJL (2017) Inverse electron demand Diels-Alder reactions in chemical biology. Chem Soc Rev 46(16):4895–4950. https://doi.org/10.1039/c7cs00184c
18. Plass T et al (2012) Amino acids for Diels-Alder reactions in living cells. Angew Chem Int Ed Engl 51(17):4166–4170. https://doi.org/10.1002/anie.201108231
19. Nikic I et al (2014) Minimal tags for rapid dual-color live-cell labeling and super-resolution microscopy. Angew Chem Int Ed Engl 53(8):2245–2249. https://doi.org/10.1002/anie.201309847
20. Borrmann A et al (2012) Genetic encoding of a bicyclo[6.1.0]nonyne-charged amino acid enables fast cellular protein imaging by metal-free ligation. Chembiochem 13(14):2094–2099. https://doi.org/10.1002/cbic.201200407
21. Lang K et al (2012) Genetic Encoding of bicyclononynes and trans-cyclooctenes for site-specific protein labeling in vitro and in live mammalian cells via rapid fluorogenic Diels-Alder reactions. J Am Chem Soc 134(25):10317–10320. https://doi.org/10.1021/ja302832g
22. Lang K (2012) Genetically encoded norbornene directs site-specific cellular protein labelling via a rapid bioorthogonal reaction. Nat Chem 4(4):298–304. https://doi.org/10.1038/nchem.1250
23. Kaya E et al (2012) A genetically encoded norbornene amino acid for the mild and selective modification of proteins in a copper-free click reaction. Angew Chem Int Ed Engl 51(18):4466–4469. https://doi.org/10.1002/anie.201109252
24. Seitchik JL et al (2012) Genetically encoded tetrazine amino acid directs rapid site-specific in vivo bioorthogonal ligation with trans-cyclooctenes. J Am Chem Soc 134(6):2898–2901. https://doi.org/10.1021/ja2109745
25. Zhou W, Wesalo JS, Liu J, Deiters A (2020) Genetic code expansion in mammalian cells: a plasmid system comparison. Bioorg Med Chem 28(24):115772. https://doi.org/10.1016/j.bmc.2020.115772
26. Ryan A et al (2022) Quantitative analysis and optimization of site-specific protein bioconjugation in mammalian cells. Bioconjug Chem 33(12):2361–2369. https://doi.org/10.1021/acs.bioconjchem.2c00451
27. Fu Y, Finney NS (2018) Small-molecule fluorescent probes and their design. RSC Adv 8(51):29051–29061. https://doi.org/10.1039/c8ra02297f
28. Murrey HE et al (2015) Systematic evaluation of bioorthogonal reactions in live cells with clickable HaloTag ligands: implications for intracellular imaging. J Am Chem Soc 137(35):11461–11475. https://doi.org/10.1021/jacs.5b06847

Chapter 7

Intelligence Disclosure on Deploying SpyTag/SpyCatcher for Nanoparticle Assembly

Rory A. Hills and Mark R. Howarth

Abstract

Immunization with antigens presented multivalently on virus-like particles elicits potent immune responses. The SpyTag/SpyCatcher system, which allows spontaneous covalent bond formation between two proteins, provides a simple route to functionalize virus-like particles for multivalent presentation. Termed SpyVLPs, these reactive nanoparticles allow modular plug-and-display of different antigens. SpyCatcher003-mi3 is a resilient SpyVLP that can be produced at high yield, allows efficient coupling, and can be applied to vaccines against a diverse range of pathogens, as well as against cancer and allergy. Here, we describe the procedure for the expression and purification of SpyCatcher003-mi3. We outline considerations related to antigen design, multivalent display, nanoparticle validation, and immunogen preparation that are critical for effective implementation of SpyCatcher003-mi3 and other SpyVLPs. The methods discussed are applicable beyond vaccines, extending to therapeutic targeting, diagnostics, and catalysis.

Keywords Bioconjugation, Synthetic biology, Nanotechnology, Nanomedicine, Outbreak pathogen, Isopeptide, VLP, Crosslink

1 Introduction

Nanoparticles decorated with functional molecules have created new possibilities in diverse areas, including imaging, catalysis, protein therapy, and drug delivery [1, 2]. Protein-based nanoparticles allow atomic-level design precision, while leveraging existing microbial fermentation infrastructure to facilitate large-scale production [3, 4]. Cargo proteins can be attached to a protein nanoparticle-scaffold by direct genetic fusion [5, 6]. However, changes to the scaffold can interfere with the sensitive assembly process, so cargo-scaffold genetic fusions often require substantial optimization and have not been tractable for many complex cargo molecules [7, 8]. Additionally, the optimal expression conditions for the cargo protein and nanoparticle are often poorly matched, requiring truncations, altered post-translational modification, or sacrifices to production efficiency

Alexander Deiters (ed.), *Bioconjugation Methods*, Methods in Molecular Biology, vol. 3009,
https://doi.org/10.1007/978-1-0716-5112-4_7,

for one component. An effective strategy that addresses these issues is to produce the cargo and nanoparticle separately, before assembling a conjugated nanoparticle [7].

This assembly can be accomplished using the SpyTag/SpyCatcher system [9, 10]. The SpyTag peptide and SpyCatcher protein are both engineered from the fibronectin binding protein (FbaB) of *Streptococcus pyogenes*. When mixed, the pair spontaneously forms an irreversible isopeptide bond (Fig. 1A) [11]. SpyTag and SpyCatcher can be genetically fused to separate proteins, to facilitate covalent conjugation between partners that can be expressed at different times, under different conditions, and in different cellular systems [11]. Fusion of SpyTag or SpyCatcher to a protein nanoparticle can create a modular assembly platform. The subsequent addition of molecules fused to the cognate Tag or Catcher allows for facile multivalent display without the need for additional optimization steps - the "plug-and-display" approach (Fig. 1B).

SpyTag/SpyCatcher functionalized nanoparticles have been employed against a wide range of targets in the field of vaccinology (Fig. 2). A common vaccine strategy is to immunize against a specific component of the targeted pathogen, such as a protein or carbohydrate [12]. However, these subunit antigens are often poorly immunogenic on their own, while efforts to enhance immunogenicity through more potent adjuvants can increase reactogenic inflammatory responses [13, 14]. Multivalent display on a nanoparticle presents antigens in a structured, highly repetitive manner that may mimic a virus surface and is an effective strategy to enhance immunogenicity [13, 15]. Nanoparticles derived from virus structural proteins and those that have been selected or engineered to mimic the structural features of a virus are termed virus-like particles (VLPs) [16]. VLP display of antigens facilitates enhanced cross-linking by B cell receptors (BCRs), uptake by antigen-presenting cells (APCs), recognition by complement, and delivery to draining lymph nodes [17].

SpyTag/SpyCatcher has been used to functionalize a variety of VLPs including AP205 phage coat [9, 10], *Aquifex aeolicus* lumazine synthase [18], hepatitis B surface antigen [19], hepatitis B core antigen [20], encapsulin [21], and ferritin [22, 23], among others [24, 25]. The modular nature of these SpyVLPs is adaptable to various antigens including monomeric proteins [9], oligomeric proteins [26], peptides [20], and carbohydrates for conjugate vaccines [27]. These antigens have been produced using a variety of expression systems, including *Escherichia coli* [28, 29], *Shigella flexneri* [30], *Komagataella phaffii* (also known as *Pichia pastoris*) [19], *Thermothelomyces heterothallica* C1 [31], baculovirus with an insect cell-line [32], Expi293 mammalian cell-line [33], *Bombyx mori* larvae [34], *Nicotiana benthamiana* [35], and direct peptide synthesis [20]. SpyVLPs have been used to target diverse diseases including cancers [36],

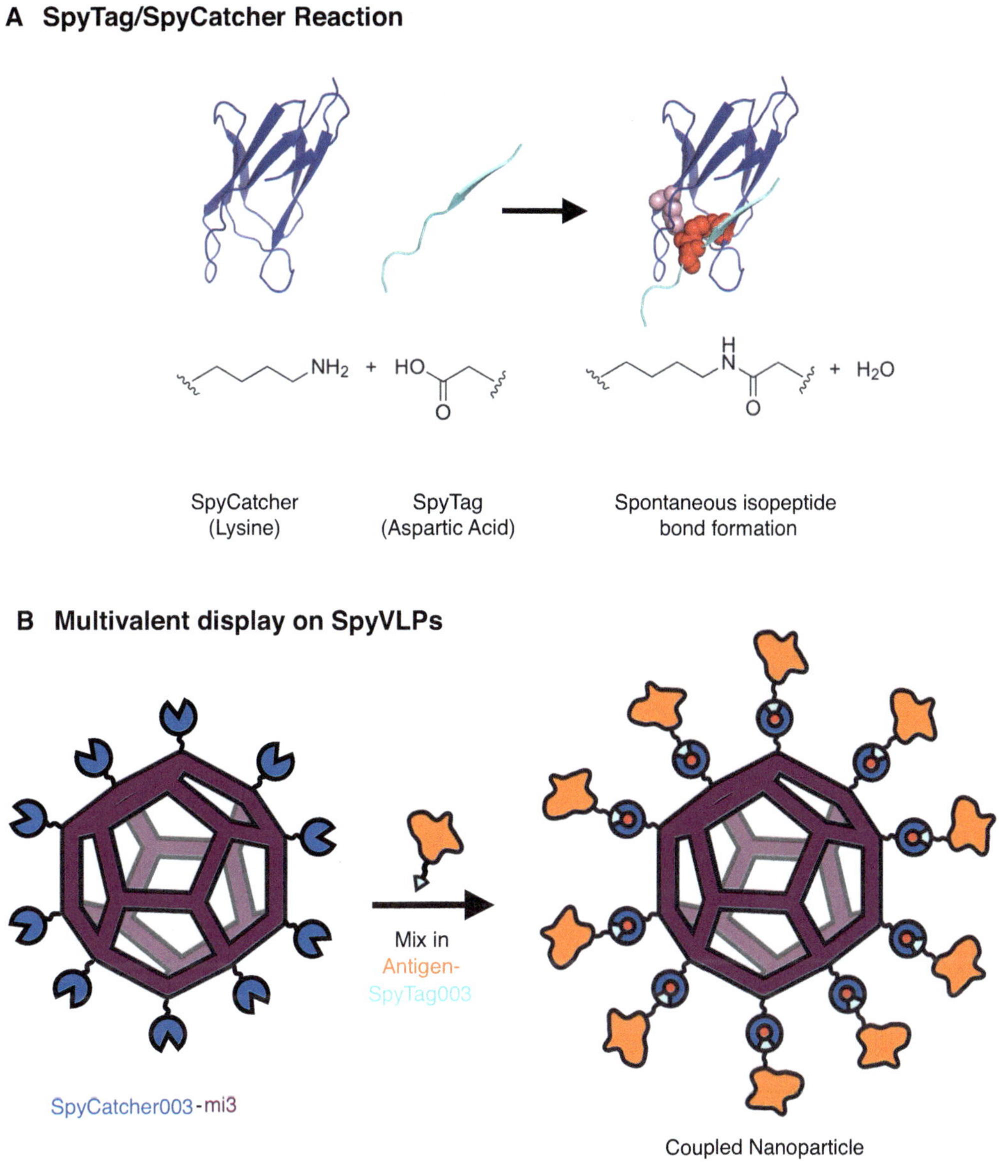

Fig. 1 SpyTag/SpyCatcher facilitates multivalent display on SpyVLPs. (**A**) Schematic of the SpyTag reaction with SpyCatcher: SpyCatcher (dark blue) spontaneously forms a covalent isopeptide bond with SpyTag (cyan). The reactive lysine (K31) and aspartic acid (D117) are shown in red, and the apposed glutamic acid (E77) is shown in pink for the final SpyTag:SpyCatcher structure (PDB 4MLI). (**B**) Cartoon representation of a SpyVLP. In this example, a protein antigen (orange) is genetically fused to SpyTag003 (cyan) and incubated with the SpyCatcher003-mi3 nanoparticle. This procedure facilitates formation of a covalent bond (red) between SpyTag003 and SpyCatcher003 (dark blue), allowing as many as 60 copies of the protein antigen to be displayed per protein nanoparticle formed by mi3 (purple). Only some copies of the antigen and SpyCatcher003 are shown for clarity

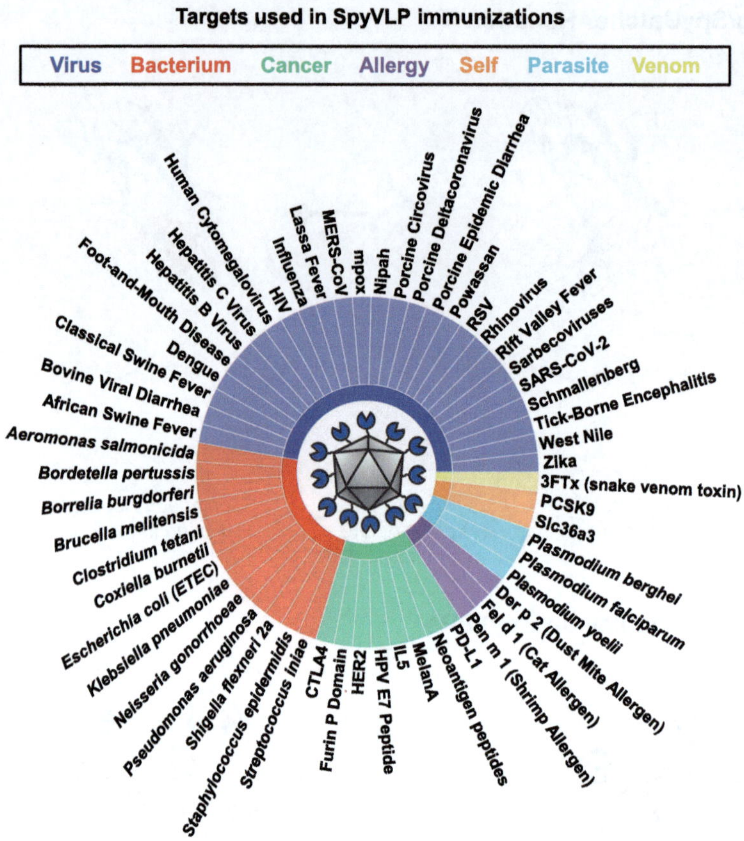

Fig. 2 Targets immunized against using SpyVLPs. A summary of 57 targets that SpyVLPs immunizations have been performed against. Targets are classified as virus (dark blue), bacteria (red-orange), cancer (green), allergy (purple), self (orange), parasite (light blue), and venom (yellow). Criteria for inclusion are a protein nanoparticle scaffold coupled to an antigen by an isopeptide bond and immunized in an appropriate clinical or preclinical model

allergies [37, 38], heart disease [39], and a wide variety of infectious diseases (Fig. 2). SpyVLP immunizations have been performed in at least 17 different animal species, including mammals, birds and ray-finned fish (Fig. 3). This evaluation includes ongoing Phase I human clinical trials for a Human Cytomegalovirus (HCMV) vaccine [40] (Trial ID NCT06145178) and a malaria vaccine [41] (NCT05357560), as well as a successfully concluded Phase I human clinical trial against SARS-CoV-2 [42] (Trial ID NCT05329220).

SpyCatcher003-mi3 (Table 1) is a SpyVLP that demonstrates several favorable qualities, including robust nanoassembly, compatibility with diverse geometries, and fast reactivity that facilitates coupling to completion [26]. SpyCatcher003-mi3's key properties are listed in Table 1. SpyCatcher003-mi3 is a genetic fusion of the SpyCatcher003 and mi3 proteins [26]. SpyCatcher003 is the most recent version of SpyCatcher and includes several mutations selected

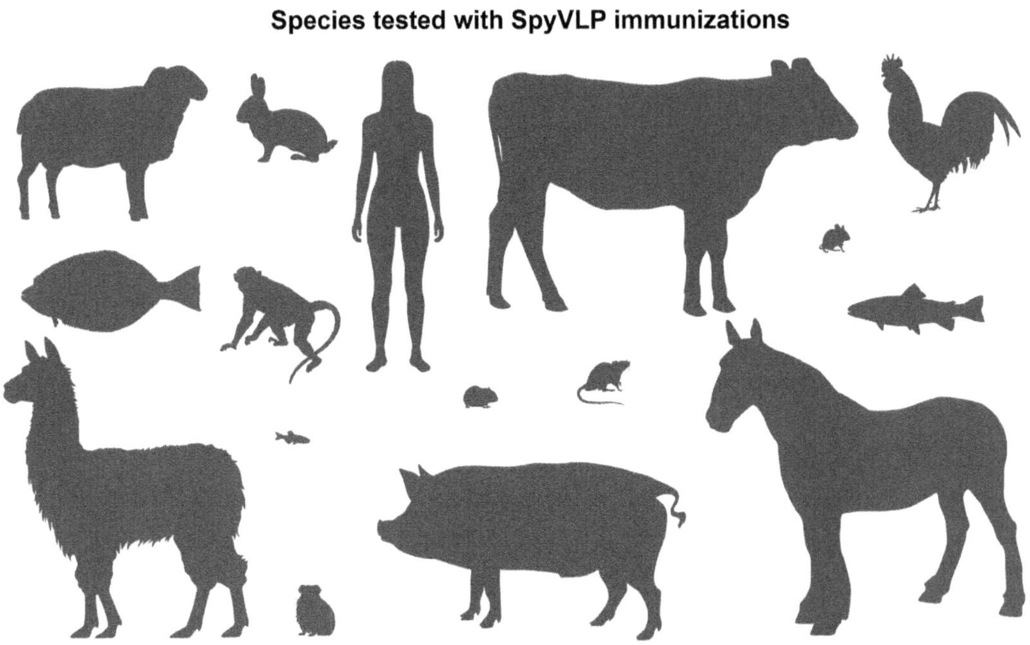

Fig. 3 Animals immunized with SpyVLPs. A representation of 17 animal species that have been immunized using SpyVLPs. Silhouettes represent mouse (*Mus musculus*), sheep (*Ovis aries*), hamster (*Mesocricetus auratus*), macaque (*Macaca mulatta* and *Macaca fascicularis*), rabbit (*Oryctolagus cuniculus*), cow (*Bos taurus*), guinea pig (*Cavia porcellus*), human (*Homo sapiens*), horse (*Equus ferus caballus*), olive flounder (*Paralichthys olivaceus*), zebrafish (*Danio rerio*), chicken (*Gallus gallus domesticus*), rat (*Rattus norvegicus*), rainbow trout (*Oncorhynchus mykiss*), pig (*Sus scrofa domesticus*), and llama (*Lama glama*)

Table 1
Key properties of SpyCatcher003-mi3. Characteristics of a SpyCatcher003-mi3 monomer, determined from the SpyCatcher003-mi3 sequence GenBank MT945417 using ExPASy ProtParam [83], assuming removal of the N-terminal formyl methionine

Feature	Value
Molecular weight	35,829 Da
Theoretical isoelectric point (pI)	5.0
Molar extinction coefficient	18,575 $M^{-1}.cm^{-1}$
Number of amino acid residues	340

through rational design and phage display evolution to enhance reaction rate and efficiency as a fusion partner [43]. SpyCatcher003 covalently couples to SpyTag003 at a rate that approaches the diffusion limit. SpyCatcher003 retains back-compatibility with SpyTag

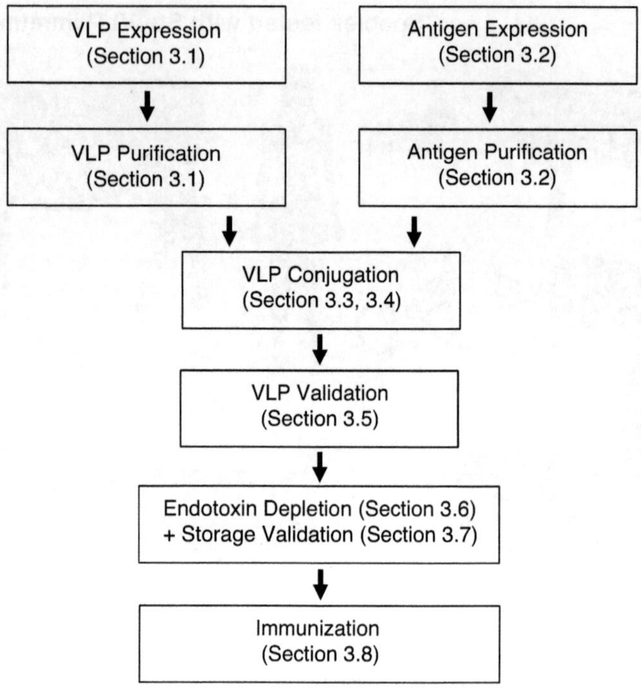

Fig. 4 Production of SpyCatcher003-mi3 vaccine. A generic workflow for production of multivalent display vaccines using SpyCatcher003-mi3. The section of this protocol that expands on each step is indicated in the figure

and reacts faster with SpyTag than the original SpyCatcher [43]. The nanoparticle scaffold is a hollow and porous structure formed by 60 copies of the mi3 protein [33]. mi3 is derived from the computationally designed I3-01 nanoparticle [3] with two cysteine to alanine mutations introduced to improve yield and reduce aggregation [33]. I3-01 itself was based on a trimeric aldolase from *Thermotoga maritima* that had four computationally derived non-polar residues introduced to generate a new interface and bridge trimers together [3]. The mi3 scaffold is robust to heating (1 h at 80°C), freeze-thaw cycling, and lyophilization [33].

This protocol will outline the production of SpyCatcher003-mi3 nanoparticles and their application as multivalent display platforms for molecules of interest. The protocol, summarized in Fig. 4, is tailored to the development of vaccine candidates, but the methods are applicable to broad applications of nanoparticles, including catalysis, therapeutic targeting, and diagnostics [44]. While focused on SpyCatcher003-mi3, the considerations outlined, including SpyTag-fusion coupling, VLP validation, endotoxin depletion, and dose selection, are important considerations for all SpyVLPs.

2 Materials

2.1 Equipment

1. Personal protective equipment (lab coat, safety glasses, gloves).
2. Pipettors (P1000, P200, P10).
3. Pipette Tips (Filtered) for manipulating samples during and after endotoxin depletion.
4. Pipette Tips (Non-Filtered), optional for manipulating samples prior to endotoxin depletion.
5. Bacterial incubator capable of shaking at 200 rpm with temperature control from 22°C to 37°C.
6. Spectrophotometer capable of measuring OD_{600}.
7. Floor-standing centrifuge with temperature control to 4°C, capable of spinning 1 L bacterial culture at 4000 × *g* or ~30 mL of cell lysate at 30,000 × *g*.
8. Bench-top microcentrifuge capable of spinning 1.5 mL microcentrifuge tubes at 16,000 × *g* with temperature control from 4°C to 37°C.
9. Rotator for end-over-end mixing.
10. Sonicator or other cell-disrupting apparatus (e.g., French press).
11. Syringe with a capacity of at least 10 mL.
12. 0.45 μm syringe filters.
13. 0.22 μm syringe filters.
14. High molecular weight cut-off (MWCO) dialysis tubing (e.g., Spectra Por 131420 Biotech-Grade CE Dialysis Tubing, 100,000 MWCO).
15. Gel filtration system (e.g., ÄKTA Pure 25 system from GE Healthcare).
16. Size-Exclusion Chromatography (SEC) column capable of separating 2.15 MDa particles from the void peak (e.g., HiPrep™ 16/60 Sephacryl® S-400 HR from GE Healthcare).
17. SDS-PAGE gels at the appropriate acrylamide percentage for resolving SpyCatcher003-mi3 monomer, the selected SpyTag antigen(s), and the pair after covalent coupling.
18. Electrophoresis apparatus for running SDS-PAGE.
19. Heating block capable of maintaining 37°C and compatible with 1.5 mL microcentrifuge tubes.
20. Pyrogen-free 96-well flat bottom microplates with a volume of at least 300 μL for bicinchoninic acid (BCA) and Limulus amebocyte lysate (LAL) assays.

21. Microplate reader for measurement of plates as part of a BCA or LAL assay.
22. Dynamic light scattering (DLS) instrument and compatible cuvettes for measurement (e.g., Zetasizer Nano S DLS and Disposable Solvent Resistant Micro Cuvette [ZEN0040] from Malvern Panalytical).
23. System for concentrating proteins (e.g., Vivaspin 20 100 kDa MWCO centrifugal concentrator from GE Healthcare).

2.2 Proteins, DNA, and Other Reagents

1. pET28a SpyCatcher003-mi3 plasmid (GenBank MT945417, Addgene 159995).
2. Plasmid for the SpyTag fusion antigen of interest. For example, an N-terminal SpyTag003 fusion with the OspC protein from *Borrelia burgdorferi* (Addgene 159998) or a C-terminal SpyTag fusion with Hemagglutinin-Neuraminidase from Newcastle Disease virus (Addgene 160000).
3. 100 mg/mL kanamcyin solution: 1 g of kanamycin sulfate in 10 mL MilliQ water. Syringe-filter and store at −20°C.
4. 20% (w/v) glucose solution: 200 g/L of D-glucose in MilliQ water. Autoclave and store at room temperature.
5. 420 mM IPTG solution: 1 g isopropyl-β-D-thiogalactopyranoside in 10 mL MilliQ water. Syringe-filter and store at −20°C.
6. Complete Protease Inhibitor Cocktail tablets (Roche) for inhibiting *Escherichia coli* proteases during purification. Store at −20°C.
7. 100 mM PMSF solution: 17.4 mg of phenylmethylsulfonyl fluoride (PMSF; beware-hazardous) in 1 mL of isopropanol. Store at −20°C.
8. 10 mg/mL lysozyme solution: 10 mg Hen Egg White Lysozyme (e.g., Lysozyme from chicken egg white [L6876] from Sigma-Aldrich) in 1 mL MilliQ water. Store at −20°C.
9. Ammonium sulfate (laboratory reagent grade).
10. Coomassie-based gel stain for SDS-PAGE gels (e.g., InstantBlue™ from Sigma-Aldrich).
11. Protein ladder marker for SDS-PAGE (e.g., PageRuler™ Plus Prestained Protein Ladder from Thermo Fisher).
12. Triton X-114.
13. Reagents for BCA assay (e.g., Pierce™ BCA Protein Assay Kit from Thermo Fisher).
14. Reagents for endotoxin quantification (e.g., Pierce™ Chromogenic Endotoxin Quant Kit from Thermo Fisher).

2.3 Buffers, Media, and Cells

1. Appropriate *E. coli* expression strain, such as BL21(DE3) RIPL (Agilent).
2. Luria Bertani broth (LB): 10 g/L tryptone, 5 g/L yeast extract, 5 g/L NaCl. Autoclave and store at room temperature.
3. Resuspension Buffer: 25 mM Tris-HCl, 300 mM NaCl, pH adjusted to 7.9 with NaOH when measured at 25°C. Store at room temperature.
4. SpyCatcher003-mi3 Buffer: 25 mM Tris-HCl, 150 mM NaCl, pH adjusted to 7.4 with NaOH when measured at 25°C. Store at room temperature.
5. 6× SDS protein gel loading buffer: 234 mM Tris-HCl pH 6.8, 24% (v/v) glycerol, 120 μM bromophenol blue, 234 mM SDS. Store at −20°C.

3 Methods

3.1 Production of SpyCatcher003-mi3

This procedure for the production of SpyCatcher003-mi3 takes at least 5 days to complete.

1. Transform pET28a SpyCatcher003-mi3 plasmid into an appropriate *E. coli* expression strain, such as BL21(DE3) RIPL.
2. Use a single colony from this transformation to inoculate 10 mL LB plus 50 μg/mL kanamycin and grow for 16 h at 37°C with shaking at 200 rpm.
3. Use this culture to inoculate 1 L LB plus 50 μg/mL kanamycin and 0.8% (w/v) glucose and grow at 37°C with shaking at 200 rpm until OD_{600} is between 0.6 and 0.8. Take a preinduction sample for analysis by SDS-PAGE (Fig. 5A).
4. Induce protein expression with 420 μM IPTG.
5. Incubate the culture at 22°C with shaking at 200 rpm for 16 h.
6. Take a post-induction sample for analysis by SDS-PAGE to confirm protein expression (Fig. 5A).
7. Centrifuge culture at 4000 × *g* at 4°C for 20 min to pellet the cells. Discard the supernatant.
8. Expression can be validated by diluting a small amount of pellet in MilliQ water and running on SDS-PAGE with coomassie staining. There should be an intense band corresponding to SpyCatcher003-mi3. Similar to other SpyCatcher-containing proteins, SpyCatcher003-mi3 tends to run larger than expected by SDS-PAGE. This discrepancy may be due to the low pI of SpyCatcher003 (Table 1) [45]. The apparent

molecular weight of SpyCatcher003-mi3 by SDS-PAGE is ~45 kDa (Fig. 5A).

Resuspend the cells in resuspension buffer using 40 mL buffer per 1 L culture, a final concentration of 0.1 mg/mL lysozyme, 1 mg/mL cOmplete mini EDTA-free protease inhibitor (Merck), and 1 mM PMSF.

9. Incubate for 30 min at room temperature with end-over-end mixing.

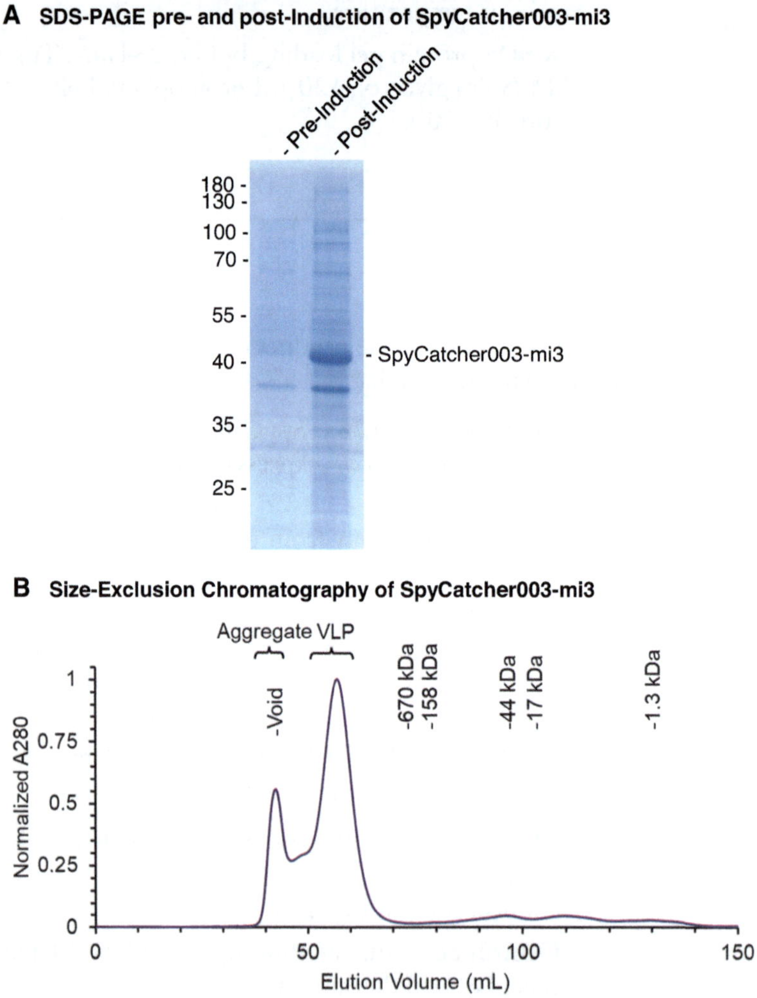

Fig. 5 Expression of SpyCatcher003-mi3. (**A**) SDS-PAGE gel showing an *E. coli* culture transformed with the SpyCatcher003-mi3 plasmid before induction and post-induction. The post-induction sample was taken immediately prior to harvest. This technique demonstrates that the protein is being expressed before the time-consuming purification process is undertaken. (**B**) A size-exclusion chromatography trace for SpyCatcher003-mi3 run on a S400 Sephacryl column in SpyCatcher003-mi3 Buffer. Markers based on gel filtration standard are shown. VLP represents the fraction that was pooled and taken forward during the purification process. There is a clear separation of nanoparticle and protein aggregate from ammonium sulfate precipitation

10. Sonicate resuspended cells on ice for 60 s at 50% duty-cycle (alternating 1 s on and 1 s off) using a sonicator equipped with a microtip. From this step onwards, the protein solution should be kept on ice or stored at or below 4°C.
11. Centrifuge for 45 min at 30,000 × *g* at 4°C.
12. Harvest the supernatant.
    Steps 13–17 describe enrichment of SpyCatcher003-mi3 by ammonium sulfate precipitation, which provides a scalable route for purification without the need for any affinity chromatography step [26]. The nanoparticle is subsequently further separated from bacterial proteins by dialysis with high MWCO dialysis tubing and SEC. Alternatively, SpyCatcher003-mi3 can be purified by affinity chromatography using its C-tag [33, 46] and the protocol can continue using pooled elution fractions at step 18.
13. For every 1 mL of supernatant, add 170 mg of ammonium sulfate as a fine powder. Any lumps that have formed should be broken up prior to being added to the supernatant. Incubate at 4°C with gentle mixing using a stir bar at 120 rpm for 1 h.
14. Centrifuge for 30 min at 30,000 × *g* at 4°C.
15. Discard the supernatant and, at 4°C or on ice, resuspend the pellet in 10 mL SpyCatcher003-mi3 Buffer by pipetting up and down using a serological pipette.
16. Filter the resuspended pellet through a 0.45 μm syringe filter and then through a 0.22 μm syringe filter.
17. Dialyze the filtrate for 16 h at 4°C against at least 1000-fold excess of SpyCatcher003-mi3 Buffer using 100,000 MWCO dialysis tubing.
18. Pre-equilibrate an appropriate SEC column using SpyCatcher003-mi3 Buffer that has been filtered (0.22-μm filter) and degassed under vacuum. Column selection is a critical step. Properly formed SpyCatcher003-mi3 nanoparticles are 2.15 MDa, and a column should be selected that has suitable resolution to separate these particles from aggregate, e.g., HiPrep™ 16/60 Sephacryl® S-400 HR (GE Healthcare) (*see* **Note 1**).
19. Centrifuge the post-dialysis protein for 30 min at 16,000 × *g* at 4°C.
20. Filter the supernatant through a 0.22-μm syringe filter.
21. Run the flow-through over the pre-equilibrated SEC column (Fig. 5B). For a HiPrep™ 16/60 Sephacryl® S-400 HR, typically 5–20 mg of SpyCatcher003-mi3 is loaded.
22. Assess fractions containing protein (based on A_{280} measurement) by SDS-PAGE with coomassie staining.

23. Pool the fractions from the peak which is determined to be the SpyCatcher003-mi3 nanoparticle. This peak should be separate from the void peak which contains protein aggregate (Fig. 5B). If there is overlap between the two peaks, do not pool protein from the overlapping region.
24. If desired, the protein can be concentrated to 1–3 mg/mL at this point, for example, using a Vivaspin 20 100 kDa MWCO centrifugal concentrator (GE Healthcare).
25. Quantification of SpyCatcher003-mi3 protein concentration can be performed by a bicinchoninic acid (BCA) assay. At this point, you would expect to have at least 25 mg SpyCatcher-mi3 per L of initial culture.

3.2 Production of Antigen

SpyCatcher003-mi3 can couple with antigens through reaction with SpyTag or SpyTag003 [26]. Antigens can be designed with the tag at either terminus [47] or included within permissive loops of the protein [48]. When deciding on tag location, consideration should be given to the impact that Tag/Catcher coupling will have on the accessibility of different antigen surfaces. A flexible linker composed of glycine and serine can be used to create space between the tag and antigen. For example, the cysteine-rich protective antigen (CyRPA) from *Plasmodium falciparum* was cloned to incorporate a C-terminal SpyTag separated from the antigen with a 9-residue $(GSG)_3$ spacer (GenBank MH425516).

SpyTagged antigens can be produced using a variety of expression systems (Sect. 1) and purification techniques, with Ni-NTA purification of His_6-tagged protein most common (*see* **Note 2**). The SpySwitch system has been developed to use SpyTag directly as the affinity purification tag. SpySwitch is a SpyCatcher variant that binds non-covalently to all generations of SpyTag. SpySwitch captures SpyTagged proteins at pH 7.5 and 4°C, while releasing proteins at a moderately acidic pH (~5.0) or at 37°C. This system enables a single tag to be used for purification and multivalent display [49].

3.3 Coupling of Antigen to SpyCatcher003-mi3

Mixing SpyCatcher003-mi3 with SpyTagged antigens is all that is required to facilitate coupling. The reaction will proceed in various buffer conditions without the need for any additional co-factors [43, 50]. Typically, the reaction is performed for 16 h at 4°C with the monomeric concentration of SpyCatcher003-mi3 and antigen between 1 and 10 μM (Fig. 6A). It is possible for the reaction to proceed at lower concentrations or for a shorter time; optimal conditions can be empirically determined for each antigen.

If one aims to achieve antigen saturation on the nanoparticle, an excess of antigen is typically required (Fig. 7). Titration between an antigen:nanoparticle subunit ratio of 1:1 and 3:1 can be performed to determine the optimal coupling conditions (Fig. 6A) (*see* **Note 3**). The degree of coupling can be determined by densitometry after

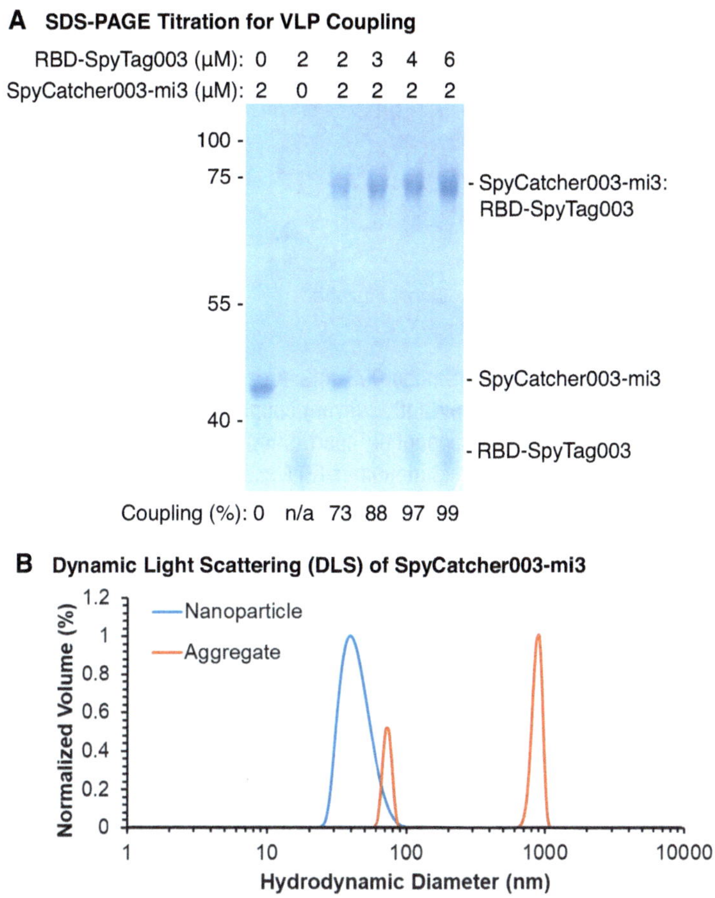

Fig. 6 Validation of SpyCatcher003-mi3. (**A**) An example titration of SpyTag antigen with SpyCatcher003-mi3. Increasing molar amounts of SARS-CoV-2 receptor-binding domain (RBD) with a C-terminal SpyTag003 are reacted for 16 h at 4°C and run on reducing SDS-PAGE with coomassie staining. n/a = not applicable. (**B**) DLS is performed on SpyCatcher003-mi3 coupled to completion with excess SARS-CoV-2 RBD-SpyTag003 (blue). To demonstrate DLS of an aggregate, the particles were vortexed at high speed for 5 min and re-measured (orange)

SDS-PAGE with coomassie staining (Sect. 3.5). Excess uncoupled antigen can be removed by dialysis or SEC (Sect. 3.4). Immunization with nanoparticles displaying antigens at sub-saturating levels can also provide potent immune enhancement and simplify the production process by eliminating the need to remove excess antigen [51–53].

3.3.1 Coupling of Multiple Antigens

SpyCatcher003-mi3 has been used to co-display several SpyTagged antigens on the same nanoparticle. For example, receptor-binding domains (RBDs) from as many as eight different sarbecoviruses have been co-displayed on SpyCatcher003-mi3. These mosaic nanoparticles elicit immune responses targeted to regions conserved across

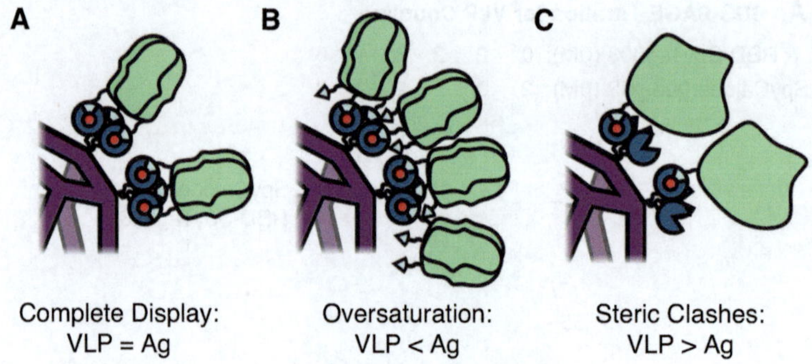

Fig. 7 Scenarios for antigen coupling to SpyCatcher003-mi3. A cartoon representation of different scenarios for an antigen (green) genetically fused to SpyTag003 (cyan) coupling with SpyCatcher003 (dark blue) on an mi3 scaffold (purple). The formation of an isopeptide bond is indicated in red. (**A**) A scenario where every SpyTag003 interacts with a SpyCatcher003-mi3 monomer. (**B**) A scenario where an oligomeric antigen couples to SpyCatcher003-mi3 in an oversaturated manner, leaving free SpyTag003-antigen, which is not covalently linked to the SpyVLP. (**C**) A scenario where steric hindrance prevents full saturation on the nanoparticle, leading to uncoupled SpyCatcher003-mi3 monomers

the coupled antigens [54]. For optimal batch-to-batch consistency of co-displayed nanoparticles, antigens should be thoroughly mixed together prior to the addition of SpyCatcher003-mi3. If antigens were added sequentially, then because of the fast kinetics of the SpyTag003/SpyCatcher003 reaction [43], the antigens added earlier may be over-represented on particles relative to antigens added later. Consideration should be given to features of the different antigens (e.g., size, oligomeric state, and SpyTag accessibility) that may impact their relative rates of coupling and efficiency of display [26].

3.4 Removal of Free Antigen

Excess uncoupled antigen that was introduced to achieve saturation can be removed by a variety of methods, typically reliant on the large size difference between soluble antigens and SpyCatcher003-mi3 nanoparticles.

SEC can be performed post-coupling to separate antigen (2-250 kDa) from the much larger nanoparticle (>2.15 MDa). Compared with SpyCatcher003-mi3 purification, a smaller column (e.g., HiPrep™ 16/60 Sephacryl® S-200 HR) can be used, since removal of antigen does not require separation of the nanoparticle and void peak. A larger column (e.g., HiPrep™ 16/60 Sephacryl® S-400 HR) can be implemented to assess any coupling-induced aggregation, while also removing excess free antigen.

Alternatively, dialysis can be performed post-coupling, using high MWCO dialysis tubing. It is critical to select dialysis tubing that is an appropriate size to allow antigen to pass through while retaining the nanoparticle. Dialysis can be performed against any buffer that is appropriate for the nanoparticle-antigen conjugate.

The presence of excess antigen can be determined by SDS-PAGE with coomassie staining. When assessing the removal of excess antigen, it is important to consider the oligomeric state of the antigen. It is possible that some SpyTags in a multimeric assembly will not couple to SpyCatcher003. For example, a trimeric protein may couple to SpyCatcher003-mi3 via only one of the three accessible SpyTags (Fig. 7). The full trimer would be attached to the SpyCatcher003-mi3 nanoparticle and be retained through dialysis or SEC. However, when assessed by SDS-PAGE, the two subunits with free SpyTags would run indistinguishably from the subunits of an unattached trimer.

3.5 VLP Validation

The coupling of antigen and SpyCatcher003-mi3 can be assessed by SDS-PAGE with coomassie staining (Fig. 6A). When SpyCatcher003-mi3 is boiled and treated with denaturing loading buffer as part of the SDS-PAGE process, the nanoparticle will disassemble into its monomeric subunits. In contrast, the covalent bond that forms between SpyTag and SpyCatcher survives the SDS-PAGE process. SpyCatcher003-mi3 monomers that have been covalently linked to an antigen will migrate more slowly through the gel than uncoupled monomers, resulting in an upper conjugated band and a lower unconjugated band (*see* **Note 4**). SDS-PAGE densitometry can be used to quantify coupling. Typically, this quantification is most reliable when the depletion of the uncoupled SpyCatcher003-mi3 band is compared to a control lane that has the same amount of SpyCatcher003-mi3 but no additional antigen (Fig. 6A). The equation to calculate coupling percentage is:

$$\text{Coupling Percentage} = 100\% - \left(\frac{\text{VLP Band Intensity After Antigen Reaction}}{\text{VLP Band Intensity Without Antigen Added}} \times 100\%\right)$$

It is valuable to assess SpyCatcher003-mi3 nanoparticles by Dynamic Light Scattering (DLS) (Fig. 6B). This technique measures particle size distribution and can give insight into proper assembly of the nanoparticles [55]. Larger particles scatter much more and so this technique is especially sensitive to particle aggregation (*see* **Note 5**). Smaller proteins, such as uncoupled antigen, are unlikely to be detected by DLS in the presence of SpyCatcher003-mi3 nanoparticle. Antigens being coupled to SpyCatcher003-mi3 increase the size of the particles that may be detectable by DLS. Typically, DLS is performed with particles at a concentration of ~30 nM (~2 μM of SpyCatcher003-mi3 monomer). Samples are generally centrifuged for 30 min at 16,000 × *g* at 4°C prior to DLS measurement, to match treatment of the particles prior to immunization.

SpyCatcher003-mi3 nanoparticles, on their own or coupled to antigen, can be imaged using negative stain transmission electron microscopy (TEM) [26]. Cryo-electron microscopy has been used to demonstrate the reasonable uniformity of the mi3 nanocage but the flexibility of the linkers between SpyCatcher003 and mi3 means that coupled antigens are often disordered and do not give well-resolved electron density [53].

3.6 Endotoxin

Endotoxins, also known as lipopolysaccharides or LPS, are components of the outer membrane of Gram-negative bacteria [56]. Endotoxins are immunostimulatory and can cause potent inflammatory responses [57]. Endotoxin levels are a vital consideration for any recombinant protein that reaches clinical use [58]. In preclinical animal models, variability in endotoxin levels can lead to different immune stimulation, independent of the components being tested, and can cause misinterpretation of differences. It is therefore critical to ensure that low levels of endotoxin are present in pre-clinical vaccine candidates. A maximum threshold of 20 endotoxin units (EU) per mL has been recommended for recombinant protein vaccines [59].

Endotoxin is introduced into solution upon lysis of a Gram-negative expression strain such as *E. coli* BL21 (DE3) [60]. The production process for SpyCatcher003-mi3 or for antigens expressed in *E. coli* is therefore likely to introduce endotoxin to the sample. Endotoxin contamination can also occur in proteins produced in different organisms, especially in settings where *E. coli* protein expression is also being undertaken. Endotoxin depletion and quantification are important for the implementation of SpyCatcher003-mi3 vaccines. Nonetheless, endotoxin can be depleted through a simple and cost-effective procedure.

3.6.1 Endotoxin Depletion

Endotoxin depletion can be performed immediately after purification of SpyCatcher003-mi3 or at a later stage in the vaccine production process. It is important to consider any steps that will be downstream of endotoxin depletion (e.g., antigen coupling or excess antigen removal) and whether these steps may contaminate the sample with endotoxin. There are a variety of methods that can be used for endotoxin depletion [61]. The following Triton X-114 phase separation [62] protocol has been used successfully for SpyCatcher003-mi3.

1. All manipulations should be performed using filtered pipette tips and all microcentrifuge tubes should be certified as endotoxin free (non-pyrogenic).
2. Clean all work surfaces and equipment with 70% (v/v) ethanol.
3. Set the ThermoMixer and microcentrifuge (both with 1.5 mL adaptor) to 37°C.

4. To avoid contamination of the stock, transfer the total required volume of Triton X-114 from the bottle to a microcentrifuge tube. Triton X-114 is viscous and moves slowly when being pipetted.
5. The recombinant protein volume will ideally be 400–1500 μL with a protein concentration below 10 mg/mL. The procedure can be performed with a lower volume, but this is likely to increase protein loss. The procedure can be performed with a higher volume but will need to be performed across multiple tubes.
6. Add 1% (v/v) Triton X-114 to the protein (e.g., 1 μL Triton X-114 per 100 μL protein). The Triton X-114 should form a thin visible line in the solution. Mix the Triton X-114 into solution using the pipette tip until homogeneous. Incubate on ice for 5 min.
7. Transfer to a heating block at 37°C and incubate for 5 min. The solution should become cloudy at this point.
8. Transfer to a microcentrifuge and spin for 2 min at 16,000 × *g* at 37°C. A visible phase separation should occur at this point. The top phase contains the protein of interest and the bottom phase contains the Triton X-114 with a higher concentration of endotoxin.
9. Transfer the top phase to a new endotoxin-free tube. It is preferable to leave a small amount of the top phase than to accidentally transfer the endotoxin to the new tube.
10. Repeat steps 6–9 twice, for a total of three rounds of endotoxin depletion.
11. Perform steps 6–9 a fourth time but do not add 1% (v/v) Triton X-114 at step 6. After centrifugation at 37°C no visible phase separation should occur. Leave ~5 μL in the tube when performing the final transfer. This step ensures that a minimal amount of Triton X-114 and endotoxin remains in the protein sample.

3.6.2 Endotoxin Quantification

Endotoxin quantification of SpyCatcher003-mi3 vaccine candidates must be performed after any step that has the potential to introduce endotoxin. There are a variety of available techniques for endotoxin quantification [63, 64]. One of these techniques that has been successfully implemented for SpyCatcher003-mi3 is a LAL assay [26, 47] using the Pierce™ Chromogenic Endotoxin Quant Kit, following the manufacturer's guidance.

3.7 Storage of SpyCatcher003-mi3

SpyCatcher003-mi3 nanoparticles are robust to lyophilization [53] and several freeze-thaw cycles [26]. For storage of 1–3 days, the nanoparticles may be stored at 4°C. For storage of more than a few days, purified particles should be maintained at −80°C. Purified

SpyCatcher003-mi3 remain viable for at least one year stored at −80°C. It is desirable to minimize freeze-thaw cycling and it is recommended to separate the protein stock into multiple aliquots. If an aliquot of SpyCatcher003-mi3 is run on SDS-PAGE/coomassie and novel bands appear, proteolysis has probably occurred and the aliquot should be discarded.

There is no certainty that the robustness of the nanoparticle will be maintained when coupled to an antigen. Freeze-thaw cycling or lyophilization and resuspension may destabilize the coupled antigen and cause particle aggregation [65]. It is important to validate appropriate storage conditions for each new antigen that is coupled to SpyCatcher003-mi3. Storage-induced particle aggregation can be assessed by DLS, as previously described (Sect. 3.5). Integrity of the coupled antigen can be determined by methods such as an enzyme-linked immunosorbent assay (ELISA), using a monoclonal antibody specific to the antigen in its desire form [53]. The ELISA procedure depends on the particulars of the antigen. If particle aggregation or antigen misfolding occurs during storage, consider the inclusion of a cryoprotectant such as sucrose or trehalose [66, 67].

3.8 Immunization

Immunizations have been performed with SpyVLPs using a variety of routes, including intramuscular [9, 10], subcutaneous [68, 69], intraperitoneal [70], intranasal [71, 72], intradermal [73], and intratracheal [74] administration. Although a large variety of animal species have been immunized with SpyVLPs (Fig. 3), mice (*Mus musculus*) are the most common animal model for SpyVLP immunizations. SpyVLP immunizations have been performed on BALB/c [33, 52], C57BL/6 [69, 75], CD-1 [76], and a variety of transgenic mice strains [32, 77].

When considering the dose of SpyVLP to administer, it is critical to consider the effect of coupling on the molarity of the antigens involved. When comparing an antigen in soluble or VLP-displayed format, it is best practice to normalize by the number of antigens that are present in both samples. For example, in an immunization testing a SpyTag SARS-CoV-2 RBD (Addgene 159999), a 1 μg dose of the soluble RBD (25.4 kDa) would require a 2.4 μg dose of SpyCatcher003-mi3 coupled to RBD (61.2 kDa per monomer), in order to achieve a molar equivalence of antigen. The dose required for immunization will vary based on the specifics of the experiment. Higher doses will generally increase the likelihood of an effective response. However, higher doses can lead to saturation of antibody responses in mice in a manner that is difficult to achieve for human clinical studies. Consequently, high doses can obscure relevant differences between the groups being tested [78, 79]. A range of doses for immunization of mice with antigen conjugated to SpyCatcher003-mi3 has been reported, ranging from 0.1 [53] to 10 μg of antigen [80].

4 Notes

1. Aggregation of SpyCatcher003-mi3 is an important consideration. Although the SpyCatcher003-mi3 nanoparticle is robust to various stresses, it is common for some aggregation to occur during the ammonium sulfate precipitation process. SEC should effectively separate the properly assembled nanoparticle from aggregated protein. If aggregate is retained in the sample after SEC, it is possible that the peaks were not properly resolved. Ensure that the SEC column has the appropriate resolution to separate the 2.15 MDa particle from aggregate in the void peak. If there is any overlap between the aggregate and nanoparticle elution peaks, do not pool fractions in the overlapping region. If the SEC chromatogram shows clear separation between the aggregate and nanoparticle peak, then consider the post-SEC storage conditions for the nanoparticle. The particle should not be stored above 4°C and, if the VLPs will be stored for longer than a few days, should be stored at −80°C or lyophilized. Effort should be made to minimize the number of freeze-thaw cycles.
2. If an antigen has previously been expressed successfully but fails to express when genetically fused to SpyTag, consider re-designing the construct with the SpyTag at the other terminus and/or alter the length of the flexible linker between tag and antigen [44]. The SpyTag may also be added internally if a permissive region of the antigen is identified [48].
3. If an antigen fails to react to completion (or react at all) with SpyCatcher003-mi3, attempt the coupling at a higher overall concentration (aiming for at least 5 μM) and consider increasing the ratio of antigen relative to SpyCatcher003-mi3. If this approach does not drive the reaction to completion, then the construct may need to be redesigned with the tag at a different location or with longer flexible linkers separating the tag and the antigen. If SpyTag has been used, it is recommended to update the construct with the faster reacting SpyTag003 [43]. As a positive control, soluble SpyCatcher003 (plasmid at Addgene 133447 or protein from Bio-Rad, TZC025) can be reacted with the antigen in question. Similarly, SpyTag003-Maltose-Binding Protein (MBP) (plasmid at Addgene 133450 or protein from Kerafast, EOX005) can be reacted with SpyCatcher003-mi3 to validate the reactivity of the nanoparticle batch. If the antigen fails to react with soluble SpyCatcher003, mass spectrometry can be performed to ensure that the SpyTag remains present on the antigen after expression and purification. It is possible that steric hindrance will prevent an antigen from reacting to completion for multivalent display on a SpyVLP (Fig. 7) [81].

If SDS-PAGE/coomassie shows complete depletion of SpyCatcher003-mi3 when reacted with excess antigen but shows free antigen with excess SpyCatcher003-mi3, consider the oligomeric state of the antigen. As discussed in Sect. 3.4, a multimeric protein may not react through all accessible SpyTags. This situation can lead to a subunit that is indirectly attached to the nanoparticle through non-covalent interactions with other multimer subunits covalently linked to SpyCatcher003 (Fig. 7). On SDS-PAGE/coomassie, the subunit that is non-covalently linked to the nanoparticle will run indistinguishably from a subunit of a fully soluble multimer.

4. If the antigen runs similarly to SpyCatcher003-mi3 on SDS-PAGE/coomassie, it may be difficult to assess what level of coupling has been accomplished. If the antigen is glycosylated, the reaction mixture can be treated with PNGase F to remove glycans and facilitate band separation [47]. If the antigen is not glycosylated or the glycan removal does not produce a sufficient downward shift, then a western blot can be performed using a commercial antibody specific to the C-tag on SpyCatcher003-mi3. This approach will allow for determination of depletion of uncoupled SpyCatcher003-mi3.
5. It is critical to monitor the solubility of SpyCatcher003-mi3 after antigen coupling. If the concentration of the antigen-coupled nanoparticle drastically declines after centrifugation (16,000 × *g*, 4°C, at least 20 min), the coupling may have induced insolubility and the pair may have crashed out of solution. Precipitation of nanoparticles does not necessarily reflect unfolding and may simply reflect insufficient repulsion between the particles [82]. In this scenario, then modifications should be made to the reaction buffer and/or pH. The SpyTag/SpyCatcher reaction is amenable to a wide range of reaction conditions [43] and so the reaction buffer should be selected based on the ideal conditions for the antigen. Consider the pI of the antigen on its own or coupled to SpyCatcher003-mi3 [83]. Avoid using a buffer at this pH, which will cause a net neutral charge for the protein that can reduce solubility. Consider whether the addition of divalent cations, sucrose, or other co-factors may improve antigen stability. A buffer panel can be constructed in order to test a variety of reaction conditions. If these buffer changes are insufficient, then perform the reactions at reduced concentrations of both the antigen and SpyCatcher003-mi3.

If modifications to the reaction conditions do not overcome the solubility issues, consider the quality control that has been performed on the antigen. Misfolding or aggregation already present in the antigen can be exacerbated when multimerized on SpyCatcher003-mi3. Quality control to address this possibility could include

SEC to identify and remove aggregate, as well as ELISA with a conformation-specific antibody to assess antigen folding [41]. Protein engineering to enhance stability and solubility may also be relevant for the most challenging antigens [41].

Acknowledgments

R.A.H. received funding from the Rhodes Trust and Townsend-Jeantet Prize Trust.

Conflict of Interest M.R.H. is an author on patents or patent applications for Tag/Catcher/Ligase technologies (UK Intellectual Property Office 1002362.0, 1509782.7, 1705750.6, 1706430.4, 1903479.2, 2104999.4). M.R.H. is a co-founder and shareholder and was formerly a consultant for SpyBiotech. R.A.H. has no conflict of interest.

References

1. Habibi N, Mauser A, Ko Y, Lahann J (2022) Protein nanoparticles: uniting the power of proteins with engineering design approaches. Adv Sci 9(8):2104012. https://doi.org/10.1002/advs.202104012
2. Kaltbeitzel J, Wich PR (2023) Protein-based nanoparticles: from drug delivery to imaging, nanocatalysis and protein therapy. Angew Chem 62(44):202216097. https://doi.org/10.1002/anie.202216097
3. Hsia Y, Bale JB, Gonen S, Shi D, Sheffler W, Fong KK, Nattermann U, Xu C, Huang PS, Ravichandran R, Yi S, Davis TN, Gonen T, King NP, Baker D (2016) Design of a hyperstable 60-subunit protein icosahedron. Nature 535(7610):136–139. https://doi.org/10.1038/nature18010
4. Wang J, Liu Y, Chen Y, Wang A, Wei Q, Liu D, Zhang G (2020) Large-scale manufacture of VP2 VLP vaccine against porcine parvovirus in *Escherichia coli* with high-density fermentation. Appl Microbiol Biotechnol 104(9):3847–3857. https://doi.org/10.1007/s00253-020-10483-5
5. Joyce MG, Chen W-H, Sankhala RS, Hajduczki A, Thomas PV, Choe M, Martinez EJ, Chang WC, Peterson CE, Morrison EB, Smith C, Chen RE, Ahmed A, Wieczorek L, Anderson A, Case JB, Li Y, Oertel T, Rosado L, Modjarrad K (2021) SARS-CoV-2 ferritin nanoparticle vaccines elicit broad SARS coronavirus immunogenicity. Cell Rep 37(12):110143. https://doi.org/10.1016/j.celrep.2021.110143
6. Ober Shepherd BL, Scott PT, Hutter JN, Lee C, McCauley MD, Guzman I, Bryant C, McGuire S, Kennedy J, Chen W-H, Hajduczki A, Mdluli T, Valencia-Ruiz A, Amare MF, Matyas GR, Rao M, Rolland M, Mascola JR, Rosa SC, Gazi M (2024) SARS-CoV-2 recombinant spike ferritin nanoparticle vaccine adjuvanted with army liposome formulation containing monophosphoryl lipid A and QS-21: a phase 1, randomised, double-blind, placebo-controlled, first-in-human clinical trial. Lancet Microbe 5(6):581–593. https://doi.org/10.1016/S2666-5247(23)00410-X
7. Brune KD, Howarth M (2018) New routes and opportunities for modular construction of particulate vaccines: stick, click, and glue. Front Immunol 9:1432. https://doi.org/10.3389/fimmu.2018.01432
8. Mateu MG (2011) Virus engineering: functionalization and stabilization. Protein Eng Des Sel 24(1–2):53–63. https://doi.org/10.1093/protein/gzq069
9. Brune KD, Leneghan DB, Brian IJ, Ishizuka AS, Bachmann MF, Draper SJ, Biswas S, Howarth M (2016) Plug-and-display: decoration of virus-like particles via isopeptide bonds for modular immunization. Sci Rep 6:19234. https://doi.org/10.1038/srep19234
10. Thrane S, Janitzek CM, Matondo S, Resende M, Gustavsson T, Jongh WA, Clemmensen S, Roeffen W, Vegte Bolmer M, Gemert GJ, Sauerwein R, Schiller JT, Nielsen MA, Theander TG, Salanti A, Sander AF (2016) Bacterial superglue enables easy development of efficient virus-like particle based vaccines. J Nanobiotechnol 14(1):30. https://doi.org/10.1186/s12951-016-0181-1
11. Zakeri B, Fierer JO, Celik E, Chittock EC, Schwarz-Linek U, Moy VT, Howarth M (2012)

Peptide tag forming a rapid covalent bond to a protein, through engineering a bacterial adhesin. Proc Natl Acad Sci 109(12):E690–E697. https://doi.org/10.1073/pnas.1115485109
12. Pollard AJ, Bijker EM (2021) A guide to vaccinology: from basic principles to new developments. Nat Rev Immunol 21(2):83–100. https://doi.org/10.1038/s41577-020-00479-7
13. Bachmann MF, Jennings GT (2010) Vaccine delivery: a matter of size, geometry, kinetics and molecular patterns. Nat Rev Immunol 10(11):787–796. https://doi.org/10.1038/nri2868
14. Hervé C, Laupèze B, Del Giudice G, Didierlaurent AM, Silva F (2019) The how's and what's of vaccine reactogenicity. NPJ Vaccines 4(1):39. https://doi.org/10.1038/s41541-019-0132-6
15. Hills RA, Howarth M (2022) Virus-like particles against infectious disease and cancer: guidance for the nano-architect. Curr Opin Biotechnol 73:346–354. https://doi.org/10.1016/j.copbio.2021.09.012
16. Heddle JG, Chakraborti S, Iwasaki K (2017) Natural and artificial protein cages: design, structure and therapeutic applications. Curr Opin Struct Biol 43:148–155. https://doi.org/10.1016/j.sbi.2017.03.007
17. Gomes A, Mohsen M, Bachmann M (2017) Harnessing nanoparticles for immunomodulation and vaccines. Vaccines 5(1):6. https://doi.org/10.3390/vaccines5010006
18. Okba NMA, Widjaja I, Dieren B, Aebischer A, Amerongen G, Waal L, Stittelaar KJ, Schipper D, Martina B, Brand JMA, Beer M, Bosch B-J, Haagmans BL (2020) Particulate multivalent presentation of the receptor binding domain induces protective immune responses against MERS-CoV. Emerg Microbes Infect 9(1):1080–1091. https://doi.org/10.1080/22221751.2020.1760735
19. Dalvie NC, Tostanoski LH, Rodriguez-Aponte SA, Kaur K, Bajoria S, Kumru OS, Martinot AJ, Chandrashekar A, McMahan K, Mercado NB, Yu J, Chang A, Giffin VM, Nampanya F, Patel S, Bowman L, Naranjo CA, Yun D, Flinchbaugh Z, Barouch DH (2022) SARS-CoV-2 receptor binding domain displayed on HBsAg virus–like particles elicits protective immunity in macaques. Sci Adv 8(11):6015. https://doi.org/10.1126/sciadv.abl6015
20. Ji M, Zhu J, Xie X, Liu D, Wang B, Yu Z, Liu R (2020) A novel rapid modularized hepatitis B core virus-like particle-based platform for personalized cancer vaccine preparation via fixed-point coupling. Nanomed Nanotechnol Biol Med 28:102223. https://doi.org/10.1016/j.nano.2020.102223
21. Corrigan AR, Duan H, Cheng C, Gonelli CA, Ou L, Xu K, DeMouth ME, Geng H, Narpala S, O'Connell S, Zhang B, Zhou T, Basappa M, Boyington JC, Chen SJ, O'Dell S, Pegu A, Stephens T, Tsybovsky Y, Mascola JR (2021) Fusion peptide priming reduces immune responses to HIV-1 envelope trimer base. Cell Rep 35(1):108937. https://doi.org/10.1016/j.celrep.2021.108937
22. Wang W, Liu Z, Zhou X, Guo Z, Zhang J, Zhu P, Yao S, Zhu M (2019) Ferritin nanoparticle-based spytag/spycatcher-enabled click vaccine for tumor immunotherapy. Nanomed Nanotechnol Biol Med 16:69–78. https://doi.org/10.1016/j.nano.2018.11.009
23. Song J, Wang M, Zhou L, Tian P, Sun Z, Sun J, Wang X, Zhuang G, Jiang D, Wu Y, Zhang G (2023) A candidate nanoparticle vaccine comprised of multiple epitopes of the African swine fever virus elicits a robust immune response. J Nanobiotechnology 21(1):424. https://doi.org/10.1186/s12951-023-02210-9
24. Liu Z-H, Deng Z-F, Lu Y, Fang W-H, He F (2022) A modular and self-adjuvanted multivalent vaccine platform based on porcine circovirus virus-like nanoparticles. J Nanobiotechnology 20(1):493. https://doi.org/10.1186/s12951-022-01710-4
25. Yang JI, Kim KH (2022) Display of *Streptococcus iniae* α-enolase on the surface of virus-like particles (VLPs) of nervous necrosis virus (NNV) using SpyTag/SpyCatcher. Mar Biotechnol 24(6):1066–1072. https://doi.org/10.1007/s10126-022-10166-4
26. Rahikainen R, Rijal P, Tan TK, Wu HJ, Andersson AMC, Barrett JR, Bowden TA, Draper SJ, Townsend AR, Howarth M (2021) Overcoming symmetry mismatch in vaccine nanoassembly through spontaneous amidation. Angew Chem 60(1):321–330. https://doi.org/10.1002/anie.202009663
27. Liu Y, Pan C, Wang K, Guo Y, Sun Y, Li X, Sun P, Wu J, Wang H, Zhu L (2023) Preparation of a *Klebsiella pneumoniae* conjugate nanovaccine using glycol-engineered *Escherichia coli*. Microb Cell Factories 22(1):95. https://doi.org/10.1186/s12934-023-02099-x
28. Govasli ML, Diaz Y, Puntervoll P (2019) Virus-like particle-display of the enterotoxigenic *Escherichia coli* heat-stable toxoid STh-A14T elicits neutralizing antibodies in mice. Vaccine 37(43):6405–6414. https://doi.org/10.1016/j.vaccine.2019.09.004
29. Belghith AA, Cotter CA, Ignacio MA, Earl PL, Hills RA, Howarth MR, Yee DS, Brenchley JM, Moss B (2025) Mpox Multiprotein virus-like nanoparticle vaccine induces neutralizing and protective antibodies in mice and non-human primates. Nat Commun 16(1):4726. https://doi.org/10.1038/s41467-025-59826-8
30. Li X, Pan C, Sun P, Peng Z, Feng E, Wu J, Wang H, Zhu L (2022) Orthogonal modular

biosynthesis of nanoscale conjugate vaccines for vaccination against infection. Nano Res 15(2):1645–1653. https://doi.org/10.1007/s12274-021-3713-4

31. Gonzalez-Hernandez M, Kaiser FK, Steffen I, Ciurkiewicz M, Amerongen G, Tchelet R, Emalfarb M, Saloheimo M, Wiebe MG, Vitikainen M, Albulescu IC, Bosch B-J, Baumgärtner W, Haagmans BL, Osterhaus ADME (2023) Preclinical immunogenicity and protective efficacy of a SARS-CoV-2 RBD-based vaccine produced with the thermophilic filamentous fungal expression system thermothelomyces heterothallica C1. Front Immunol 14:1204834. https://doi.org/10.3389/fimmu.2023.1204834
32. Oosten L, Altenburg JJ, Fougeroux C, Geertsema C, End F, Evers WAC, Westphal AH, Lindhoud S, Berg W, Swarts DC, Deurhof L, Suhrbier A, Le TT, Torres Morales S, Myeni SK, Kikkert M, Sander AF, Jongh WA, Dagil R, Pijlman GP (2021) Two-component nanoparticle vaccine displaying glycosylated spike s1 domain induces neutralizing antibody response against SARS-CoV-2 variants. mBio 12(5):0181321. https://doi.org/10.1128/mBio.01813-21
33. Bruun TUJ, Andersson AMC, Draper SJ, Howarth M (2018) Engineering a rugged nanoscaffold to enhance plug-and-display vaccination. ACS Nano 12(9):8855–8866. https://doi.org/10.1021/acsnano.8b02805
34. Xu J, Sekiguchi T, Boonyakida J, Kato T, Park EY (2023) Display of multiple proteins on engineered canine parvovirus-like particles expressed in cultured silkworm cells and silkworm larvae. Front Bioeng Biotechnol 11:1096363. https://doi.org/10.3389/fbioe.2023.1096363
35. Stander J, Chabeda A, Rybicki EP, Meyers AE (2021) A plant-produced virus-like particle displaying envelope protein domain III elicits an immune response against West Nile virus in mice. Front Plant Sci 12:738619. https://doi.org/10.3389/fpls.2021.738619
36. Zheng P, Yang Y, Fu Y, He J, Hu Y, Zheng X, Duan B, Wang M, Liu Q, Li W, Li D, Yang Y, Yang Z, Yang X, Huang W, Ma Y (2023) Engineered norovirus-derived nanoparticles as a plug-and-play cancer vaccine platform. ACS Nano 17(4):3412–3429. https://doi.org/10.1021/acsnano.2c08840
37. Chanasit S, Johnston E, Thanasarnthungcharoen C, Kamath SD, Bohle B, Lopata AL, Jacquet A (2024) Hypoallergenic chimeric virus-like particles for the development of shrimp tropomyosin allergen Pen m 1-specific blocking antibodies. Allergy 79(4):1052–1055. https://doi.org/10.1111/all.15892
38. Ogrina A, Skrastina D, Balke I, Kalnciema I, Jansons J, Bachmann MF, Zeltins A (2022) Comparison of bacterial expression systems based on potato virus Y-like particles for vaccine generation. Vaccines 10(4):485. https://doi.org/10.3390/vaccines10040485
39. Goksøyr L, Skrzypczak M, Sampson M, Nielsen MA, Salanti A, Theander TG, Remaley AT, Jongh WA, Sander AF (2022) A cVLP-based vaccine displaying full-length PCSK9 elicits a higher reduction in plasma PCSK9 than similar peptide-based cVLP vaccines. Vaccines 11(1):2. https://doi.org/10.3390/vaccines11010002
40. SpyBiotech Limited (2014) A study of human cytomegalovirus (HCMV) vaccine SPYVLP01 with and without adjuvants. SPYVAC01. https://clinicaltrials.gov/study/NCT06145178
41. King LDW, Pulido D, Barrett JR, Davies H, Quinkert D, Lias AM, Silk SE, Pattinson DJ, Diouf A, Williams BG, McHugh K, Rodrigues A, Rigby CA, Strazza V, Suurbaar J, Rees-Spear C, Dabbs RA, Ishizuka AS, Zhou Y, Draper SJ (2024) Preclinical development of a stabilized RH5 virus-like particle vaccine that induces improved antimalarial antibodies. Cell Rep Med 5(7):101654. https://doi.org/10.1016/j.xcrm.2024.101654
42. Smit MJ, Sander AF, Ariaans MBPA, Fougeroux C, Heinzel C, Fendel R, Esen M, Kremsner PG, Heine R, Wertheim HF, Idorn M, Paludan SR, Underwood AP, Binderup A, Ramirez S, Bukh J, Soegaard M, Erdogan SM, Gustavsson T, Vidal-Calvo E (2023) First-in-human use of a modular capsid virus-like vaccine platform: an open-label, non-randomised, phase 1 clinical trial of the SARS-CoV-2 vaccine ABNCoV2. Lancet Microbe 4(3):140–148. https://doi.org/10.1016/S2666-5247(22)00337-8
43. Keeble AH, Turkki P, Stokes S, Khairil Anuar INA, Rahikainen R, Hytönen VP, Howarth M (2019) Approaching infinite affinity through engineering of peptide–protein interaction. Proc Natl Acad Sci 116(52):26523–26533. https://doi.org/10.1073/pnas.1909653116
44. Keeble AH, Howarth M (2020) Power to the protein: enhancing and combining activities using the spy toolbox. Chem Sci 11(28):7281–7291. https://doi.org/10.1039/d0sc01878c
45. Tiwari P, Kaila P, Guptasarma P (2019) Understanding anomalous mobility of proteins on SDS-PAGE with special reference to the highly acidic extracellular domains of human E- and N-cadherins. Electrophoresis 40(9):1273–1281. https://doi.org/10.1002/elps.201800219
46. Huang J, Nagy SS, Koide A, Rock RS, Koide S (2009) A peptide tag system for facile purification and single-molecule immobilization. Biochemistry 48(50):11834–11836. https://doi.org/10.1021/bi901756n
47. Hills RA, Tan TK, Cohen AA, Keeffe JR, Keeble AH, Gnanapragasam PNP, Storm KN, Rorick AV, West AP, Hill ML, Liu S, Gilbert-Jaramillo J, Afzal M, Napier A, Admans G, James WS,

Bjorkman PJ, Townsend AR, Howarth MR (2024) Proactive vaccination using multiviral quartet nanocages to elicit broad anti-coronavirus responses. Nat Nanotechnol 19(8):1216–1223. https://doi.org/10.1038/s41565-024-01655-9

48. Zhu H, Li X, Ren X, Chen H, Qian P (2022) Improving cross-protection against influenza virus in mice using a nanoparticle vaccine of mini-HA. Vaccine 40(44):6352–6361. https://doi.org/10.1016/j.vaccine.2022.09.058
49. Vester SK, Rahikainen R, Khairil Anuar INA, Hills RA, Tan TK, Howarth M (2022) SpySwitch enables pH- or heat-responsive capture and release for plug-and-display nanoassembly. Nat Commun 13(1):1–16. https://doi.org/10.1038/s41467-022-31193-8
50. Keeble AH, Yadav VK, Ferla MP, Bauer CC, Chuntharpursat-Bon E, Huang J, Bon RS, Howarth M (2022) DogCatcher allows loop-friendly protein-protein ligation. Cell Chem Biol 29(2):339–350.e10. https://doi.org/10.1016/j.chembiol.2021.07.005
51. Aebischer A, Wernike K, König P, Franzke K, Wichgers Schreur PJ, Kortekaas J, Vitikainen M, Wiebe M, Saloheimo M, Tchelet R, Audonnet J-C, Beer M (2021) Development of a modular vaccine platform for multimeric antigen display using an orthobunyavirus model. Vaccines 9(6):651. https://doi.org/10.3390/vaccines9060651
52. Marini A, Zhou Y, Li Y, Taylor IJ, Leneghan DB, Jin J, Zaric M, Mekhaiel D, Long CA, Miura K, Biswas S (2019) A universal plug-and-display vaccine carrier based on HBsAg VLP to maximize effective antibody response. Front Immunol 10:2931. https://doi.org/10.3389/fimmu.2019.02931
53. Tan TK, Rijal P, Rahikainen R, Keeble AH, Schimanski L, Hussain S, Harvey R, Hayes JWP, Edwards JC, McLean RK, Martini V, Pedrera M, Thakur N, Conceicao C, Dietrich I, Shelton H, Ludi A, Wilsden G, Browning C, Townsend AR (2021) A COVID-19 vaccine candidate using spycatcher multimerization of the SARS-CoV-2 spike protein receptor-binding domain induces potent neutralising antibody responses. Nat Commun 12(1):542. https://doi.org/10.1038/s41467-020-20654-7
54. Cohen AA, Gnanapragasam PNP, Lee YE, Hoffman PR, Ou S, Kakutani LM, Keeffe JR, Wu H-J, Howarth M, West AP, Barnes CO, Nussenzweig MC, Bjorkman PJ (2021) Mosaic nanoparticles elicit cross-reactive immune responses to zoonotic coronaviruses in mice. Science 371(6530):735–741. https://doi.org/10.1126/science.abf6840
55. Huang CQ, Hills RA, Carnell GW, Vishwanath S, Aguinam ET, Chan ACY, Palmer P, O'Reilly L, Tonks P, Temperton N, Frost SDW, Tiley LS, Howarth MR, Heeney JL (2025) Computationally designed haemagglutinin with nanocage plug-and-display elicits pan-H5 influenza vaccine responses. Emerg Microbes Infect 14(1):2511132. https://doi.org/10.1080/22221751.2025.2511132
56. Raetz CRH, Whitfield C (2002) Lipopolysaccharide endotoxins. Annu Rev Biochem 71(1):635–700. https://doi.org/10.1146/annurev.biochem.71.110601.135414
57. Cavaillon J-M (2018) Exotoxins and endotoxins: inducers of inflammatory cytokines. Toxicon 149:45–53. https://doi.org/10.1016/j.toxicon.2017.10.016
58. Schwarz H, Schmittner M, Duschl A, Horejs-Hoeck J (2014) Residual endotoxin contaminations in recombinant proteins are sufficient to activate human CD1c+ dendritic cells. PLoS One 9(12):113840. https://doi.org/10.1371/journal.pone.0113840
59. Brito LA, Singh M (2011) Acceptable levels of endotoxin in vaccine formulations during preclinical research. J Pharm Sci 100(1):34–37. https://doi.org/10.1002/jps.22267
60. Mamat U, Wilke K, Bramhill D, Schromm AB, Lindner B, Kohl TA, Corchero JL, Villaverde A, Schaffer L, Head SR, Souvignier C, Meredith TC, Woodard RW (2015) Detoxifying *Escherichia coli* for endotoxin-free production of recombinant proteins. Microb Cell Factories 14(1):57. https://doi.org/10.1186/s12934-015-0241-5
61. Schneier M, Razdan S, Miller AM, Briceno ME, Barua S (2020) Current technologies to endotoxin detection and removal for biopharmaceutical purification. Biotechnol Bioeng 117(8):2588–2609. https://doi.org/10.1002/bit.27362
62. Aida Y, Pabst MJ (1990) Removal of endotoxin from protein solutions by phase separation using triton X-114. J Immunol Methods 132(2):191–195. https://doi.org/10.1016/0022-1759(90)90029-U
63. Sondhi P, Adeniji T, Lingden D, Stine KJ (2024) Advances in endotoxin analysis. Adv Clin Chem 118:1–34. https://doi.org/10.1016/bs.acc.2023.11.001
64. Su W, Ding X (2015) Methods of endotoxin detection. SLAS Technol 20(4):354–364. https://doi.org/10.1177/2211068215572136
65. Dash R, Rathore AS (2021) Freeze thaw and lyophilization induced alteration in mAb therapeutics: trastuzumab as a case study. J Pharm Biomed Anal 201:114122. https://doi.org/10.1016/j.jpba.2021.114122
66. Fayter AER, Hasan M, Congdon TR, Kontopoulou I, Gibson MI (2020) Ice recrystallisation inhibiting polymers prevent

irreversible protein aggregation during solvent-free cryopreservation as additives and as covalent polymer-protein conjugates. Eur Polym J 140:110036. https://doi.org/10.1016/j.eurpolymj.2020.110036

67. Kommineni N, Butreddy A, Sainaga Jyothi VGS, Angsantikul P (2022) Freeze-drying for the preservation of immunoengineering products. iScience 25(10):105127. https://doi.org/10.1016/j.isci.2022.105127
68. Escolano A, Gristick HB, Gautam R, DeLaitsch AT, Abernathy ME, Yang Z, Wang H, Hoffmann MAG, Nishimura Y, Wang Z, Koranda N, Kakutani LM, Gao H, Gnanapragasam PNP, Raina H, Gazumyan A, Cipolla M, Oliveira TY, Ramos V, Bjorkman PJ (2021) Sequential immunization of macaques elicits heterologous neutralizing antibodies targeting the V3-glycan patch of HIV-1 Env. Sci Transl Med 13(621):1533. https://doi.org/10.1126/scitranslmed.abk1533
69. Zheng P, He J, Yang Z, Fu Y, Yang Y, Li W, Ding Y, Yang X, Ma Y (2023) Neoantigen-based nanovaccine in combination with immune checkpoint inhibitors abolish postsurgical tumor recurrence and metastasis. Small 19(50):2302922. https://doi.org/10.1002/smll.202302922
70. Cohen AA, Yang Z, Gnanapragasam PNP, Ou S, Dam K-MA, Wang H, Bjorkman PJ (2021) Construction, characterization, and immunization of nanoparticles that display a diverse array of influenza HA trimers. PLoS One 16(3):0247963. https://doi.org/10.1371/journal.pone.0247963
71. Lee KS, Rader NA, Miller-Stump OA, Cooper M, Wong TY, Amin S, M B, M B, R J, Ernst RK, Heath Damron F (2023) Intranasal VLP-RBD vaccine adjuvanted with BECC470 confers immunity against delta SARS-CoV-2 challenge in K18-hACE2-mice. Vaccine 41(34):5003–5017. https://doi.org/10.1016/j.vaccine.2023.06.080
72. Patel DR, Minns AM, Sim DG, Field CJ, Kerr AE, Heinly TA, Luley EH, Rossi RM, Bator CM, Moustafa IM, Norton EB, Hafenstein SL, Lindner SE, Sutton TC (2024) Intranasal SARS-CoV-2 RBD decorated nanoparticle vaccine enhances viral clearance in the Syrian hamster model. Microbiol Spectr 12(3):04998–22. https://doi.org/10.1128/spectrum.04998-22
73. Janitzek CM, Carlsen PHR, Thrane S, Khanna VM, Jakob V, Barnier-Quer C, Collin N, Theander TG, Salanti A, Nielsen MA, Sander AF (2021) The immunogenicity of capsid-like particle vaccines in combination with different adjuvants using different routes of administration. Vaccines 9(2):131. https://doi.org/10.3390/vaccines9020131
74. Sharma J, Shepardson K, Johns LL, Wellham J, Avera J, Schwarz B, Rynda-Apple A, Douglas T (2020) A self-adjuvanted, modular, antigenic VLP for rapid response to influenza virus variability. ACS Appl Mater Interfaces 12(16):18211–18224. https://doi.org/10.1021/acsami.9b21776
75. Wamhoff E-C, Ronsard L, Feldman J, Knappe GA, Hauser BM, Romanov A, Case JB, Sanapala S, Lam EC, Denis KJS, Boucau J, Barczak AK, Balazs AB, Diamond MS, Schmidt AG, Lingwood D, Bathe M (2024) Enhancing antibody responses by multivalent antigen display on thymus-independent DNA origami scaffolds. Nat Commun 15(1):795. https://doi.org/10.1038/s41467-024-44869-0
76. Pyles GM, Huckaby AB, Gutierrez M, de la P, Witt WT, Mateu-Borrás M, Dublin SR, Rocuskie-Marker C, Sesti BN, Peasak K, Bitzer GJ, Rader N, Weaver KL, Boehm DT, Fitzgerald N, Chapman J, Ulicny S, Damron FH, Barbier M (2024) Virus-like particles displaying the mature C-terminal domain of filamentous hemagglutinin are immunogenic and protective against *Bordetella pertussis* respiratory infection in mice. Infect Immun 92(8):e0027024. https://doi.org/10.1128/iai.00270-24
77. Wang W, Zhou X, Bian Y, Wang S, Chai Q, Guo Z, Wang Z, Zhu P, Peng H, Yan X, Li W, Fu Y-X, Zhu M (2020) Dual-targeting nanoparticle vaccine elicits a therapeutic antibody response against chronic hepatitis B. Nat Nanotechnol 15(5):406–416. https://doi.org/10.1038/s41565-020-0648-y
78. Davis HL (2008) Novel vaccines and adjuvant systems: the utility of animal models for predicting immunogenicity in humans. Hum Vaccin 4(3):246–250. https://doi.org/10.4161/hv.4.3.5318
79. Graham SP, McLean RK, Spencer AJ, Belij-Rammerstorfer S, Wright D, Ulaszewska M, Edwards JC, Hayes JWP, Martini V, Thakur N, Conceicao C, Dietrich I, Shelton H, Waters R, Ludi A, Wilsden G, Browning C, Bialy D, Bhat S, Lambe T (2020) Evaluation of the immunogenicity of prime-boost vaccination with the replication-deficient viral vectored COVID-19 vaccine candidate ChAdOx1 nCoV-19. NPJ Vaccines 5:69. https://doi.org/10.1038/s41541-020-00221-3
80. Gristick HB, Hartweger H, Loewe M, Schooten J, Ramos V, Oliveira TY, Nishimura Y, Koranda NS, Wall A, Yao KH, Poston D, Gazumyan A, Wiatr M, Horning M, Keeffe JR, Hoffmann MAG, Yang Z, Abernathy ME, Dam KMA, Bjorkman PJ (2023) CD4 binding site immunogens elicit heterologous anti–HIV-1

neutralizing antibodies in transgenic and wild-type animals. Sci Immunol 8(80):6364. https://doi.org/10.1126/sciimmunol.ade6364
81. Malebo K, Woodward J, Ximba P, Mkhize Q, Cingo S, Moyo-Gwete T, Moore PL, Williamson AL, Chapman R (2024) Development of a two-component nanoparticle vaccine displaying an HIV-1 envelope glycoprotein that elicits tier 2 neutralising antibodies. Vaccines 12(9):1063. https://doi.org/10.3390/vaccines12091063
82. Mejía-Méndez JL, Vazquez-Duhalt R, Hernández LR, Sánchez-Arreola E, Bach H (2022) Virus-like particles: fundamentals and biomedical applications. Int J Mol Sci 23(15):8579. https://doi.org/10.3390/ijms23158579
83. Gasteiger E, Hoogland C, Gattiker A, Duvaud S, Wilkins MR, Appel RD, Bairoch A (2005) Protein identification and analysis tools on the ExPASy server. In: The proteomics protocols handbook. Humana Press, New York, 571–607. https://doi.org/10.1385/1-59259-890-0:571pp

Check for updates

Chapter 8

Rapid and Wash-free Live Cell Labeling with BenzoTag

Bryan J. Lampkin

Abstract

Self-labeling proteins (SLPs) enable the site-specific labeling of genetically expressed proteins with synthetic tags. BenzoTag was evolved from the commonly use self-labeling protein HaloTag7, for improved labeling kinetics with the fluorogenic dye, **Bz1**. This developed **Bz1**•BenzoTag conjugate was successfully applied in wash-free live cell imaging of intracellularly expressed BenzoTag and shown to have orthogonal reactivity against the HaloTag7•**CA-JF**$_{635}$ system.

Keywords Fluorescent labeling, Biorthogonal, Fluorogenic, Self-labeling protein, HaloTag, Wash-free

1 Introduction

Self-labeling proteins (SLPs) have proved to be valuable bioconjugation tools in the chemical and biological sciences [1]. SLPs are enzymes that have been engineered to be covalently labeled by their native substrate instead of turning over a reaction. The most used SLP is HaloTag7, is a modified bacterial alkyl dehalogenase that is covalently labeled by a linear chloroalkane ligand (Fig. 1a) [2, 3]. HaloTag7 contains a H272N mutation, which prevents water-mediated hydrolysis of the intermediate ester, thus trapping the covalently labeled adduct. The linear chloroalkane ligand can be functionalized by conjugation to a large variety of synthetic substrates, usually fluorescent dyes [4–6]. Therefore, HaloTag7 and other SLPs enable the colocalization of a synthetically derived substrate to a protein of interest fused to HaloTag7 with chemical and genetic specificity.

Despite HaloTag7's popularity, it preferentially reacts with chloroalkane-tagged rhodamine derived fluorescent dyes. In fact, the prototypical HaloTag7 substrate, chloroalkane tetramethyl rhodamine (**CA-TMR**, Fig. 1b) conjugates to HaloTag7 with a second order rate constant of $2 \times 10^7 M^{-1}s^{-1}$ [7]. However, when modifications are made to the rhodamine substrate, or when nonrhodamine dyes are

Alexander Deiters (ed.), *Bioconjugation Methods*, Methods in Molecular Biology, vol. 3009,
https://doi.org/10.1007/978-1-0716-5112-4_8,

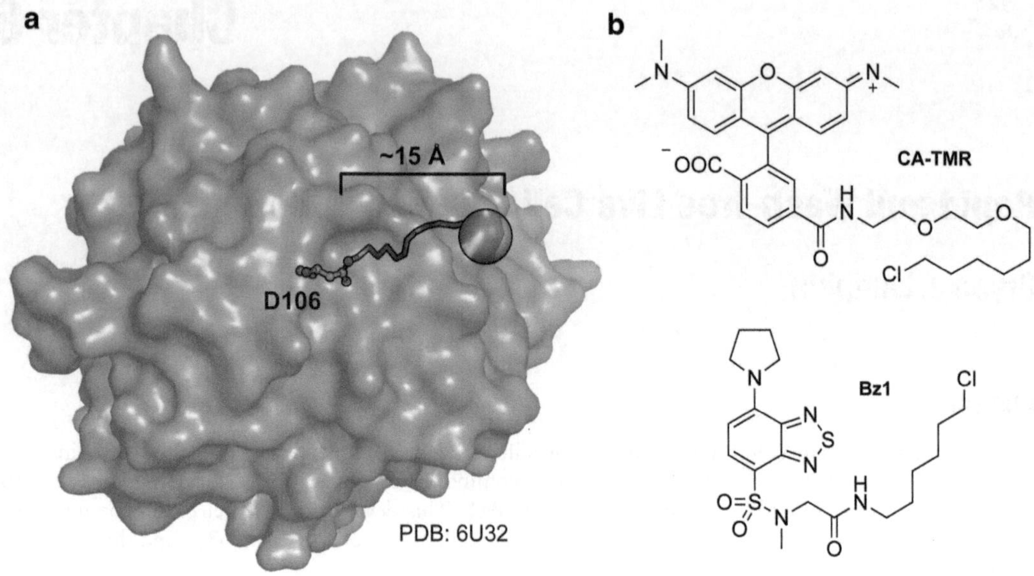

Fig. 1 Self-labeling protein HaloTag7. (**a**) Structure of HaloTag7 covalently labeled with a chloroalkane ligand at the base of a ~15 Å hydrophobic channel. The pink sphere represents the attachment point for a synthetic substrate. (**b**) Chemical structure of the commonly used chloroalkane tetramethylrhodamine (**CA-TMR**) ligand and the fluorogenic dye **Bz1** that BenzoTag was evolved for improved labeling kinetics

used, like the fluorogenic benzothiadiazole, these conjugation rates plumet to 10^3–$10^5 M^{-1}s^{-1}$ [6, 8, 9]. Given this inherent substrate specificity of HaloTag7, we developed a directed evolution platform to rapidly evolve HaloTag7 for fast conjugation kinetics with novel substrates by using yeast surface display [10].

From this system, we developed BenzoTag, a triple mutant of HaloTag7 that provides a 60-fold conjugation rate enhancement with the novel fluorogenic benzothiadiazole **Bz1** (Fig. 2) [6, 10]. The BenzoTag•**Bz1** system enables rapid and wash-free labeling of intracellularly tagged proteins at low concentrations of dye. Because **Bz1** produces a fluorescent signal only after conjugated to BenzoTag, cells can be imaged directly without needing to wash cells of excess dye. Further, BenzoTag•**Bz1** was found to react orthogonally with the commonly used HaloTag7•$\mathbf{JF}_{635}$ system, where $\mathbf{JF}_{635}$ is a fluorogenic rhodamine derivative [5]. This capability allowed for wash-free multiplexed labeling of cells coexpressing BenzoTag and HaloTag7 with a single cocktail of the two fluorogenic dyes. Because of the high signal-over-background afforded by the BenzoTag•**Bz1** system, coupled with its ease of use, we anticipate that this bioconjugation tool will be broadly used in dynamic cellular applications where rapid intracellular labeling of target proteins is necessary, such as measuring protein turnover, folding, or trafficking.

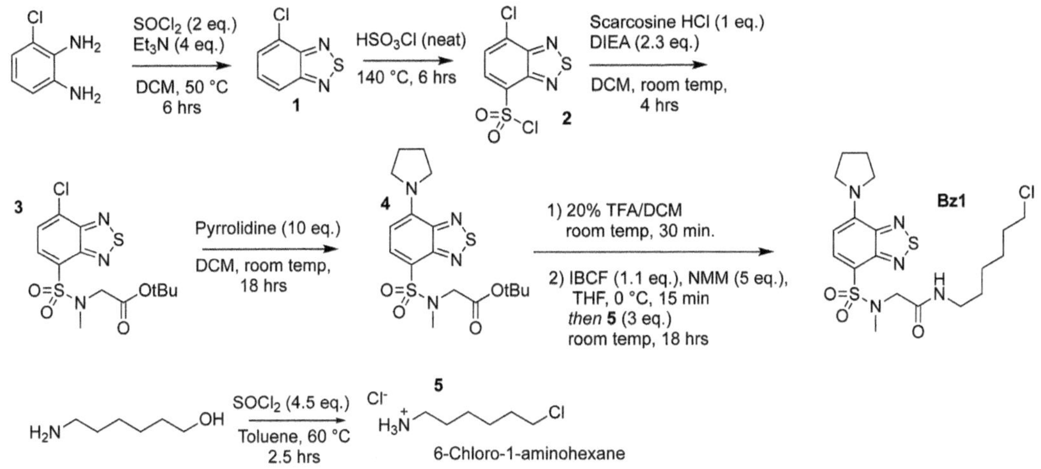

Fig. 2 Synthetic scheme for **Bz1**

2 Materials

2.1 Synthesis of Bz1

Chemicals and Reagents

1. 1,2-diamino-3-chlorobenzene
2. Thionyl chloride
3. Triethylamine
4. Chlorosulfonic acid
5. Sarcosine•HCl
6. Diisopropyl ethylamine
7. Pyrrolidine
8. Trifluoroacetic acid
9. Isobutylchloroformate
10. N-methylmorpholine
11. 6-aminohexanol
12. Sodium sulfate
13. 1 M hydrochloric acid
14. Dichloromethane (DCM)
15. Tetrahydrofuran (THF)
16. Toluene
17. Dimethylsulfoxide (DMSO)
18. Potassium chloride
19. Sodium chloride
20. Sodium phosphate dibasic

21. Potassium phosphate monobasic
22. Sodium hydroxide
23. Saturated NaCl solution (Brine)

Equipment

1. Stirring hotplate
2. Silicon oil bath
3. Silica coated glass plates for thin layer chromatography
4. Standard synthetic glassware (i.e., round bottom flasks, separatory funnel, chromatography columns, etc.)
5. Rotary evaporator

2.2 Live Cell Labeling

2.2.1 Materials and Reagents

1. 1 mM **Bz1** stock in DMSO diluted to appropriate 2× concentration in reduced serum media.
2. Sterile nontreated flat bottomed 96-well plates. For imaging applications, black plates with clear optical bottoms are required.
3. McCoy's 5 A Medium.
4. Fetal bovine serum (FBS).
5. U-2 OS cells (ATCC HTB-96) (*see* **Note 1**).
6. X-tremeGene™ 9 DNA transfection reagent (Roche).
7. Mammalian expression plasmid for BenzoTag: pcDNA™5/FRT vector encoding for BenzoTag expressed as a fusion Histone 2B for nuclear localization (*see* **Note 2**).
8. Gibco Opti-MEM™ reduced serum media.
9. Gibco Trypsin-EDTA (0.5%) without phenol red diluted to 0.05% working stock.
10. 1× Phosphate buffered saline (pH = 7.4): Dissolve 8 g NaCl, 0.2 g KCl, 1.44 g of NA_2HPO_4, and KH_2PO_4 into 800 mL of distilled water and adjust the pH to 7.4 using 0.1 M HCl or 0.1 M NaOH. Dilute up to 1 L of total volume using distilled water and then filter through a 0.22-micron filter.
11. 1 mM **CA-JF$_{635}$** stock in DMSO diluted to appropriate 2× concentration in reduced serum media (*see* **Note 3**).

Equipment

1. Biosafety cabinet with HEPA filter
2. CO_2 incubator
3. Benchtop incubator

4. Microliter and serological pipettes
5. Cytek Guava easyCyte Flow Cytometer
6. Leica TCS SP2 Microscope

3 Methods

3.1 Synthesis of Bz1

1. Dissolve 1,2-diamino-3-chlorobenzene (19.7 mmol, 2.8 g) in dichloromethane (200 mL) and dropwise add triethylamine (4 eq.) followed by thionyl chloride (2 eq.). Reflux the reaction at 50°C for 6 h.
2. Add 1 M HCl (100 mL) to quench the reaction. The organic layer is then collected, washed with 50 mL brine, dried with granular sodium sulfate (Na_2SO_4), and concentrated in vacuo to yield **1** as a light brown solid. Purify the target compound by silica gel chromatography to yield a white solid (1.87 g, 56% yield).
3. Dissolve **1** (8.8 mmol, 1.5 g) in neat chlorosulfonic acid (10 mL) and heat to 140°C for 6 h.
4. Cool the reaction back to room temperature and add 80 mL of DI H_2O. Extract the product from the resulting aqueous solution by treating with ethyl acetate (20 mL) and collect the organic layer. Repeat three times. Combine the organic layers, wash with 30 mL brine, dry with Na_2SO_4, and concentrate in vacuo to yield **2** as an off-white solid (1.02 g, 43% yield).
5. Dissolve **2** (3.0 mmol, 0.81 g) in dichloromethane (30 mL) and add diisopropyl amine (2.3 eq., 1.2 mL) and sarcosine• HCl (1 eq., 0.27 g). Stir the reaction at room temperature for 4 h.
6. When the reaction is confirmed complete by TLC, add 30 mL of 1 M HCl to quench the reaction. Collect the organic layer and back extract the aqueous layer three times with dichloromethane (10 mL). Combine the organic layers, wash with brine, dry with Na_2SO_4, and concentrated in vacuo. Purify the target compound by silica gel chromatography to yield **3** as a white solid (0.96 g, 85% yield).
7. Dissolve **3** (0.2 mmol, 75 mg) in dichloromethane (10 mL), add pyrrolidine (10 eq., 164 μL) and stir the reaction for 18 h at room temperature.
8. When the reaction is confirmed complete by TLC, add 10 mL of 1 M HCl to quench the reaction. Collect the organic layer and back extract the aqueous layer three times with dichloromethane (10 mL). Combine the organic layers, wash with

15 mL of brine, dry with Na_2SO_4, and concentration in vacuo. Purify the target compound by silica gel chromatography to yield **4** as an orange solid (81 mg, quantitative yield).

9. Dissolve **4** (0.12 mmol, 50 mg) in 1 mL of 20% trifluoroacetic acid in dichloromethane and stir at room temperature for 30 min. Remove solvent in vacuo, redissolve in 1 mL of dichloromethane, and concentrate again. Repeat two more times.
10. Dissolve the resulting orange gel in dry tetrahydrofuruan (2.5 mL) and cool the solution to 0°C using an ice bath.
11. Add N-methylmorpholine (5 eq., 66 μL) followed by isobutylchloroformate (1.1 eq., 17 μL) dropwise. Allow the solution to stir for 15 min. A white precipitate should form.
12. Add **5** (3 eq., 62 mg) in one portion and remove the solution from ice. Allow the reaction to stir at room temperature for 18 h.
13. When the reaction is confirmed complete via TLC analysis, filter the solution and concentrate the filtrate in vacuo. Purify the resulting orange gel by silica gel chromatography to yield the target compound, **Bz1** (40 mg, 71% yield), as an orange solid.
14. Make a 1 mM stock solution of **Bz1** in DMSO using an extinction coefficient of 10,000 $M^{-1}cm^{-1}$ or by weighing out >5 mg of **Bz1** powder on an analytical balance and dissolving in the appropriate amount of DMSO. Store this stock solution at 4°C.

3.2 Live Cell Labeling with BenzoTag and Bz1

1. Seed 1 × 10^4 U-2 OS cells per well in a total volume of 100 μL of high glucose MyCoy's 5 A media supplemented with 10% fetal bovine serum in a 96-well microplate (~20% confluency, *see* **Note 4**) approximately 48 h prior to the labeling experiment. Incubate at 37°C with 5% carbon dioxide.
2. Approximately 24 h prior to the labeling experiment, cells should be transfected with H2B-BenzoTag expression plasmid using the X-tremeGene™ 9 DNA transfection reagent (Roche) following manufacturers protocols (*see* **Note 5**). Briefly, 1 μg of plasmid and 3 μL of transfection reagent were diluted to 100 μL with Gibco Opti-MEM™ reduced serum media. The resulting mixture was left to incubate at room temperature for 15–30 min. Following incubation, the transfection mix was diluted to 1000 μL with full serum growth media. Aspirate existing media from each well and add 100 μL of the transfection mix to each well. Incubate the cells at 37°C for another 18–24 h.
3. On the day of the labeling experiment, aspirate transfection media and gently wash and aspirate cells once with 50 μL

Opti-MEM™, followed by an additional 50 μL of Opti-MEM™. The transfected cells are ready for labeling by **Bz1**.

3.2.1 Labeling for Flow Cytometry Applications *(Fig. 3a)*

4. Treat cells with 50 μL of 2× **Bz1** prepared in Opti-MEM™ and incubate at room temperature for 10 min.
5. Aspirate dye containing media and add 40 μL of 0.05% Trypsin per well to dissociate cells from the microplate. Incubate at 37°C until complete dissociation (typically 5–10 min).
6. Dilute cell suspension with 160 μL of PBS to a total volume of 200 μL and analyze BenzoTag labeling by **Bz1** using a Cytek Guava easyCyte Flow Cytometer. The spectral properties of **Bz1** allow the use of the common 488 blue laser and a green 525/50 detection filter.
7. Quantification of cell labeling is normalized to transfected, but untreated controls (*see* **Note 6**).

3.2.2 Labeling for Fluorescent Microscopy applications *(Fig. 3b)*

8. Treat cells with 50 μL of 20 nM **Bz1** (final concentration 10 nM) prepared in Opti-MEM™ and incubate for necessary time to see adequate signal (*see* **Notes 7–9**).
9. Acquire image(s) directly on a Leica TCS SP2 Microscope using the 488 nm blue laser without needing to replace media or washing cells of excess dye.

3.2.3 Orthogonal Wash-free Labeling Using BenzoTag•Bz1 and HaloTag•JF$_{635}$ *(Fig. 4)*

1. Seed 1 × 10^4 cells per well in a 96-well microplate as described above.
2. Co-transfect cells with mammalian expression plasmids for H2B-BenzoTag and Tomm20-HaloTag as described above (*see* **Note 10**).
3. 16–18 h after transfection, aspirate transfection media and gently wash and aspirate cells once with 50 μL Opti-MEM™,

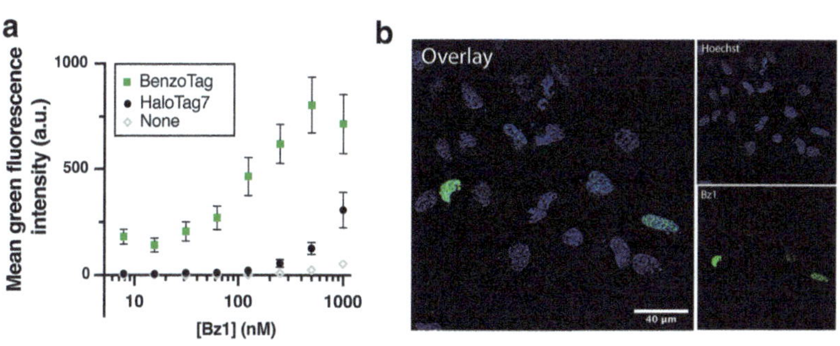

Fig. 3 **Bz1** rapidly and selectively labels cells expressing BenzoTag. (**a**) Dose-dependent **Bz1** labeling of U-2 OS cells that were transiently transfected with H2B-BenzoTag, H2B-HaloTag7, and non-expressing cells. Fluorescence was measured by flow cytometry. (**b**) Confocal microscopy images of U-2 OS cells transiently transfected with H2B-Benzotag that were imaged directly after being labeled with 10 nM **Bz1**. Figure adapted from [10]

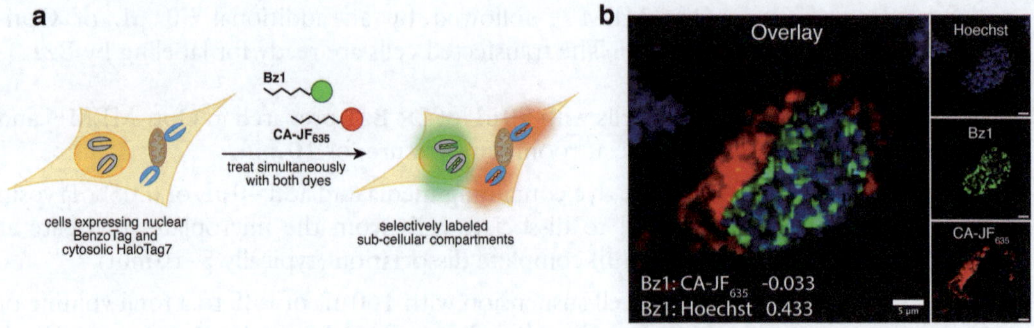

Fig. 4 Multiplexed wash-free labeling of live cells. (**a**) Labeling scheme of cells co-transfected with H2B-BenzoTag (nuclear) and Tomm20-HaloTag7 (cytosol) treated with a single solution of **Bz1** and **CA-JF**$_{635}$. (**b**) Confocal microscopy images of multiplexed labeling. CA-JF$_{635}$ was imaged using the 647 nm red laser. Pearson correlation analysis demonstrates selective compartmental labeling between **Bz1**•BenzoTag and **CA-JF**$_{635}$•HaloTag7. Figure is adapted from [10]

followed by an additional 50 μL of Opti-MEM™. The transfected cells are ready for colabeling by **Bz1** and **JF**$_{635}$.

4. Co-treat cells with 50 μL of a cocktail of 100 nM **Bz1** and 250 nM **JF**$_{635}$ (*see* **Note 11**; final concentration 50 and 125 nM, respectively). Prepare cells for either flow cytometry analysis or microscopy using methods described above.

4 Notes

1. U-2 OS cells were used in this study. However, we anticipate any immortalized cell lines may be suitable. Transfection and treatment conditions should be optimized for the cell line used.
2. The Histone 2B fusion was used in this study. However, much like the versatility of HaloTag7, we anticipate that BenzoTag can be expressed as fusions to other organelles or proteins of interest.
3. Chloroalkane tagged **CA-JF**$_{635}$ can be purchased from Promega Corporation. Other nonconjugated Janelia Fluor® dyes can purchased from Tocris® and Hello Bio®.
4. Adjust cell seeding density relative to the cell type being used and when transfection will occur for a givenexperiment.
5. Transfection conditions may need to be optimized. Screening different plasmid and transfection reagent is recommended. Transfection efficiency can be determined by treating

transfected cells with 5 μM of chloroalke-tagged tetramethyl rhodamine (TMR) for 15 min, followed by washing, and measuring the percentage of labeled cells by flow cytometry. Transfection efficiencies in our hands were approximately 25%.

6. The mean fluorescence intensity for each population of cells treated was normalized by subtracting the autofluorescence value of untreated transfected cells. The degree of background labeling was determined by treated nontransfected cells with the same concentration of dye and subtracting the autofluorescence of untreated nontransfected cells.
7. Label cells with required organelle stains (Hoechst, LysoTracker™, etc.) for colocalization quantification, if necessary, by following manufacturer protocols.
8. **Bz1** appears to be very cell permeable when treating U-2 OS cells and appreciable signal was observed in as little as 10 min when using 10 nM of dye. This time window may need to be optimized for a specific experimental setup.
9. At higher concentrations of **Bz1** (>500 nM) and longer incubations times (>60 min), small puncta were observed. While the identity of these puncta were not confirmed, related benzothiadiazole dyes have been reported to selectively label lipid droplets when labeled at 20 μM for 25 min [11].
10. Again, different proteins or organelles of interest may be used, and cotransfection conditions should be optimized for the specific cell line being used. The Tomm20-HaloTag7 construct, which expresses HaloTag7 in the cytosol as an outer mitochondrial fusion, was provided by Kai Johnsson (addgene: 169330).
11. The final concentrations of **Bz1** and **JF$_{635}$** were 50 nM and 125 nM, respectively. These were empirically determined for optimal *selective* labeling. Higher concentrations of **Bz1** reacted with cytosolic expressed HaloTag7. Lower concentrations of **CA-JF$_{635}$** did not produce a strong enough signal to be imaged.

Acknowledgments

This work was supported by NIH 5F32GM139371. The author wishes to thank Professor Joshua Kritzer and Benjamin Goldberg for their assistance in the original BenzoTag work. **CA-JF$_{635}$** was generously donated by the lab of Luke Lavis at HHMI-Janelia.

References

1. Hoelzel CA, Zhang X (2020) Visualizing and manipulating biological processes using HaloTag and SNAP-Tag technologies. Chembiochem 21:1935–1946
2. Encell LP, Friedman Ohana R, Zimmerman K et al (2012) Development of a dehalogenase-based protein fusion tag capable of rapid, selective and covalent attachment to customizable ligands. Curr Chem Genomics 6:55-71
3. Los GV, Encell LP, McDougall MG et al (2008) HaloTag: a novel protein labeling technology for cell imaging and protein analysis. ACS Chem Biol 3:373–382
4. Cook A, Walterspiel F, Deo C (2023) HaloTag-based reporters for fluorescence imaging and biosensing. ChemBioChem 24:e202300022
5. Grimm JB, Muthusamy AK, Liang Y et al (2017) A general method to fine-tune fluorophores for live-cell and in vivo imaging. Nat Methods 14:987–994
6. Lampkin BJ, Kritzer JA (2023) Engineered fluorogenic HaloTag ligands for turn-on labelling in live cells. Chem Commun 60:200–203
7. Wilhelm J, Kühn S, Tarnawski M et al (2021) Kinetic and structural characterization of the self-labeling protein tags HaloTag7, SNAP-tag, and CLIP-tag. Biochemistry 60:2560–2575
8. Clark SA, Singh V, Vega Mendoza D et al (2016) Light-Up "Channel Dyes" for haloalkane-based protein labeling in vitro and in bacterial cells. Bioconjugate Chem 27:2839–2843
9. Liu Y, Miao K, Dunham NP et al (2017) The cation-π interaction enables a Halo-Tag fluorogenic probe for fast no-wash live cell imaging and gel-free protein quantification. Biochemistry 56:1585–1595
10. Lampkin BJ, Goldberg BJ, Kritzer JA (2024) Multiplexed no-wash cellular imaging using BenzoTag, an evolved self-labeling protein. Chem Sci 15:17337–17347
11. Suarez SI, Warner CC, Brown-Harding H et al (2020) Highly selective staining and quantification of intracellular lipid droplets with a compact push–pull fluorophore based on benzothiadiazole. Org Biomol Chem 18:495–499

Check for updates

Chapter 9

Antibody Conjugation to Cell Surface Receptors

Avani Parikh, Elisa Ruffo, Alexander Deiters and Jason Lohmueller

Abstract

Here, we describe the covalent modification of cell surfaces with antibodies through bioconjugation to the self-labeling SNAP-tag enzyme. We apply this approach to the reprogramming of T cell signaling through the universal chimeric antigen receptor (CAR), "SNAP-CAR." Universal CARs are a highly programmable class of engineered receptors that interact with co-administered "adaptor" antibodies to recognize one or more antigens of interest to trigger receptor signaling. Universal CARs have promise for use in cell therapeutics and as research tools. SNAP-CAR covalently attaches to adaptor antibodies containing a benzyl guanine (BG) motif. The protocols presented here include methods for SNAP-CAR T cell and antibody adaptor generation using gamma retroviral transduction and BG-NHS ester conjugation, respectively. The chapter also describes methods to evaluate cell surface bioconjugate formation, including quantification of BG motifs on antibodies and co-incubation experiments to assay for antigen-directed SNAP-CAR T cell functions of target cell killing and SNAP-CAR T cell activation.

Keywords Universal CAR T cells, Antibody therapeutics, Antibody conjugation, CAR T cell therapy, Synthetic biology

1 Introduction

Bioconjugation to proteins on the surfaces of living mammalian cells is a powerful approach for probing and manipulating cell behavior [1]. Various molecules have been successfully conjugated, including functional small molecules [2], dyes [3], DNAs [4], synthetic coatings and polymers [5], nanoparticles [6], and binding proteins [7]. These modifications have allowed for applications such as cell tracking [8], re-wiring of cellular signaling [9], altering cell migration [10], and delivery of therapeutic agents [11], among others. Bioconjugation strategies have been developed both for broadly labeling cell surface proteins as well as conjugating to specific proteins of interest. Approaches include direct chemical covalent bonding [10], in which payloads are directly conjugated to reactive chemical groups on the surface of proteins (such as thiols and amines) by

Alexander Deiters (ed.), *Bioconjugation Methods*, Methods in Molecular Biology, vol. 3009,
https://doi.org/10.1007/978-1-0716-5112-4_9,

reactive chemicals like NHS esters or maleimides; metabolic glycan labeling [12, 13], in which synthetic sugars with click-chemistry handles such as *N*-azidoacetylmannosamine-tetraacetate are fed to cells and incorporated onto proteoglycans for subsequent click-chemistry conjugation; and enzymatic modification [14, 15], in which enzymes such as oxidoreductases or glycosyl-transferases are applied to cells to create reactive groups for further chemical conjugation. Overall, these techniques open numerous exciting avenues for biotechnological applications.

One therapeutic application for cell surface bioconjugation approaches is universal chimeric antigen receptor (CAR) therapy. CAR T cell therapy is an adoptive cell therapy in which T lymphocytes are genetically modified to express a CAR, an artificial receptor that binds to a specific target antigen on diseased cells to trigger immune activation and target cell lysis [16]. Universal CARs are a highly programmable, next-generation system variants that do not bind directly to disease antigens but instead bind to co-administered "adaptor" molecules that recognize the target antigens [17]. Separating the target antigen binding from the signaling receptor allows for several capabilities, including (1) simultaneous targeting of multiple antigens via multiple adaptors to avoid therapeutic resistance, (2) targeting multiple diseases through one receptor by using different adaptors, and (3) tuning CAR T cell activity by the adaptor dose to avoid toxicity. Several universal CARs have been developed, differing primarily in the CAR and adaptor interacting domains, including: scFv:FITC [15], streptavidin:biotin [18, 19], scFv:peptide neoepitope [20], and leucine zippers [21]. Recent approaches have generated systems in which the adaptors covalently attach to the universal CAR. These receptors have shown the advantages of enhanced potency and the ability to be pre-loaded with adaptors to be "armed" for antigen attack [9, 22–24].

In this chapter, we describe methods to arm the covalent universal SNAP-CAR for antigen targeting by conjugating it with antibody adaptors (Fig. 1) [9]. The adaptor binding domain for this receptor is SNAP-tag, a modified human O6-methylguanine-DNA-methyltransferase (MGMT) protein, that interacts with O^6-benzyl guanine (BG) to form a covalent bond [25]. We generated SNAP-CAR T cells by transducing primary human T cells with a gamma retrovirus construct encoding SNAP-CAR expression. To generate the antibody adaptors consisting of monoclonal antibodies (mAbs) conjugated to BG, we used BG-GLA-NHS. This NHS ester conjugates BG to random lysine residues or the N-termini of proteins (Fig. 2). Adaptors targeting several tumor antigens, including CD20, EGFR, CD19, and HER2 were assembled. These adaptors included both full-length immunoglobulins and Fab antibody fragments. Quantification of BG conjugation via

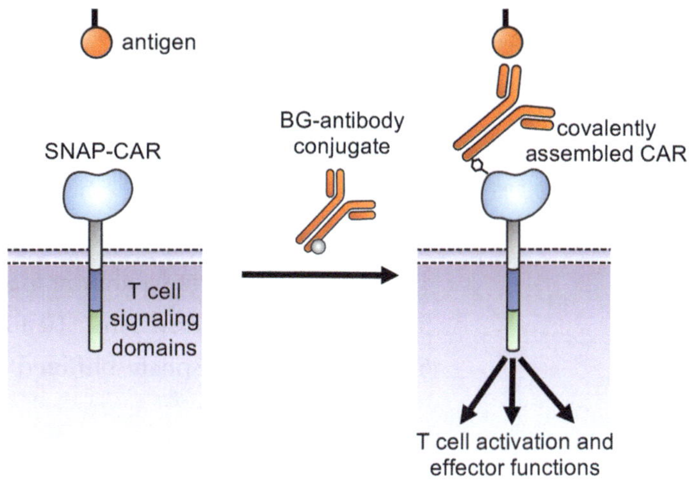

Fig. 1 Arming universal SNAP-CAR with antibody adaptors. The SNAP-CAR is targeted by a BG-conjugated antibody to activate T-cell signaling and effector functions upon antigen recognition. (Reproduced from [9], with permission from Springer Nature)

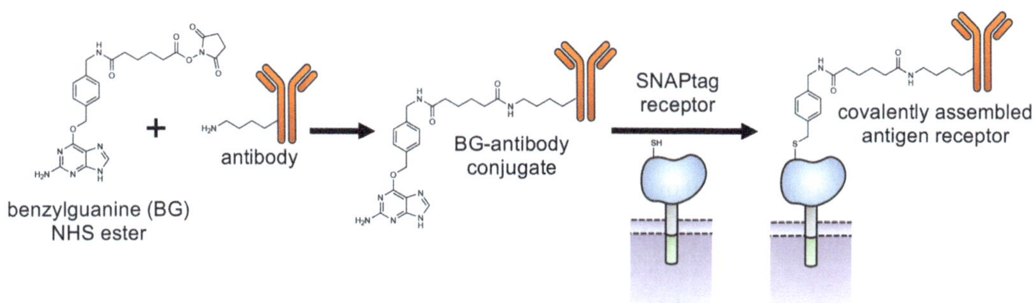

Fig. 2 Covalent labeling of monoclonal antibodies with BG-GLA-NHS ester. A benzylguanine motif (BG) is chemically conjugated to an antibody using a benzylguanine NHS ester, which recognizes lysine residues on the monoclonal antibodies. The BG-antibody conjugate then covalently binds to the extracellular SNAP-tag enzyme through a self-labeling reaction. (Reproduced from [9], with permission from Springer Nature)

an SDS-PAGE gel assay showed that this process yields an average of 2.6 BG molecules per antibody. SNAP-CAR T cells could be readily armed with these antibody adaptors by a simple hour-long co-incubation and wash. Co-incubation of these armed cells with antigen-positive target cells led to robust antigen-specific T cell activation and target cell lysis by flow cytometry analysis. Similar arming approaches have been conducted for other covalent universal CAR platforms [22, 24, 26].

2 Materials

2.1 Isolation of Peripheral Blood Mononuclear Cells and Primary Human T Cells

1. Peripheral Blood Mononuclear Cells (PBMCs) isolated from a half leukopack (Miltenyi Biotec) or buffy coat (Vitalant).
2. Scalpel.
3. Conical tubes (50 mL and 15 mL).
4. LSM, lymphocyte separation medium (MP Biomedicals).
5. Serological pipettes: 5 mL, 10 mL, 25 mL volumes.
6. Dulbecco's phosphate-buffered saline, no calcium, no magnesium (DPBS).
7. RBC lysis buffer.
8. Complete RPMI Human AB (HAB) media: RPMI supplemented with heat-inactivated 10% human AB serum (Gemini Bio), 1% each of pen strep, non-essential amino acids, L-glutamine, sodium pyruvate, and 4 mM L-arginine. Weigh L-arginine and add to RPMI media alone (no other components added). After complete dissolution of L-arginine, adjust pH to 7.2 using 1 M HCl. Add Human AB serum, all other components. Sterilize using a 0.22 μM PES vacuum filter.
9. Trypan Blue.
10. Hemocytometer.
11. Bambanker Freezing Media (Bulldog-Bio).
12. 2 mL cryogenic tube.
13. Deoxyribonuclease I (DNase I), bovine pancreas, purified (stock concentration, 1 mg/mL).
14. MACS buffer: Dulbecco's PBS (DPBS) (no magnesium, no calcium), add 0.5% BSA. Dissolve thoroughly in DPBS and then add 2 mM ethylenediaminetetraacetic acid (EDTA).
15. Human Pan T-cell isolation kit (Miltenyi Biotec).
16. LS columns (Miltenyi Biotec).
17. QuadroMACS Separator (Miltenyi Biotec).
18. Human T Cell TransAct (Miltenyi Biotec).
19. Human IL-2 IS (Miltenyi Biotec), stock solution 100,000 IU.
20. Human IL-15 (Miltenyi Biotec), stock solution 1000 ng/mL.
21. 24-well tissue culture-coated plate or a T-75 tissue culture flask.

2.2 Flow Cytometry Staining and Reagents

1. Flow cytometry buffer recipe (FACS buffer): DPBS + 1% Penicillin Streptomycin + 2% fetal bovine serum (FBS).
2. 96-well V-bottom plate.
3. 96-well U-bottom plate.
4. 1.2 mL clear microtiter tubes.

5. Antibodies for flow cytometry: anti-CD3 FITC, UCHT1 clone (Biolegend), anti-CD271-BV421, C40-1457 clone (BD Biosciences), Alexa Fluor® 647 AffiniPure® Fab Fragment Goat Anti-Human IgG (H+L), (109-607-003, Jackson ImmunoResearch), anti-CD69 BV711, FN50 clone (Biolegend), anti-CD62L, SK11 clone (BD Biosciences).
6. Flow cytometer: LSR Fortessa (BD Biosciences) or equivalent.

2.3 Retrovirus Production

1. Complete Dulbecco's Modified Eagle Media (DMEM) supplemented with 1× MEM amino acids solution, 10 mM sodium pyruvate, 10% FBS, and 1% Penicillin-Streptomycin.
2. HEK293-GP cells (gift from U. Kammula, University of Pittsburgh) [27].
3. 100 mm poly-D-lysine-coated plate.
4. Microcentrifuge tubes, 1.5 mL.
5. Opti-MEM (Gibco).
6. Lipofectamine 3000.
7. MSGV1-SNAP-CAR-T2A-LNGFR, SNAP-CAR retroviral expression plasmid (Fig. 3) [9].
8. RD114 envelope plasmid (gift from Dr. U. Kammula, University of Pittsburgh) [27].
9. Vortex-Genie 2 Lab Mixer.
10. Cell culture incubator with 37°C and 5% CO_2.
11. Temperature-regulated centrifuge.
12. 50 mL conical tubes.
13. 0.45 μM PES membrane filters.
14. 10 mL sterile syringes.
15. 2 mL volume cryovials.

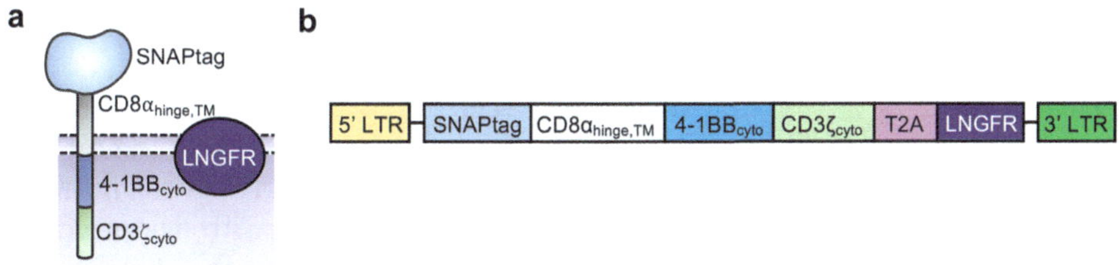

Fig. 3 Design of the Universal SNAP-CAR T cells construct and vector design. (**a**) Diagram of the SNAP-CAR with T2A-LNGFR receptor. (**b**) Schematic of SNAP-CAR construct design with T2A-LNGFR tag. CD8a hinge, TM, transmembrane domain, 4-1BB, cytoplasmic activation domain from CD137, CD3ξ, cytoplasmic activation domain of CD3ξ. (Reproduced from [9], with permission from Springer Nature)

2.4 Generation and Maintenance of Primary Human T Cells Encoding SNAP-CAR

1. Complete RPMI Human AB (HAB) media (see Sect. 2.1, step 8).
2. T Cell TransAct (Miltenyi Biotec).
3. Human IL-2 IS (Miltenyi Biotec), stock solution 100,000 IU.
4. Human IL-15 (Miltenyi Biotec), stock solution 1000 ng/mL.
5. Retronectin: (Takara), diluted with sterile DPBS to a 10 μg/mL solution.
6. 6-well non-tissue culture-treated polystyrene plate.
7. Temperature-regulated centrifuge
8. 6-well tissue culture-treated plate.

2.5 Conjugation of BG to mAbs to Create Antibody Adaptors

1. Monoclonal antibody of interest 0.1–1.5 mg, e.g., rituximab, anti-human CD20 targeting antibody. Recommended volume of the antibody is between 200 μL and 700 μL for a 2 mL column size.
2. 2 mL Zeba spin column (7 K MWKO) (Thermo Scientific), store at 4°C.
3. 15 mL conical tube.
4. Dulbecco's phosphate-buffered saline (DPBS), no calcium, no magnesium for BG-conjugation to monoclonal antibody.
5. BG-GLA-NHS-ester (New England Biolabs). Store at −20°C. To create the 10.3482 mM stock working solution, immediately prior to conjugation, weigh out 500 μg of BG-GLA-NHS ester and dissolve it in 100 μL of DMSO.
6. Nanodrop One C or equivalent.
7. Dimethyl sulfoxide (DMSO).
8. SNAP-tag protein (New England Biolabs).
9. Dulbecco's phosphate-buffered saline (DPBS) for gel: Add 100 mg of $CaCl_2$, 200 mg of KCl, 100 mg of $MgCl_2.6H_20$, 2.16 grams of Na_2HPO_4 in 800 mL of MilliQ water. Adjust to pH 7.4 with 65% [v/v] aq. HCl while monitoring with a pH meter. Adjust volume to 1 L with milliQ water.
10. Dithiothreitol (DTT) 1 mM in PBS, pH 7.4.

2.6 Quantification of the BG Conjugation to Antibody Adaptors

1. Laemmli buffer: Tris buffer 1 M, pH 6.8, sodium dodecyl sulphate (SDS) 6 g, glycerol, 2-mercaptoethanol, bromophenol blue. Take 15 mL of Tris buffer in a 50 mL conical tube and add 6 g of SDS, 30 mL of glycerol, 15 mL of bromophenol blue. Make up the remaining volume with deionized water to reach a final volume of 50 mL. Vortex the solution until the solids dissolve. Store at −20°C.
2. 8% SDS-PAGE gel: for one gel but can be scaled up linearly for multiple gels at once. 7.95 mL of water, 3 mL of 40% acrylamide, 3.75 mL of 1.5 M Tris at pH 8.8, 150 μL of SDS, 150 μL

of 10% ammonium persulfate, 9 μL of tetramethylenediamine TEMED.

3. Protein ladder for SDS-PAGE, 10–250 kDa.
4. Gel staining solution: 200 mM imidazole aqueous solution containing 0.1% SDS in a 50 mL volume of milliQ water.
5. Zinc sulfate developing gel buffer: 200 mM zinc sulphate solution in a 50 mL volume of milliQ water.
6. ChemiDoc.
7. Image J image analysis software.

2.7 Conjugation of Adaptors to SNAP-CAR on the Surface of Primary Human T Cells and Assessment of Conjugation Efficiency

1. Primary human SNAP-CAR T cells cultured in HAB media supplemented with 100 IU/mL of IL-2 and 1 ng/mL of IL-15 (see Sect. 3.5).
2. Monoclonal antibody adaptor (see Sect. 3.6).
3. Microcentrifuge tubes, 1.5 mL.
4. 96-well V-bottom plate.
5. Temperature-regulated centrifuge.

2.8 Functional Assessment of SNAP-CAR Function

1. Complete RPMI 1640 supplemented with 1× MEM amino acids solution, 10 mM sodium pyruvate, 10% fetal bovine serum (FBS), and Penicillin-Streptomycin.
2. Armed primary human SNAP-CAR T cells (generated in Sect. 3.5 and armed in Sect. 3.8).
3. Antigen positive target cells (e.g., NALM6-CD20) [9]. Antigen negative cells can also be used as an additional negative control (e.g., NALM6, ATCC).
4. Trypan Blue.
5. Hemocytometer.
6. 15 mL conical tube.
7. Dulbecco's Phosphate Buffered Saline, no magnesium, no calcium.
8. Cell Trace Yellow Proliferation Kit (ThermoFisher Scientific), 1:1000×. Resuspend the lyophilized pellet per tube in 40 μL DMSO (*see* **Note 1**).
9. Water bath maintained at 37°C.
10. Ghost Dye Red Viability Dye (Tonbo Biosciences), 1:1000× in DPBS.
11. Antibodies for flow cytometry: anti-CD3 FITC, UCHT1 clone (Biolegend), anti-CD271-BV421, C40-1457 clone (BD Biosciences), Alexa Fluor® 647 AffiniPure® Fab Fragment Goat Anti-Human IgG (H+L), (109-607-003,

Jackson ImmunoResearch), anti-CD69 BV711, FN50 clone (Biolegend), anti-CD62L, SK11 clone (BD Biosciences).

12. Flow cytometer: LSR Fortessa (BD Biosciences) or equivalent.

3 Methods

3.1 Isolation of Peripheral Blood Mononuclear Cells (PBMCs)

1. Open the leukopack bag(s) in a sterile tissue culture hood using a scalpel (*see* **Note 2**).
2. Transfer the blood into a 50 mL conical tube; scale up if needed.
3. Add 15 mL of Lymphocyte Separation Medium (LSM) to each 50 mL conical tube containing blood.
4. Transfer blood on top of the LSM-containing tubes carefully with a serological pipette (*see* **Note 3**).
5. Spin the tubes at 400 × *g* for 30 min with one acceleration and zero deceleration speed.
6. Carefully aspirate the top layer which contains residual plasma.
7. Collect the second layer of mononuclear cells into a new 50 mL conical tube using a serological pipette, without taking any volume from the LSM layer below.
8. Add 40 mL of DPBS to wash the cells.
9. Spin the cells at 400 × *g* for 15 min with full acceleration and deceleration speed.
10. Aspirate DPBS and incubate cells with 2 mL of RBC lysis buffer.
11. Incubate RBC lysis buffer at room temperature for 2 min and add 40 mL of PBS to the tube.
12. Spin the cells at 300 × *g* for 10 min with full acceleration and deceleration speed.
13. Aspirate PBS and resuspend the cell pellet in complete HAB media.
14. Count the cells using Trypan Blue on a hemocytometer using the appropriate dilution factor.
15. Freeze 50 × 10^6 cells/mL in a 2 mL cryogenic tube using Bambanker Freezing Media.
16. Transfer the cryogenic tubes into −80°C immediately.
17. After 24 h, transfer the cryogenic tubes into LN_2 maintained at −140°C or lower.

3.2 Isolation of Primary Human Pan T Cells

1. The process is applicable for freshly isolated or frozen PBMCs as described in Sect. 3.1.

2. Prepare a mixture of pre-warmed 9 mL of HAB media + 1 mL of deoxyribonuclease I, DNase if using frozen PBMCs.
3. Carefully pipette PBMCs into a 15 mL conical tube.
4. Add HAB with DNase as described in step 2 dropwise to the PBMCs.
5. Spin the tube at 300 × *g* for 5 min.
6. Aspirate the supernatant.
7. Resuspend the cell pellet in 10 mL of HAB media (no DNase).
8. Count the cells using Trypan Blue on a hemocytometer and proceed to the isolation of T cells. The following steps 9–15 consider a total of 10 × 10^6 PBMCs isolated. The protocol can be scaled to larger cell numbers as needed.
9. Spin the tube and resuspend cells in 40 μL of MACS buffer.
10. Add 10 of μL of Pan T-cell biotin-antibody cocktail.
11. Gently mix and incubate at 4°C for 5 min.
12. After 5 min add 30 μL of MACS buffer.
13. Add 20 μL of Pan T-cell microbead cocktail.
14. Gently mix and incubate at 4°C for 10 min.
15. Add MACS buffer to raise volume to a minimum of 500 μL (*see* **Note 4**).
16. Use LS columns with QuadroMACS separator for isolating Pan T cells (*see* **Note 5**).
17. Place a conical tube such that the tip of the LS column goes right into the opening of the conical tube.
18. Wash each LS column with 3 mL of MACS buffer.
19. Discard the conical tube that has the flow-through in it.
20. Place a new conical tube under the column and add the cell suspension into the LS column.
21. Collect the flow-through which contains unlabeled isolated Pan T cells.
22. Wash the column with 3 mL of MACS buffer.
23. Count the cells from the flow-through.
24. Collect 100 μL of cells, of which 50 μL will be stained for checking the purity of CD3 on isolated cells. The remaining 50 μL will be unstained.
25. Stain cells with anti-human CD3-FITC antibody, using the flow cytometry staining protocol as mentioned below in Sect. 3.3.
26. Culture cells in a 24-well tissue culture-coated plate, with 2 × 10^6 cells in a well in 2 of HAB media (*see* **Note 6**).

27. Maintain isolated primary human T cells at a concentration of 1×10^6 cells/mL.
28. Supplement cells with 2 μL of 100,000 IU/mL IL-2 and 2 μL of 1000 ng/mL IL-15 cytokines per well, considering a 2 mL volume per well. Scale up the volume of cytokines if using a different culture volume plate.

3.3 Flow Cytometric Surface Staining of Cells

1. Harvest an appropriate number of cells (5×10^4–1×10^6 cells per sample) by spinning at 300 × *g* for 5 min in a round or V-bottom 96-well plate.
2. Discard the supernatant and add 200 μL of FACS buffer per well (*see* **Note 7**).
3. Spin the plate at 300 × *g* for 5 min
4. Discard the supernatant.
5. Stain cells by fully resuspending them in 50 μL of antibody mixture per sample (*see* **Note 8**). As a negative control also include a well with unstained cells.
6. Incubate the plate on ice in the dark for 30 min.
7. Add 150 μL of FACS buffer to each sample.
8. Spin the plate at 300 × *g* for 5 min.
9. Discard the supernatant.
10. Repeat washing by adding 200 μL of FACS buffer per sample.
11. Discard the supernatant and collect the cell pellet.
12. If staining with a secondary antibody, repeat steps 4–10.
13. Resuspend the cell pellet in 100 μL of FACS buffer.
14. Transfer into the 1.2 mL clear microtiter tubes and proceed to run samples on the flow cytometer.

3.4 Production of SNAP-CAR Encoding Retrovirus

1. Plate 3×10^6 293-GP cells in a final volume of 10 mL complete DMEM onto a 100 mm poly-D-lysine-coated plate one day before the transfection. Thaw 3×10^6 293-GP cells in a T-75 tissue culture flask and check the confluency before plating the cells in 100 mm poly-D-lysine-coated plate (*see* **Note 9**).
2. Twenty-four hours later, perform a media exchange by aspirating entire media from the plate (*see* **Note 10**). Add 5 mL of complete DMEM, drop-wise to the edge of the plate, being careful not to detach the cells.
3. Two hours later, prepare two microcentrifuge tubes for transfection: *Tube A* with DNA, Opti-MEM, and P3000, and *Tube B* with Opti-MEM and Lipofectamine 3000. For *Tube A*, add the following materials in order: 1.7 μg of RD114, 3.4 μg of

pMSGV1-SNAP-CAR, 650 μL of Opti-MEM, and 25 μL of P3000, mixing gently but thoroughly with a pipette. For *Tube B* add 650 μL of Opti-MEM and 40 μL of Lipofectamine 3000 and vortex for 3 s.

4. Let *Tube A* and *Tube B* incubate separately for 5 min at room temperature.
5. After 5 min, transfer dropwise the contents of *Tube A* into *Tube B*.
6. Incubate the mixture for 30–60 min.
7. Gently add the mixture to the 293-GP cells, dispensing it dropwise and distributing it evenly across the plate.
8. Gently slide the plate in upward and downward directions and avoid circular motion to further disperse the plasmid mixture evenly across the plate.
9. Move the plate to a tissue culture incubator maintained at 37°C and 5% CO_2 levels.
10. After 24 h, perform a media exchange by carefully aspirating the media and tilting the plate to gently add 11 mL of fresh DMEM media to the plate edge.
11. Move the plate to a tissue culture incubator maintained at 37°C and 5% CO_2 levels.
12. After 24 h, perform the first viral harvest by carefully collecting the supernatant into a 50 mL conical tube. Seal and store the conical tube in the 4°C fridge overnight.
13. Replenish the plate with 11 mL of fresh DMEM media as in step 10 and put the plate back into the incubator.
14. After 24 h, perform the second viral harvest, collecting the viral supernatant from the plate and combining it with the first collection. Discard the 293-GP cells into viral waste.
15. Spin the supernatant tube in a centrifuge cooled to 4°C at 1000 × *g* for 10 min to remove cellular debris.
16. Filter the supernatant via 0.45 μM PES membrane filters using 10 mL sterile syringes.
17. Aliquot 2 mL of supernatant into each 2 mL cryovial.
18. Store the virus at −80°C.

3.5 Generation of Primary Human SNAP-CAR T Cells by Retroviral Transduction and Subsequent Culture and Expansion

1. Add 1:100 volume per well of TransAct Human T cell activation reagent to T cells from Sect. 3.2 (20 μL for 2 mL well) to stimulate the T cells.
2. Gently mix the well contents by pipetting.
3. After 24 h make a Retronectin transduction plate by adding 2 mL of Retronectin solution to each well of a new 6-well

non-tissue culture-treated plate. Seal the plate and incubate for 24 h at 4°C (*see* **Note 11**).

4. Two days after the isolation of primary human T cells and one day after coating the Retronectin plate, aspirate the Retronectin solution and add 2–4 mL of purified SNAP-CAR retrovirus (generated in Sect. 3.4) to each well. Then add 0–2 mL of complete HAB RPMI media for a total volume (virus plus media) of 4 mL.
5. Spin the plate at 2000 × *g* for 2 h in a centrifuge heated to 32°C.
6. Remove 2 mL of the viral supernatant/media mix and add 1×10^6 of the TransAct-stimulated primary human T cells per well.
7. Bring the final volume to 6 mL by further adding complete HAB RPMI media.
8. Spin the plate at 32°C for 10 min at 1000 × *g* (*see* **Note 12**).
9. Place the plate into the tissue culture incubator maintained at 37°C and 5% CO_2.
10. Twenty-four hours later add 1 μL of 100,000 IU/mL IL-2 and 1 μL of 1000 ng/μL IL-15 cytokines per mL of media and place the plate back into the incubator.
11. Seventy-two hours later, transfer the cells to a 6-well tissue-culture treated plate, splitting them at a concentration of 1×10^6 cells/mL with a maximum of 6 mL per well. Again, supplement the media with 1 μL of 100,000 IU/mL IL-2 and 1 μL of 1000 ng/μL IL-15 per mL of media.
12. Place the plate back into the incubator.
13. Continue to split the cells and supplement with fresh IL-2 and IL-15 cytokines as mentioned in step 11 every 2–3 days.
14. After 7 days of stimulation and expansion stain the cells for the LNGFR marker gene of SNAP-CAR expression using the anti-CD271 BV421 antibody as described in Sect. 3.3.
15. Analyze LNGFR expression by flow cytometry. Figure 4 shows an example of LNGFR staining of primary human T cells transduced with SNAP-CAR-T2A-LNGFR retrovirus and gating to quantify the SNAP-CAR+ cell population.

3.6 Benzyl guanine (BG) Conjugation to Monoclonal Antibodies

BG conjugation to antibodies is carried out using the BG-GLA-NHS-ester reagent (Fig. 2). All steps are carried out at room temperature unless otherwise noted. This method considers the size of the Zeba Column to be 2 mL.

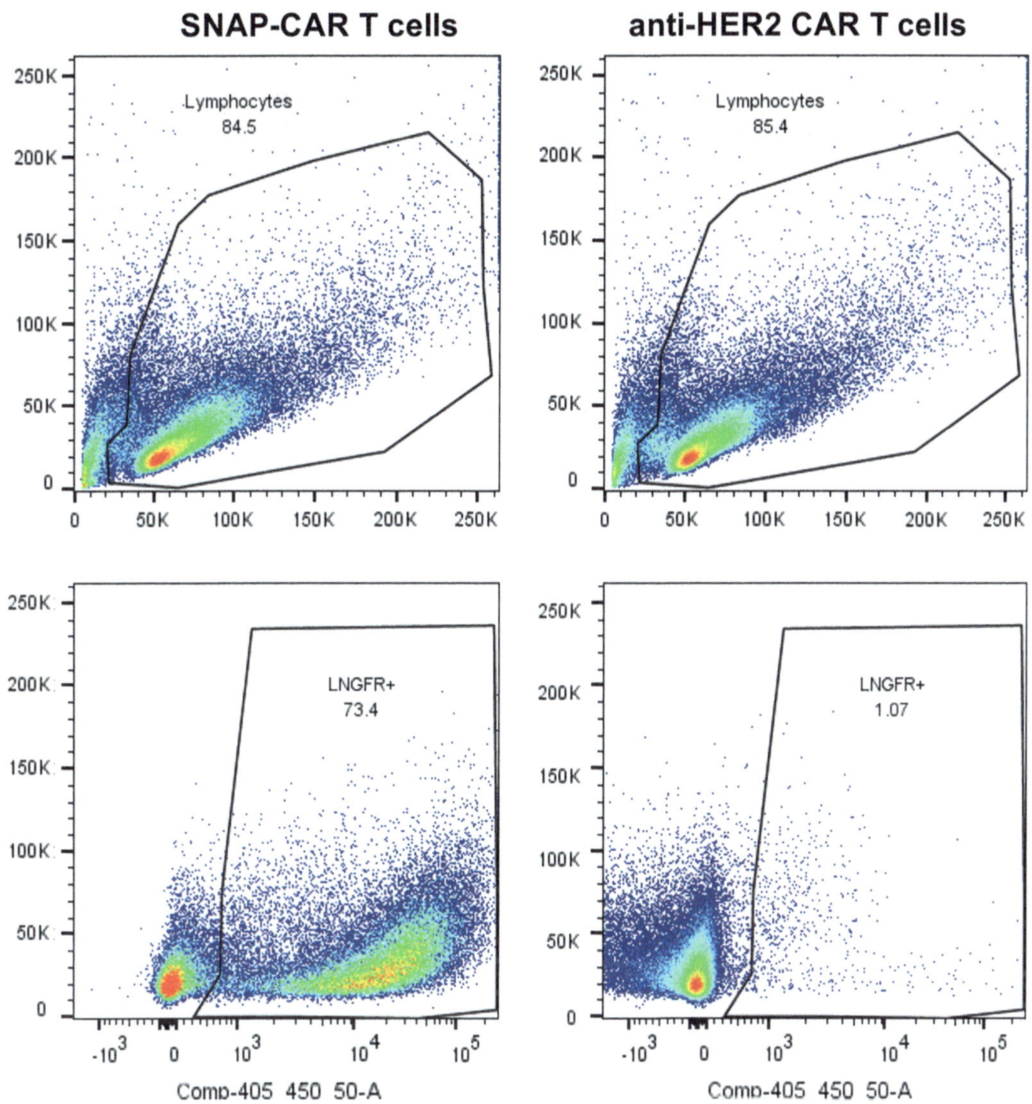

Fig. 4 Phenotypic characterization of primary human T cells expressing SNAP-CAR-T2A-LNGFR and a traditional CAR targeting the HER2 protein. Flow cytometry analysis in a two-dimensional dot plot showing expression and enzymatic functionality of the SNAP-CAR receptor on transduced cells vs. anti-HER2 CAR T cells, assessed by staining with anti-human LNGFR BV421 antibody on the x-axis. (Reproduced from [9], with permission from Springer Nature)

1. Exchange the buffer of the antibody into DPBS before proceeding with conjugation by using 7K MWCO Zeba spin desalting columns (steps 2–5).
2. Remove the bottom of the desalting column and loosen the top cap slightly. Place the column into a 15 mL conical tube.
3. Centrifuge at 1000 × *g* for 2 min to remove storage buffer.

4. Wash the columns three additional times by adding 1 mL of DPBS to each column, centrifuging as in step 3, and discarding the flow-through for every wash. For conjugation purposes, maintain volume between 200 μL and 700 μL. If the volume is less than 350 μL, add 40 μL stacker volume with water or appropriate buffer to ensure maximal protein recovery.
5. Transfer the washed column into a new 15 mL conical tube and add the antibody. Allow the antibody to wet the resin bed and then proceed with centrifugation at 1000 × *g* for 2 min. Once the antibody has passed through the column, discard the column and measure antibody concentration using a Nanodrop One C or equivalent using the "Protein A260" setting and selecting "IgG" from the drop-down menu options (*see* **Note 13**). Proceed with conjugation as described below, or store the antibody at 4°C.
6. Calculate the amount of BG-GLA-NHS ester to be added to the antibody. For a typical conjugation, aim for 20 molar equivalents of BG-GLA-NHS ester to 1 molar equivalent of antibody (*see* **Note 14**). Molecular weight of BG-GLA-NHS ester: 481.5 g/mol. Use the following formula to calculate the volume of BG-GLA-NHS ester: Volume of BG-GLA-NHS ester = 20 * volume of antibody to be conjugated * molarity of the antibody/ molarity of the ester (*see* **Note 15**). The following steps (7–10) are written for the conjugation of 500 μg of antibody. The molarity of the ester will change according to the amount of concentrated antibody.
7. Prepare the BG-GLA-NHS stock solution as described in Sect. 2.5, step 5.
   Add the calculated volume of dissolved NHS-ester to the antibody from step 6 and incubate at room temperature in the dark for 30 min (*see* **Note 16**).
8. Proceed with buffer exchange of the BG-antibody conjugate using fresh Zeba columns as previously described in steps 2–5 (*see* **Note 17**).
9. Store the antibody-BG conjugate at 4°C for future usage.

3.7 Quantification of BG Motifs on Conjugated Antibodies

The average number of BG's per antibody will be quantified by an SDS-PAGE gel shift assay in which the antibody-BG adaptor is mixed with an excess of recombinant SNAP-tag protein. The SNAP-tag conjugation causes a shift in the protein band size which can be quantified by gel analysis software.

1. Combine 0.5 μg of the BG-conjugated antibody adaptor with 2 μg of the recombinant SNAP-tag protein in a 10 μL volume of DPBS at pH 7.4. Also include an antibody BG sample with

no SNAP-tag protein added as a negative control to aid in quantification.

2. Add 10 μL of 1 mM dithiothreitol (DTT).
3. Mix and incubate the tubes at 37°C for 2 h.
4. Dilute the reactions with 3.3 μL of Laemmli buffer and boil for 5 min.
5. Run the samples and protein ladder on an 8% SDS-PAGE gel at 120 V for 90 min.
6. Stain the gel with a 200 mM imidazole aqueous solution containing 0.1% SDS for 15 min with light agitation in a 50 mL volume of milliQ water.
7. Decant the staining solution and replace it with 50 mL of water.
8. Incubate for 30 s.
9. Decant the water and develop the gel for 45 s with zinc sulphate aqueous solution with light agitation in 50 mL volume of milliQ water.
10. Rinse the gel with running water for 10 s.
11. Image the gel on a ChemiDoc using epi white light on a black background. Figure 5 shows an example of the stained gel.
12. Identify the bands for the unlabeled antibody heavy chain (~50 kDa) and additional bands corresponding to different numbers of SNAP-tag molecules conjugated to the heavy

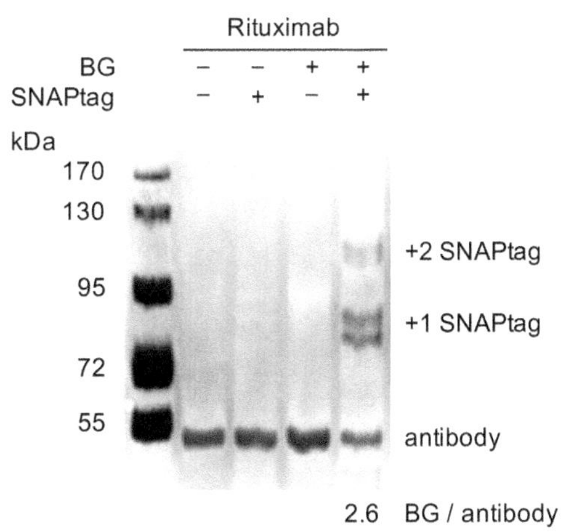

Fig. 5 Quantification of BG on conjugated monoclonal antibody. SDS-PAGE quantification of rituximab anti-human CD20 targeting monoclonal antibody-conjugated benzyl guanines by SNAP conjugation reaction with an average of 2.6 BG-labeled antibodies. The average number of BGs per antibody was calculated based on relative band intensities. (Reproduced from [9], with permission from Springer Nature)

chain (1, 2, 3...n) with the anticipated band sizes increasing by 19.4 kDa for each SNAP-tag protein. Measure the band intensities with ImageJ software.

13. Calculate the average number of BG's per full-length IgG adaptor using the band intensities.

3.8 Conjugation of BG-Antibody Adaptors to Primary Human SNAP-CAR T Cells and Quantification

1. Add 1 μg/mL of adaptor to 5 × 10^4 SNAP-CAR T cells in a microcentrifuge tube in 50 μL of complete RPMI media. While this concentration represents the recommended conditions, the adaptor loading and resulting SNAP-CAR activity can be tuned by titrating the adaptor or SNAP-CAR cell concentrations, as shown in Fig. 6.
2. Place the microcentrifuge tube in an incubator for a 1 h incubation at 37°C and 5% CO_2.
3. Following the incubation, wash away excess adaptor by adding 0.5 mL of complete RPMI media and centrifuging cells at 300 × *g* for 5 min.
4. Wash again by aspirating the media and adding 0.5 mL of FACS buffer.
5. Centrifuge the cells at 300 × *g* for 5 min and resuspend the cells in FACS buffer (for flow cytometry analysis) or complete RPMI media (for functional assays).
6. Store cells on ice and proceed immediately to flow cytometry staining (step 7) or functional assay (Sect. 3.9).
7. Stain the cell pellet with a fluorescently labeled anti-human IgG Fab such as AF647-conjugated anti-human IgG H+L following the flow cytometry staining protocol as described in Sect. 3.3. It is crucial to include a staining control sample

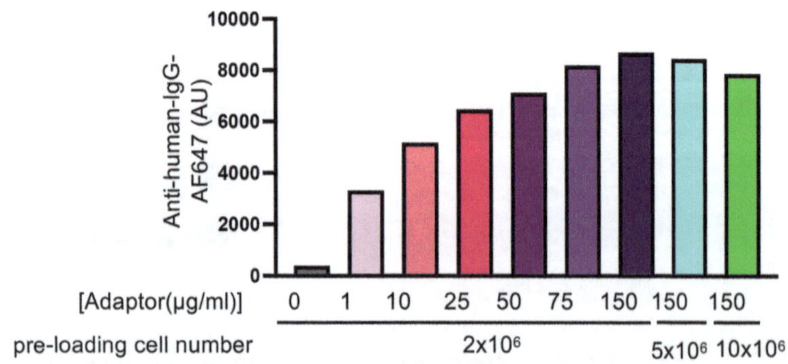

Fig. 6 Tuning SNAP-CAR T-cell labeling by antibody adaptor dose. Flow cytometric analysis of the indicated number of primary human SNAP-CAR T cells co-incubated with varying doses of rituximab-BG adaptor for 1 h at 37°C, stained with fluorescently labeled anti-human IgG-AF647 secondary antibody. (Reproduced from [9], with permission from Springer Nature)

of secondary antibody-only to account for background staining of that antibody and to accurately quantify binding to the SNAP-CAR T cells.

8. (Optional) Cells can also be stained for LNGFR expression using the anti-CD271 antibody. This staining must be performed after the anti-human IgG Fab to eliminate chance of the Fab binding to the anti-CD271 antibody (Sect. 3.3).
9. Transfer the stained cells into a 96-well V-bottom plate.
10. Analyze expression by flow cytometry. Figure 6 shows an example of staining of primary human T cells transduced with SNAP-CAR-T2A-LNGFR retrovirus, and gating to quantify the SNAP-CAR+ cell population.

3.9 Functional Testing of Armed SNAP-CAR T-Cell Activation and Target Cell Lysis In Vitro

The following steps (2–8) use 1×10^4 target cells (NALM6-CD20), per well. It is suggested performing the experiment in triplicate and with three conditions, namely, SNAP-CAR T cells armed with antibody adaptor, unarmed SNAP-CAR T cells, and target cells-only (*see* **Note 18**).

1. Count SNAP-CAR T cells and target cells using Trypan blue and a hemocytometer.
2. Harvest 2×10^5 target cells for a total of 20 wells.
3. Spin at $300 \times g$ for 5 min.
4. Aspirate the supernatant and resuspend the pellet in 500 µL of Cell Trace Yellow dye prepared in DPBS.
5. Incubate at 37°C for 8 min in a water bath.
6. Add prewarmed 10 mL of complete RPMI medium to quench the dye and mix thoroughly.
7. Incubate the cells at 37°C for 10 min in the water bath.
8. Centrifuge the cells at $300 \times g$ for 5 min.
9. Aspirate media carefully and resuspend in complete RPMI. In this case, add 100 µL of labeled target cells per well. Considering 20 wells, the final volume of resuspension is 2000 µL of complete RPMI media.
10. Plate 1×10^4 Cell Trace Yellow-labeled target cells per well in a sterile 96-well V-bottom plate.
11. Count adaptor-armed (Sect. 3.8) and unarmed primary human SNAP-CAR T cells using Trypan Blue and a hemocytometer.
12. In this case, for an effector: target cell ratio (5:1), 5×10^4 SNAP-CAR T cells per well were plated.
13. Spin the cells at $300 \times g$ for 5 min to collect the cell pellet.
14. Resuspend the cell pellet in complete RPMI media to be at 5×10^5 cells/mL.

15. Add 100 μL of SNAP-CAR T cells to each well to bring the total volume to 200 μL. Important control groups include target cells-only, target cells with effector cells and antibody, and target cells with effector cells (no antibody). Media alone should be added to the target cell-only wells.
16. Incubate the cells at 37°C for 24–48 h in the tissue culture incubator maintained at 37°C and 5% CO_2.
17. After the desired incubation time, generate a cell pellet by spinning the plate at 300 × *g* for 5 min.
18. Discard the supernatant.
19. Stain the cell pellet with Ghost Dye Red Viability Dye by adding 60 μL per well and incubating in the fridge for 15 min.
20. Wash the cells by adding 150 μL FACS buffer.
21. Spin the plate at 300 × *g* for 5 min.
22. Discard the supernatant.
23. Wash cells with 200 μL FACS buffer.
24. Spin the plate at 300 × *g* for 5 min.
25. Stain cells with a mixture of antibodies to identify SNAP CAR T cells and quantify activation: LNGFR BV421, CD69 BV711, CD62L FITC. Perform staining as described in Sect. 3.3.
26. For the gating strategy on FlowJo, refer to Fig. 7. Briefly, identify target cells by Cell Trace Yellow, and lysed target cells by Cell Trace Yellow and positivity for Ghost Dye staining.
27. Calculate specific lysis using this equation: 100*(experimental target cell lysis – % target cells-only lysis)/ (100 – % target cells-only lysis). Figure 8 shows an example of specific lysis data in addition to corresponding activation marker staining.

4 Notes

1. Cell Trace Dye working stock solutions are to be prepared as per Thermofisher's instructions. Avoid repeated freeze-thaw of the dissolved dye in DMSO.
2. Buffy coats and leukopacks contain a variable amount of blood volume; hence, consider scaling the number of 50 mL conical tubes accordingly.
3. The addition of blood into Lymphocyte Separation Medium should be done dropwise and slowly, with minimal interference of blood into the Lymphocyte Separation Medium layer.
4. If the volume of the MACS buffer and antibodies is more than 500 μL, then there is no need to add extra volume of MACS buffer.

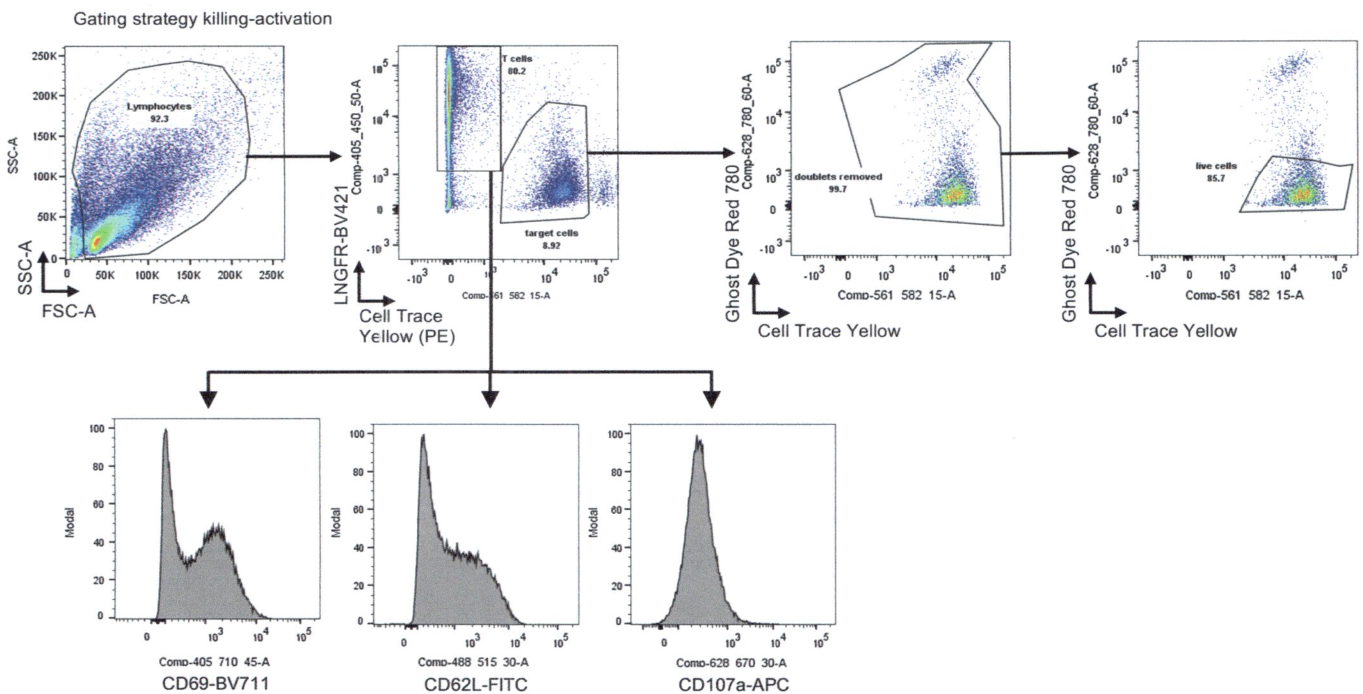

Fig. 7 Gating strategy for determining % specific cell lysis of target cells. Flow cytometry gating strategy of the co-incubation experiment to determine the % specific lysis of target cells in vitro in the presence of antibody and SNAP-CAR T cells. Target cells are labeled with Cell Trace Yellow, and SNAP-CAR T cells are stained for LNGFR antibody. Using Ghost Dye staining, we can distinguish between live and dead cells and gate out live target cells, as shown in the last panel. For activation markers, LNGFR-gated cells are quantified for mean fluorescent intensities of CD69, CD62L, and CD107a. (Reproduced from [9], with permission from Springer Nature)

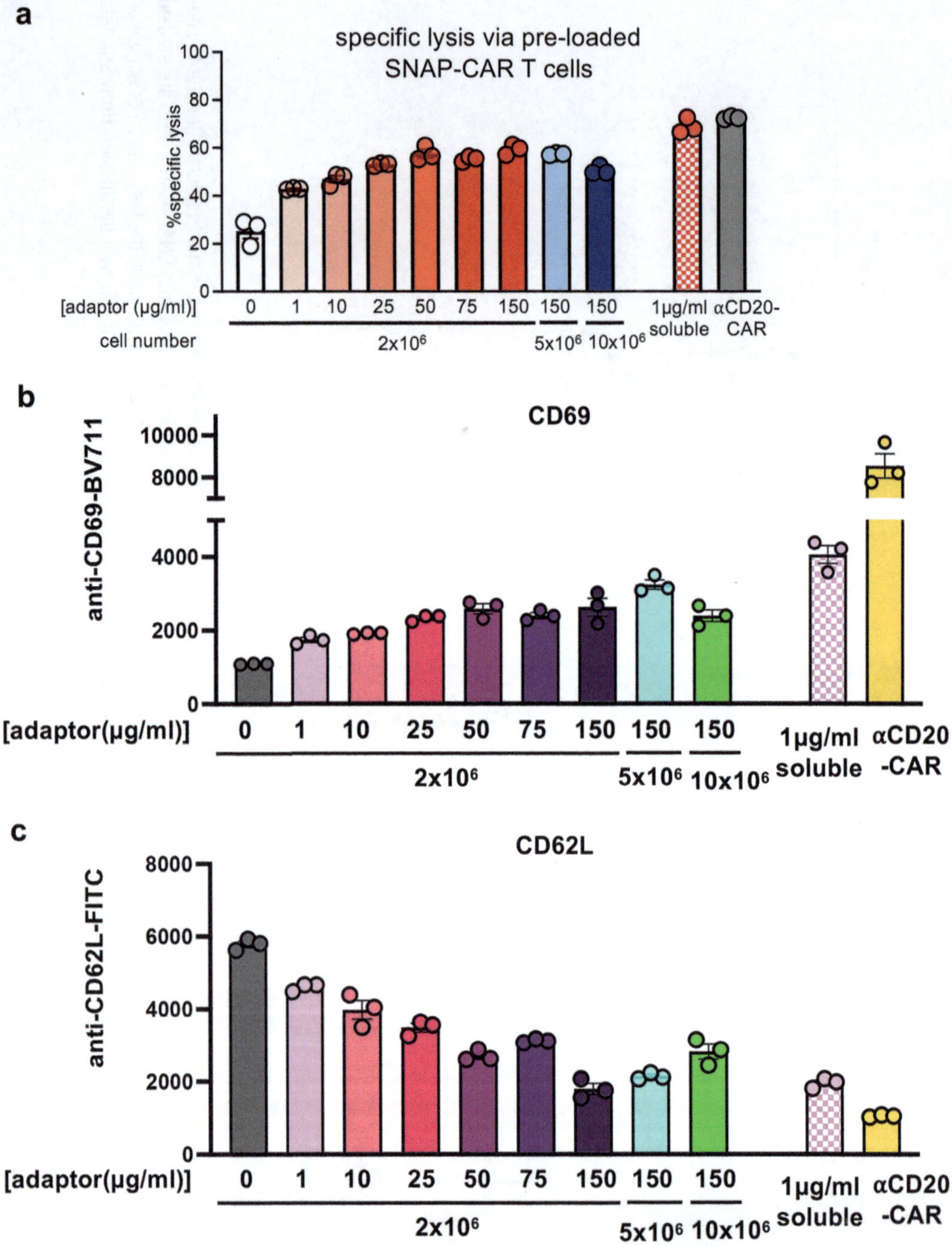

Fig. 8 Functional activity of antibody-adaptor with SNAP-CAR T cells on target cells in vitro. (**a**) Specific lysis of target cells by co-incubating primary human SNAP-CAR T cells that were pre-incubated with the indicated concentration of BG conjugated rituximab anti-human CD20 targeting antibody adaptor at the indicated cell concentration as compared to SNAP-CAR T cells incubated with 1.0 μg/mL of soluble adaptor or a positive control anti-CD20 CAR. (**b**) Flow cytometric analysis of primary human SNAP-CAR T cells at the indicated cell concentrations co-incubated with the indicated rituximab-BG adaptor dose for 1 h at 37 °C, followed by co-incubation with NALM-CD20 target cells for 24 h and staining for CD69, and (**c**) CD62L markers on the SNAP-CAR T cells (LNGFR+ cell population). Values in (**b**) and (**c**) are reported as mean fluorescent intensities (MFI). (Reproduced from [9], with permission from Springer Nature)

5. Each LS column can hold 1×10^9 labeled cells; use multiple columns if more than 1×10^9 cells are labeled.
6. When culturing cells in a 24-well plate, ensure that wells surrounding the T cells are filled with PBS to avoid loss due to evaporation in the incubator. Alternatively, cells can be cultured in a T-75 tissue culture flask.
7. When staining for flow cytometry, after every spin, ensure that a cell pellet is observed; failure to see a cell pellet indicates that the cells are not pelleted either due to a insufficient speed or time of spinning.
8. Refer to the manufacturer's recommended dilution for each staining antibody. If using more than one antibody for staining, prepare a mixed solution of different antibodies and stain cells using that master mix.
9. 293-GP cells tend to grow on top of each other once they reach confluency. It is crucial to monitor the 293-GP cells not to let them overgrow. Poor quality and/or clumped 293-GP cells will ultimately affect the quality of the virus produced, leading to lower multiplicity of infection (MOI) and lower transduction values. Avoid a higher passage number of 293-GP cells, generally above 8 is not recommended. Plating 3×10^6 cells in T-75 tissue culture flask is the recommended density.
10. When aspirating the media, carefully tilt the plate such that the media collects on one edge, and use an aspirating pipette to collect the media.
11. It is also possible to coat the Retronectin plate on the day of transduction. Use the same concentration of Retronectin and let it incubate at room temperature for 2 h, which should also be sufficient. Proceed with spinfection afterwards. For example: using a 6-well plate, add 2 mL per well; using a 24-well plate, add 500 μL per well.
12. It is crucial to maintain the temperature of the centrifuge at 32°C for the spinfection process, along with the speed when adding the virus, and when T cells are added. T cells are fragile and can undergo cell death if spun at a higher speed of $1000 \times g$ for 10 min.
13. Measure the concentration of the antibody before and after spinning through the column. This is crucial before adding the BG-GLA-NHS ester, as higher amounts can lead to precipitation and ruin the conjugation process.
14. Conjugating at lower molar equivalents results in a lower level of BG conjugation. Potentially optimize the molar ratio for different levels of BG conjugation.
15. BG-BLA-NHS ester is highly hygroscopic and stored at −20°C. If the entire 2 mg quantity is not required, aliquot

the required amount as accurately as possible, and store the unused hygroscopic powder back in −20°C.

16. Since DMSO is light sensitive, the reaction should be conducted in the dark, away from direct sunlight.
17. Buffer exchange of the columns can be started when 15 min for the conjugation step remain.
18. Scale up or down the target cells based on the design of the experiment including titrating the adaptor or testing multiple adaptors.

References

1. Yang H, Yao L, Wang Y, Chen G, Chen H (2023) Advancing cell surface modification in mammalian cells with synthetic molecules. Chem Sci 14(46):13325–13345. https://doi.org/10.1039/d3sc04597h
2. Capicciotti CJ, Zong C, Sheikh MO, Sun T, Wells L, Boons GJ (2017) Cell-surface glyco-engineering by exogenous enzymatic transfer using a bifunctional CMP-Neu5Ac derivative. J Am Chem Soc 139(38):13342–13348. https://doi.org/10.1021/jacs.7b05358
3. Streit M, Budiarta M, Jungblut M, Beliu G (2025) Fluorescent labeling strategies for molecular bioimaging. Biophys Rep (N Y) 5((1):100200. https://doi.org/10.1016/j.bpr.2025.100200
4. Dong R, Aksel T, Chan W, Germain RN, Vale RD, Douglas SM (2021) DNA origami patterning of synthetic T cell receptors reveals spatial control of the sensitivity and kinetics of signal activation. Proc Natl Acad Sci U S A 118(40):e2109057118. https://doi.org/10.1073/pnas.2109057118
5. Yu L et al (2020) Promoting the activation of T cells with glycopolymer-modified dendritic cells by enhancing cell interactions. Sci Adv 6(47):eabb6595. https://doi.org/10.1126/sciadv.abb6595
6. Stephan MT, Moon JJ, Um SH, Bershteyn A, Irvine DJ (2010) Therapeutic cell engineering with surface-conjugated synthetic nanoparticles. Nat Med 16(9):1035–1041. https://doi.org/10.1038/nm.2198
7. Hu Q et al (2018) Conjugation of haematopoietic stem cells and platelets decorated with anti-PD-1 antibodies augments anti-leukaemia efficacy. Nat Biomed Eng 2(11):831–840. https://doi.org/10.1038/s41551-018-0310-2
8. Kim H et al (2018) General and facile coating of single cells via mild reduction. J Am Chem Soc 140(4):1199–1202. https://doi.org/10.1021/jacs.7b08440
9. Ruffo E et al (2023) Post-translational covalent assembly of CAR and synNotch receptors for programmable antigen targeting. Nat Commun 14(1):2463. https://doi.org/10.1038/s41467-023-37863-5
10. Cai F, Ren Y, Dai J, Yang J, Shi X (2023) Effects of various cell surface engineering reactions on the biological behavior of mammalian cells. Macromol Biosci 23(3):e2200379. https://doi.org/10.1002/mabi.202200379
11. Zhou H, Zhang Y, Pei P, Shen W, Yi X, Yang K (2023) Liposome-anchored mesenchymal stem cells for radiation pneumonia/fibrosis treatment. Biomaterials 300:122202. https://doi.org/10.1016/j.biomaterials.2023.122202
12. Mahal LK, Yarema KJ, Bertozzi CR (1997) Engineering chemical reactivity on cell surfaces through oligosaccharide biosynthesis. Science 276(5315):1125–1128. https://doi.org/10.1126/science.276.5315.1125
13. Tomas RMF, Gibson MI (2019) Optimization and stability of cell-polymer hybrids obtained by "clicking" synthetic polymers to metabolically labeled cell surface glycans. Biomacromolecules 20(7):2726–2736. https://doi.org/10.1021/acs.biomac.9b00478
14. Ramya TN, Weerapana E, Cravatt BF, Paulson JC (2013) Glycoproteomics enabled by tagging sialic acid- or galactose-terminated glycans. Glycobiology23(2):211–221.https://doi.org/10.1093/glycob/cws144
15. Lee YG et al (2019) Use of a single CAR T cell and several bispecific adapters facilitates eradication of multiple antigenically different solid tumors. Cancer Res 79(2):387–396. https://doi.org/10.1158/0008-5472.CAN-18-1834
16. Labanieh L, Mackall CL (2023) CAR immune cells: design principles, resistance and the next generation. Nature 614(7949):635–648. https://doi.org/10.1038/s41586-023-05707-3

17. Minutolo NG, Hollander EE, Powell JrDJ (2019) The emergence of universal immune receptor T cell therapy for cancer. Front Oncol 9:176. https://doi.org/10.3389/fonc.2019.00176
18. Urbanska K et al (2012) A universal strategy for adoptive immunotherapy of cancer through use of a novel T-cell antigen receptor. Cancer Res 72(7):1844–1852. https://doi.org/10.1158/0008-5472.CAN-11-3890
19. Lohmueller JJ, Ham JD, Kvorjak M, Finn OJ (2017) mSA2 affinity-enhanced biotin-binding CAR T cells for universal tumor targeting. Oncoimmunology 7(1):e1368604. https://doi.org/10.1080/2162402X.2017.1368604
20. Benton A et al (2025) Mutant KRAS peptide targeted CAR-T cells engineered for cancer therapy. Cancer Cell 43(7):1365–1376.e5. https://doi.org/10.1016/j.ccell.2025.05.006
21. Cho JH, Collins JJ, Wong WW (2018) Universal chimeric antigen receptors for multiplexed and logical control of T cell responses. Cell 173(6):1426–1438.e11. https://doi.org/10.1016/j.cell.2018.03.038
22. Minutolo NG et al (2020) Quantitative control of gene-engineered t-cell activity through the covalent attachment of targeting ligands to a universal immune receptor. J Am Chem Soc 142(14):6554–6568. https://doi.org/10.1021/jacs.9b11622
23. Serniuck NJ et al (2024) Electrophilic proximity-inducing synthetic adapters enhance universal T cell function by covalently enforcing immune receptor signaling. Mol Ther Oncol 32(3):200842. https://doi.org/10.1016/j.omton.2024.200842
24. Stepanov AV et al (2024) Control of the antitumour activity and specificity of CAR T cells via organic adapters covalently tethering the CAR to tumour cells. Nat Biomed Eng 8(5):529–543. https://doi.org/10.1038/s41551-023-01102-5
25. Gautier A et al (2008) An engineered protein tag for multiprotein labeling in living cells. Chem Biol 15(2):128–136. https://doi.org/10.1016/j.chembiol.2008.01.007
26. Chen T et al (2024) The construction of modular universal chimeric antigen receptor T (MU-CAR-T) cells by covalent linkage of allogeneic T cells and various antibody fragments. Mol Cancer 23(1):53. https://doi.org/10.1186/s12943-024-01938-8
27. Kochenderfer JN et al (2009) Construction and preclinical evaluation of an anti-CD19 chimeric antigen receptor. J Immunother 32(7):689–702. https://doi.org/10.1097/CJI.0b013e3181ac6138

Check for updates

Chapter 10

Cysteine Re-bridging Using Divinyl Pyrimidine Reagents for the Generation of Antibody-Drug Conjugates

Thomas Wharton, Sona Krajcovicova and David R. Spring

Abstract

Antibody-drug conjugates (ADCs) are a powerful form of targeted therapy that can deliver drugs with a high level of selectivity towards a specific cell type, reducing off-target effects and increasing the therapeutic window compared to small-molecule therapeutics. However, creating ADCs that are stable, homogeneous, and with a controlled drug-to-antibody ratio (DAR) remains a significant challenge. Divinyl pyrimidine (DVP) reagents were developed as a method to irreversibly conjugate linker-payloads onto native antibodies to generate stable ADCs and antibody-probe systems with a precise drug-to-antibody ratio (DAR) of 4. This protocol describes how to conjugate DVPs to antibodies and methods for ADC characterization.

Keywords Antibody-drug conjugates, Antibody characterization, Cysteine conjugation, Disulfide re-bridging, Drug-to-antibody ratio

1 Introduction

Antibody-drug conjugates (ADCs) have emerged as powerful therapeutics over the last three decades, with interest in these biologics continuing to grow [1, 2]. ADCs are typically composed of three components: a targeting antibody, a drug, and a linker system [2]. In the conjugate, the targeting nature of the antibody is utilized to bring the therapeutic to the site in the body where it is required. Once there, the drug then acts upon the desired target. Many ADC programs have been highly successful, with, as of 2024, 13 FDA-approved therapies and more than 350 currently in clinical trials [3, 4]. However, current ADC synthetic methods lead to reversible conjugation and highly heterogenous mixtures with widely varying drug-to-antibody ratios (DARs). This heterogeneity is widely recognized to cause poor pharmacokinetic and pharmacodynamic properties [1, 5]. Furthermore, the number of payload molecules appended to each antibody is crucial for its activity, binding, toxicity,

Alexander Deiters (ed.), *Bioconjugation Methods*, Methods in Molecular Biology, vol. 3009, https://doi.org/10.1007/978-1-0716-5112-4_10, and propensity for aggregation [6, 7]; therefore, methods to precisely attain a specific DAR are of high importance.

If a conjugating agent reacts with both cysteines of an interchain disulfide bridge at once, then the antibody is said to have been chemically "re-bridged" with the conjugator covalently holding together the light and heavy antibody chains. Disulfide re-bridging has shown to be an efficient bioconjugation strategy allowing for four conjugations to take place in a site-specific way without the need for genetic or enzymic engineering of the antibody [8–10].

To add to this toolbox, the Spring group developed divinyl pyrimidine (DVP) linkers that have been shown to exhibit efficient, site-selective, and irreversible disulfide re-bridging, allowing access to DAR 4 ADCs and antibody-probe constructs from native antibodies (Fig. 1) [11–13]. DVP scaffolds have also been used in generating DAR 1 ADCs [14–16] and to staple peptides with the aim of improving stability [17, 18]. Herein is described a guide to their utilization in generating DAR 4 ADCs and an explanation of various analytical techniques used for ADC characterization. Each characterization technique gives important information about the ADC: sodium dodecyl sulfate polyacrylamide gel electrophoresis (SDS-PAGE), the extent of conjugation; mass spectroscopy (MS), the presence of the desired DAR 4 conjugate; hydrophobic interaction chromatography (HIC), the ratio between different DAR species; and size exclusion chromatography (SEC), the presence of aggregation.

Carboxylic acid functionalized DVPs (Fig. 2) are recommended for general ADC generation as they exhibit high stability, efficient bioconjugation, and facile functionalization with a desired payload. The synthesis of these DVPs and other variants has been described previously [11–13, 16], and are therefore not covered in this chapter.

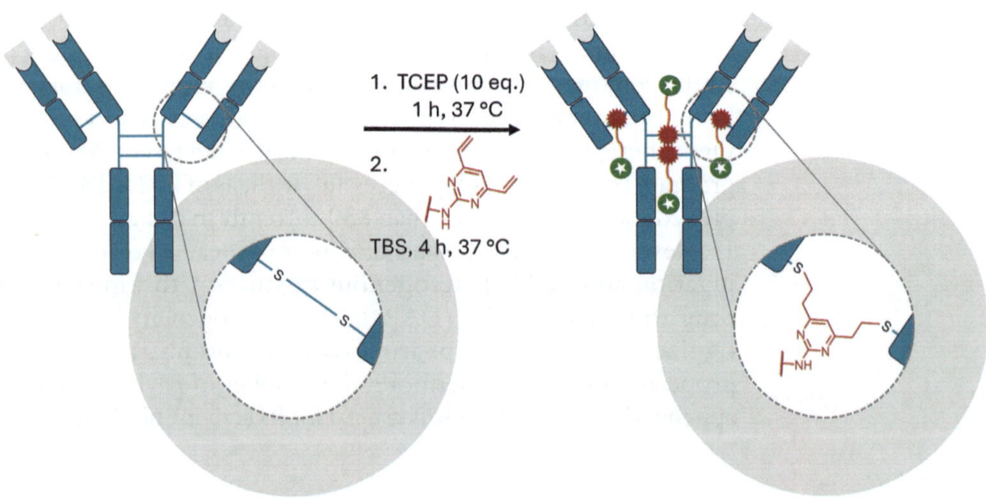

Fig. 1 Antibody schematic showing the relative locations of the interchain disulfide bridges and how DVP linkers conjugate to the antibody

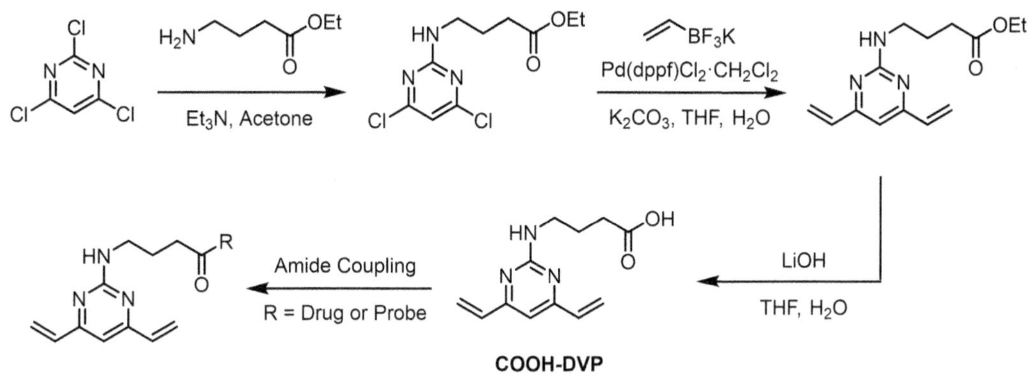

Fig. 2 Synthesis of a carboxylic acid DVP (COOH-DVP) and its functionalization [11, 12]

2 Materials

All buffers are made using ultrapure water (i.e., Milli-Q) unless otherwise stated. A selection of adjustable pipettes are required for transferring liquids between 0.4–1000 μL.

2.1 Antibody Preparation and Conjugation

1. IgG1 antibody (i.e., Trastuzumab; lyophilized solid) (*see* **Note 1**).
2. 500 μL Centrifuge tubes (*see* **Note 2**).
3. Tris-buffered saline (TBS) solution: 25 mM tris(hydroxymethyl)aminomethane (Tris), 25 mM sodium chloride, 0.5 mM ethylenediaminetetraacetic acid (EDTA), adjusted to pH 8.0 with 1 M aq. NaOH.
4. Tris(2-carboxyethyl)phosphine hydrochloride (TCEP).
5. Thermal shaker unit (such as a Grant-bio PCMT Thermo-Shaker) capable of agitating a sample whilst maintaining 37°C.
6. UV–Vis spectrometer (such as a ThermoFisher NanoDrop One) for antibody concentration determination.
7. DVP linker; 5 mM in DMSO (*see* **Note 3**) [12].
8. 0.5 mL 40k Molecular weight cut-off (MWCO) Zeba™Spin Desalting columns.
9. 0.5 mL 10k MWCO Amicon® Ultra 10k MWCO Diafiltration columns.
10. Centrifuge capable of holding 0.5–2 mL centrifuge tubes and running at up to 14,000 × *g* (such as VWR Micro Star 17).
11. Phosphate Buffer Saline (PBS) solution: 137 mM NaCl, 2.7 mM KCl, 8 mM Na_2HPO_4, and 2 mM KH_2PO_4 (*see* **Note 4**).

2.2 SDS-PAGE Analysis

1. Heating block/sand bath capable of heating 0.5 mL vials to 90°C.
2. 0.5 mL vials.
3. PBS solution (as above).
4. Broad range unstained protein standard, molecular weight (MW) ladder (10–200 kDa).
5. Non-reducing sample stain: 0.125 M 2-amino-2-(hydroxymethyl)-1,3-propanediol (Tris-HCl), 20% glycerol, 4% SDS, 0.02% bromophenol blue, adjusted to pH 6.8.
6. Reducing sample stain: 0.125 M Tris-HCl, 20% glycerol, 4% SDS, 0.02% bromophenol blue, 1.25 mM β-mercaptoethanol, adjusted to pH 6.8.
7. Laemmli running buffer (LRB): For 1 L; 3 g 2-amino-2-(hydroxymethyl)-1,3-propanediol (Tris base), 14.4 g glycine, 1 g SDS.
8. 12% Acrylamide gel including 4% stacking gel (*see* **Note 5**).
9. Gel electrophoresis tank and power supply (such as Bio-Rad mini-PROTEAN™ Tetra System).
10. Coomassie brilliant blue (CBB) stain: For 500 mL, dissolve in order: 0.5 g CBB R250 dye, 50 mL EtOH, 400 mL water, then 50 mL AcOH.

2.3 Mass Spectrometry Analysis

1. PNGase F for deglycosylation.
2. Protein LC-MS system (such as Waters Xevo G2-S TOF mass spectrometer coupled to an Acquity UPLC system using an Acquity UPLC BEH300 C4 column (1.7 μm, 2.1 × 50 mm)) with MS deconvolution software (such as Waters MaxEnt1 algorithm).
3. Mobile phase: A: 0.1% aq. formic acid. B: 95% MeCN with 5% aq. formic acid (0.1%).

2.4 HIC Analysis

1. HPLC (such as Agilent 1260 Infinity system) with analytical HIC column (such as Tosoh TSKgel Butyl-NPR column (3.5 cm × 4.6 mm, 2.5 μm)).
2. Mobile phase: A: 1.5 M ammonium sulfate, 25 mM Na_3PO_4, pH 7. B: 25% (v/v) isopropyl alcohol in 25 mM Na_3PO_4, pH 7.

2.5 SEC Analysis

1. HPLC (such as Agilent 1260 Infinity system) with analytical SEC column (such as Tosoh TSKgel G3000SWXL column (30 cm × 7.8 mm, 5 μm)).
2. Mobile phase: 50 mM Na_3PO_4, 100 mM NaCl, 0.2% (w/v) sodium azide, pH 7.

3 Methods

All procedures are carried out at ambient temperature and in air unless otherwise stated.

3.1 Antibody Preparation and Conjugation

1. Make up an antibody solution in TBS (100 μL, c. 10 μM) (*see* **Note 6**). Record the UV-vis absorbance at 280 nm (using TBS for the background reading) and calculate the concentration using the extinction coefficient of the antibody (*see* **Note 7**) and the Beer–Lambert law. For trastuzumab $\varepsilon = 215{,}380$ (*see* **Note 8**).
2. Make up a 5 mM solution of TCEP in TBS (*see* **Note 9**).
3. Add 10 eq. of TCEP (5 mM) to the antibody solution (*see* **Note 10**) and incubate on a thermal shaker with agitation (1 h, 37°C, 400 rpm).
4. Add 10 eq. of DVP (5 mM) to the antibody solution (*see* **Note 10**) and incubate on a thermal shakerwith agitation (4 h, 37°C, 400 rpm).
5. Equilibrate a 0.5 mL 40k MWCO desalting column into PBS (*see* **Note 11**) according to the manufacturer's specification.
6. Pipette the antibody solution into the desalting column and centrifuge according to the manufacturer's specifications.
7. Transfer the filtrate into a 0.5 mL 10k Amicon® diafiltration column and dilute with 300 μL PBS (*see* **Note 12**). Spin-concentrate the sample for 2 min at 14,000 × *g* using a centrifuge. Add a further 300 μL PBS and spin-concentrate again. Repeat this process of dilution and concentration four times. Dilute a fifth time and centrifuge for 10 min to concentrate the sample to ~30 μL.
8. Recover the ADC by inverting the filter section of the Amicon® into a clean tube and centrifuging for 2 min at 4000 × *g*.
9. Record the final concentration via UV–vis (*see* **Note 13**).

3.2 SDS-PAGE: Assessing Extent of Conversion

1. Preheat a heating block to 90°C.
2. Make up solutions (10 μL, 1 μM. *see* **Note 14**) of the analyte and any controls (e.g., unfunctionalized antibody) in PBS buffer in 0.5 mL vials.
3. Add 10 μL sample stain (reducing or nonreducing) to all samples.
4. Aliquot a 5 μL sample of MW Ladder into a 0.5 mL vial.
5. Heat all samples to 90°C for 5 min to denature the protein.
6. Prepare the gel electrophoresis equipment in LRB according to the manufacturer's specifications, placing the gel in the holder, and wetting the components and gaskets to ensure a good seal is made.

7. Load the individual wells of the acrylamide gel with the MW ladder followed by the analyte and control samples (*see* **Note 15**).
8. Run the gel at 200 V for 65 min.
9. Remove the gel from between the glass plates, rinse with deionized water, cut off the staking gel, and submerge in CBB stain for at least 2 h (*see* **Note 16**).
10. Holding the gel gently down with a finger, carefully decant off the CBB stain and rinse the gel gently with water. To destain, submerge the gel in 10% aq. AcOH. Replace the aqueous acid periodically until the main face of the gel is colorless (*see* **Note 17**).
11. Remove the gel from the aq. AcOH and photograph it (*see* **Note 18**).

Running the SDS-PAGE gel with unfunctionalized antibody as a control ensures that the method has been run correctly and that comparisons can be made. An efficiently re-bridged antibody will appear as high molecular weight bands in reducing conditions as there are no remaining disulfide bonds to cleave (Fig. 3, ADC 1 and ADC 2). Reduced unmodified antibody will split into the

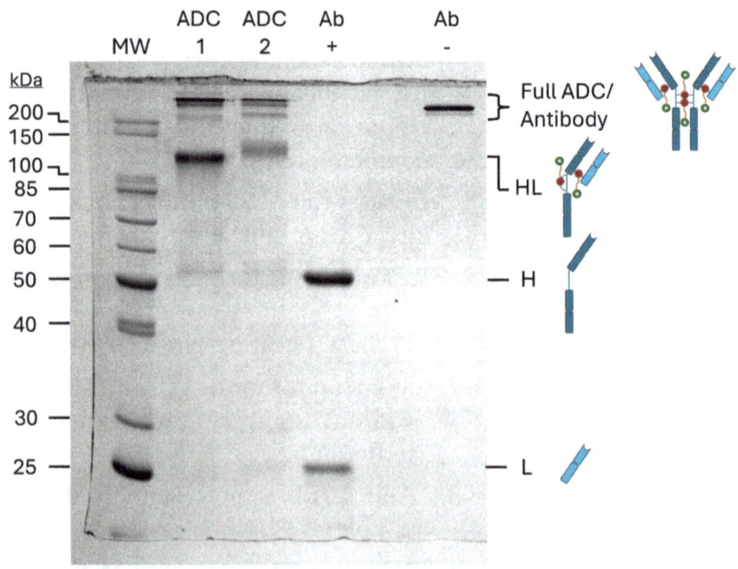

Fig. 3 SDS-PAGE gel showing the extent of conjugation for two ADCs. The ADC analytes were run in a reducing stain, therefore, bands of higher molecular weight show that the heavy and light chains have been chemically re-bridged. Ab +/− = Native antibody with/without reducing stain, MW = Molecular weight ladder, H = Heavy chain, L = Light chain

constituent heavy and light chains on the gel (Fig. 3, Ab +) whilst unreduced unmodified antibody will appear as a single high MW band on the gel (Fig. 3, Ab −). The presence of re-bridged heavy-light chain (Fig. 3, HL) is caused by intra-chain cysteine re-bridging competing with the inter-chain re-bridging. This is common for disulfide re-bridging systems and has no significant effect on ADC properties as the full ADC is held together via noncovalent interactions [8–12].

3.3 Mass Spectrometry: Assessing Most Prevalent ADC Species

1. Make up analytical samples (25 μL, 3–10 μM) in PBS (*see* **Note 19**).
2. Add 0.2 μL PNGase F and incubate the sample at 37°C for 2 h to remove sugars on the antibody surface that interfere with the MS.
3. Inject 3 μL into the MS. Run the analysis at a flow rate of 0.2 mL/min with gradient: 95% A for 0.93 min, then a gradient to 100% B over 4.28 min, then 100% B for 1.04 min, then a gradient to 95% A over 1.04 min. Capillary voltage of 2.0 kV and a cone voltage of 190 V.
4. Combine the ion count of the raw spectra and run the spectra deconvolution algorithm for m/z ions between 1400 and 3000, resolving for masses between 40,000 and 160,000 (Fig. 4). Due to the high mass values being extrapolated, inaccuracies of up to +/− 5 Da are considered reasonable.

3.4 HIC Analysis: Calculating Conversion and Average DAR

1. Make up analytical samples (30 μL, 3–10 μM) in PBS.
2. Inject 15 μL into HPLC system. Run with a linear gradient of 100% solvent A for 3 min, then 0%–100% solvent B over 17 min at a flow rate of 0.6 mL/min.
3. Monitor UV absorbance at 280 nm, with conversion calculated by peak area compared to unmodified antibody. Coupled with MS, the DAR can be calculated via comparison of the peak positions (*see* **Note 20**) and sizes (see Fig. 5a).
4. Once all analyses are complete, flush the HPLC system and column thoroughly with deionized water to prevent the high salt concentration buffers blocking the system.

3.5 SEC Analysis: Checking for Aggregation

1. Make up analytical samples (30 μL, 3–10 μM) in PBS.
2. Inject 6 μL into HPLC system. Run for 30 min at a flow rate of 0.5 mL/min.
3. Monitor UV absorbance at 280 nm, with aggregation calculated by peak area compared to the antibody. Aggregation appears as a peak of higher molecular weight (Fig. 5b) where a relative integration of <1.5% is considered ideal.
4. Once all analyses are complete, flush the HPLC system and column thoroughly with deionized water to prevent the high salt concentration buffers blocking the system.

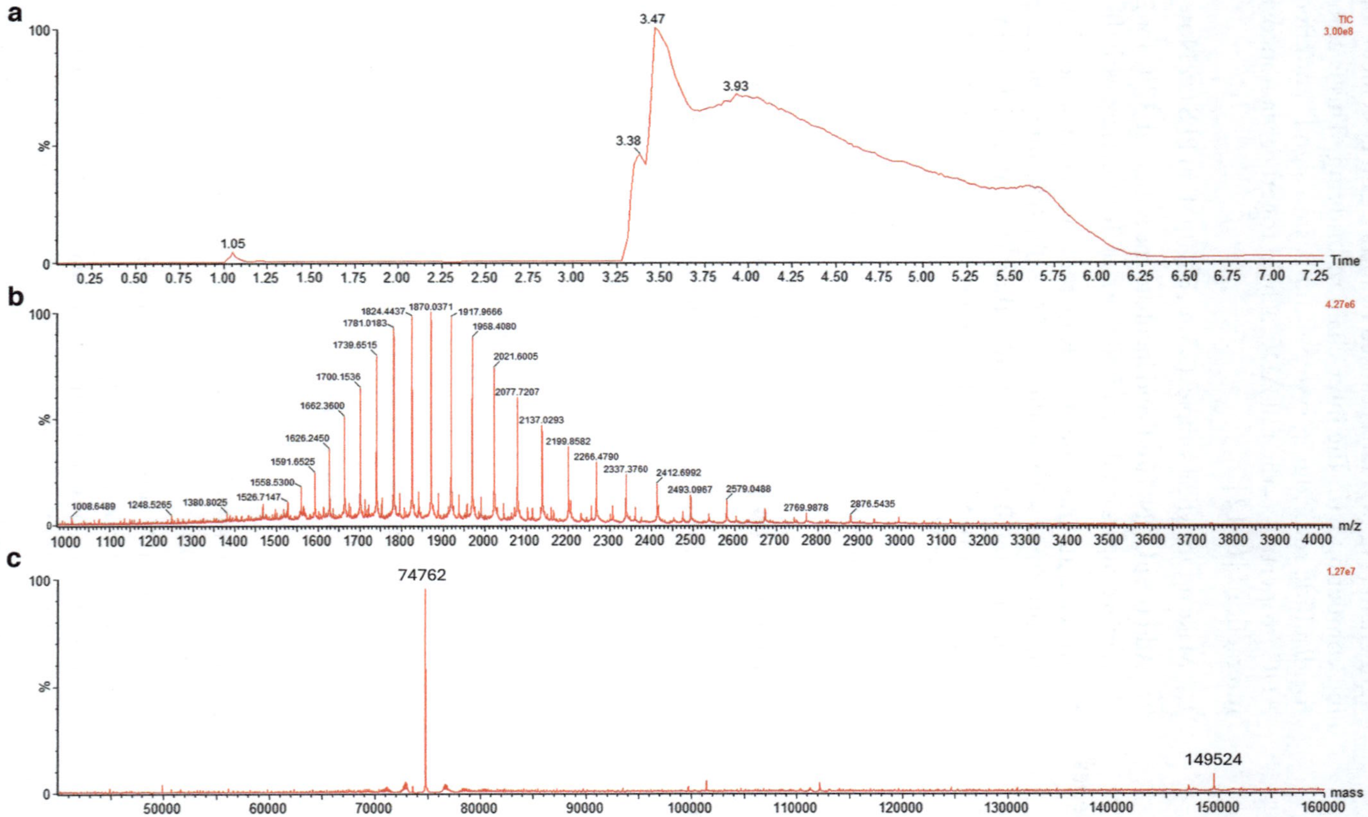

Fig. 4 (**a**) Raw total ion count, (**b**) raw *m*/*z* data from combining raw total ion count, (**c**) deconvoluted spectrum after running MaxEnt1 algorithm data between 1400 and 3000 *m*/*z* searching for masses between 40,000 and 160,000. Major peaks are the expected DAR 4 mass (exp. 149,522) and the half mass [ADC+2 H]$^{+2}$ (exp. 74,762)

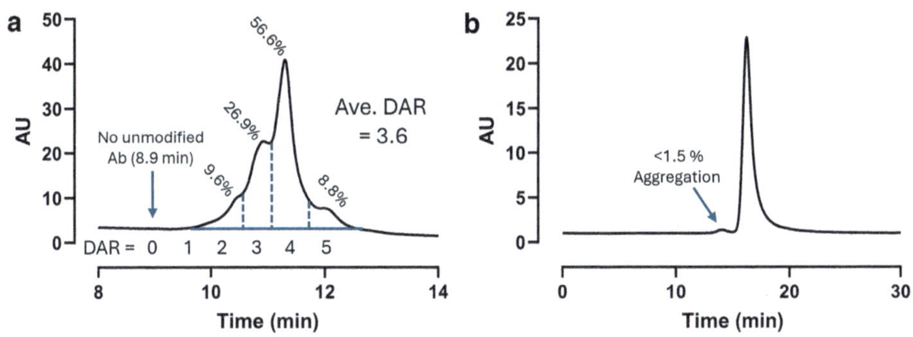

Fig. 5 (**a**) HIC spectrum showing how the peaks are assigned and integrated to calculate an average DAR. (**b**) SEC spectrum showing the presence of aggregation to the left of the antibody peak. Ab = Antibody, AU = Arbitrary units, Ave. = Average

4 Notes

1. Antibodies bought in solution can be used but require buffer exchange into TBS using a 10k MWCO Amicon® diafiltration column.
2. A 500 μL centrifuge tube is ideal for test reactions and reaction optimization. 2 mL Centrifuge tubes can be used for scale-up as long as the thermal shaker can accommodate their shape to provide even heating to the sample.
3. DVP solutions in DMSO should be stored at −4°C (or below) and can remain viable for up to 2 years.
4. PBS can be purchased as a concentrated solution or as dissolvable pellets for simple buffer preparation.
5. SDS-PAGE gels are commercially available but can be self-cast if relevant moulds are available; these should be used within 2 weeks of casting.
6. If using a lyophilized powder, about 1 mg is weighed out and diluted to 2.5 mg/mL with TBS. This gives a concentration of c. 10 μM. Antibodies acquired in solution must be buffer exchanged into TBS using an Amicon® diafiltration column via repeated dilution and concentration with TBS.
7. Concentrations between 5 μM and 20 μM are recommended. The exact concentration should be noted to allow accurate calculation of the TCEP and DVP reagents. Bioconjugation has been successfully carried out up to 2 mL in a single vial (20 μM). Due to size restraints on many thermal shakers, for larger scales multiple vials should be used.
8. If the amino acid sequence of the antibody is known then the extinction coefficient can be estimated using the Expasy ProtParam online tool [19]. If the amino acid sequence is

unknown, then use the extinction coefficient of a similar known antibody or trastuzumab (215,380). The differences in extinction coefficient for IgG1 antibodies tend to be less than 10% and so inaccuracies here are not critical and will be offset by the excess of TCEP and DVP used.

9. TCEP oxidizes in solution over time and so stocks must be made fresh directly before use to ensure complete disulfide reduction.
10. When adding TCEP or DVP to the antibody solution gently mix the solutions by repeatedly withdrawing and readmitting the solution with the pipette. This is especially key for the DVP solution to prevent the DMSO settling and creating a high local concentration, which can lead to antibody aggregation and precipitation.
11. Desalting columns help to remove excess salts (TCEP, Copper, etc.), preventing further reactivity. The need for desalting can be assessed on a case-by-case basis; whilst diafiltration should remove TCEP, experience has shown that copper salts (after CuAAC) are not removed and so desalting is required. The desalting column should be equilibrated into an appropriate buffer system for any post-purification analysis or further study. PBS is commonly used as it is compatible with copper-catalyzed "click" chemistry and biological studies.
12. Repeated PBS dilution and concentration through a 10k MWCO filter affords buffer exchange (reducing DMSO and TCEP concentration) whilst allowing excess DVP linker to be filtered off, purifying the sample. The cellulose filters are delicate so care must be taken not to pierce the membrane with the tip of a pipette. Once wetted, it is recommended not to let the membrane dry out for high sample recovery. Occasionally, agitate the solution in the filter with the tip of a pipette when adding further PBS to ensure the ADC is in solution and not stuck to the cellulose.
13. Concentrations far exceeding 100 μM are not recommended due to increased aggregation propensity. If above 100 μM, dilute with buffer.
14. The following equation can be used to easily generate 10 μL solutions at 1 μM:

$$\frac{10}{x} = y$$

where x = original sample concentration (in μM) and y = volume of sample (in μL) that needs to be diluted to 10 μL to afford a 1 μM solution (e.g., if your sample is 5 μM then 10/5 μM = 2 μL, diluted up to 10 μL to be a 1 μM solution).

15. Aiming to pipette 10% less volume than was originally in the vial (e.g., 18 µL of a 20 µL sample) will ensure good sample loading whilst preventing bubbles forming in the pipette tip; some of the solution will have evaporated during the denaturing step. It is recommended to leave an empty well between reducing and non-reducing stain samples as the mercaptoethanol can leach between wells causing partial reduction to other samples thus effecting band shape.
16. The stacking gel section can be removed at any stage before imaging but removing it early reduces the chance for parts of the stacking gel to fold over, effecting the efficiency of de/staining. Plastic pipette tip boxes are an ideal size for gel staining. A gel rocker (such as a Stuart SSM4 see-saw rocker) can be used to gently agitate the submerged gel and ensure that the gel is stained and destained evenly.
17. The gel is destained when only the protein bands are visibly blue. The protein will not be easily destained by the aqueous acid so it is fine to leave the gel in 10% aq. AcOH for several days.
18. Due to the staining, a standard camera can be used to record the bands on the gel. Gel imagers (such as Syngene ingenious 3 with GeneSys software) are useful if fluorescent components are bioconjugated to the antibody.
19. For ease, a 50 µL sample can be made up and used for HIC and SEC analyses. The remaining material can then by deglycosylated with PNGase F for MS analysis.
20. As increasing the number of payloads conjugated to an antibody usually increases the hydrophobicity of an ADC in an additive manner, different DAR species are identifiable by HIC retention time with fairly even spacing between species as the number of linkers increases. MS data is not quantitative but should show a DAR 4 species as the major product of a successful conjugation; therefore, the largest HIC peak is thus assigned "DAR 4" with the peaks around it assigned consecutively (e.g., 3, 2, 1).

References

1. Beck A, Goetsch L, Dumontet C, Corvaïa N (2017) Strategies and challenges for the next generation of antibody–drug conjugates. Nat Rev Drug Discov 16:315–337. https://doi.org/10.1038/nrd.2016.268
2. Riccardi F, Dal Bo M, Macor P, Toffoli G (2023) A comprehensive overview on antibody-drug conjugates: from the conceptualization to cancer therapy. Front Pharmacol 14:1274088. https://doi.org/10.3389/fphar.2023.1274088
3. Shen L, Sun X, Chen Z et al (2024) ADCdb: the database of antibody–drug conjugates. Nucleic Acids Res 52:D1097–D1109. https://doi.org/10.1093/nar/gkad831
4. Maecker H, Jonnalagadda V, Bhakta S et al (2023) Exploration of the antibody-drug conjugate clinical landscape. Mabs 15:2229101. https://doi.org/10.1080/19420862.2023.2229101
5. Dong W, Wang W, Cao C (2024) The evolution of antibody-drug conjugates: toward accurate DAR and multi-specificity. ChemMedChem 19(17):e202400109. https://doi.org/10.1002/cmdc.202400109

6. Hughes B (2010) Antibody–drug conjugates for cancer: poised to deliver? Nat Rev Drug Discov 9:665–667. https://doi.org/10.1038/nrd3270
7. Hamblett KJ, Senter PD, Chace DF et al (2004) Effects of drug loading on the antitumor activity of a monoclonal antibody drug conjugate. Clin Cancer Res 10:7063–7070. https://doi.org/10.1158/1078-0432.CCR-04-0789
8. Nunes JPM, Morais M, Vassileva V et al (2015) Functional native disulfide bridging enables delivery of a potent, stable and targeted antibody–drug conjugate (ADC). Chem Commun 51:10624–10627. https://doi.org/10.1039/C5CC03557K
9. Maruani A, Smith MEB, Miranda E et al (2015) A plug-and-play approach to antibody-based therapeutics via a chemoselective dual click strategy. Nat Commun 6:6645. https://doi.org/10.1038/ncomms7645
10. Kuan SL, Wang T, Weil T (2016) Site-selective disulfide modification of proteins: expanding diversity beyond the proteome. Chem Eur J 22:17112–17129
11. Walsh SJ, Omarjee S, Galloway WRJD et al (2019) A general approach for the site-selective modification of native proteins, enabling the generation of stable and functional antibody-drug conjugates. Chem Sci 10:694–700
12. Walsh SJ, Iegre J, Seki H et al (2020) General dual functionalisation of biomacromolecules via a cysteine bridging strategy. Org Biomol Chem 18:4224–4230. https://doi.org/10.1039/d0ob00907e
13. Walsh SJ, Omarjee S, Dannheim FM et al (2022) Divinylpyrimidine reagents generate antibody–drug conjugates with excellent in vivo efficacy and tolerability. Chem Commun 58:1962–1965. https://doi.org/10.1039/D1CC06766D
14. Krajcovicova S, Wharton T, Driscoll CL et al (2025) A platform for SpyCatcher conjugation to native antibodies. Chem Sci 16:10602–10609. https://doi.org/10.1039/D5SC02286J
15. Dannheim FM, Walsh SJ, Orozco CT et al (2022) All-in-one disulfide bridging enables the generation of antibody conjugates with modular cargo loading. Chem Sci 13:8781–8790. https://doi.org/10.1039/D2SC02198F
16. King TA, Walsh SJ, Kapun M et al (2023) Disulfide re-bridging reagents for single-payload antibody-drug conjugates. Chem Commun 59:9868–9871. https://doi.org/10.1039/d3cc02980h
17. Robertson NS, Walsh SJ, Fowler E et al (2019) Macrocyclisation and functionalisation of unprotected peptides: via divinyltriazine cysteine stapling. Chem Commun 55: 9499–9502
18. Iegre J, Krajcovicova S, Gunnarsson A et al (2023) A cell-active cyclic peptide targeting the Nrf2/Keap1 protein–protein interaction. Chem Sci 14:10800–10805. https://doi.org/10.1039/D3SC04083F
19. Expasy ProtParam. https://web.expasy.org/protparam/. Accessed 25 Jan 2025

Chapter 11

Sortase A Catalyzed Transpeptidation for Protein Bioengineering

Floris J. van Dalen and Hidde L. Ploegh

Abstract

Sortase A, originating from *Staphylococcus Aureus*, has become the gold standard for site-specific transpeptidation of protein substrates. Wild-type and engineered variants of sortase A catalyze the recognition and cleavage of a five-amino acid substrate motif (LPxTG), followed by transpeptidation with oligoglycine nucleophiles (peptides that contain one or more glycines at their N-terminus). This LPxTG sortagging motif can be introduced by genetic engineering of recombinantly expressed proteins of interest. Oligoglycine nucleophiles can be readily synthesized on solid-phase resin and allow the introduction of a wide range of functional groups. Here, we describe the transpeptidation of two nanobodies (VHHs, variable region of the heavy chain of heavy-chain only antibodies) with various GGG-nucleophiles.

Keywords Sortase A, Oligoglycine nucleophile, Protein bioengineering, Nanobody conjugates

1 Introduction

Sortase A, isolated from *Staphylococcus Aureus*, recognizes and cleaves a five-amino acid sequence (LPxTG) to yield an acyl-enzyme intermediate, followed by transpeptidation with polyglycine nucleophiles [1]. The wild-type version of sortase A has limited reactivity and promiscuity. Several mutants have been developed to improve the properties of wild-type sortase, for example to increase reactivity at room temperature and 4 °C, expand the range of accepted oligoglycine nucleophiles, and to remove the calcium(II)-dependence [2, 3]. Other sortase A variants have been developed to recognize complementary five-amino acid substrate motifs, such as sortase 2A9 which recognizes LAxTG or sortase 4S9 to catalyze transpeptidation of LPxSG [4]. These cross-compatible sortase A variants facilitate the site-specific modification of two sortase-motifs in-tandem, for instance, on both the heavy- and light-chain of Fab′ fragments [5]. The sortase reaction is fully (bio-)orthogonal to other protein ligation methods, such as click chemistry (both strain-promoted and

Alexander Deiters (ed.), *Bioconjugation Methods*, Methods in Molecular Biology, vol. 3009,
https://doi.org/10.1007/978-1-0716-5112-4_11,

copper(I)-catalyzed [6–8], biotin-(strept)avidin ligation [9], thiol-maleimide conjugation, and tyrosine-oxidation strain-promoted click [10]. The LPxTG sortase-recognition motif can be introduced by standard genetic engineering at the C-terminus to enable the site-specific modification of a particular protein substrate. Conversely, the N-terminus can be engineered with an oligoglycine acceptor to be modified with synthetic peptides that contain a sortase recognition motif. Then, both N- and C-terminus can be site-specifically modified through the combined use of complementary sortase motifs. The combination of the enzyme specificity of the sortase reaction with the chemical versatility of peptide synthesis enables a vast array of applications. Some examples include the development of nanobody-antigen conjugates for targeted vaccination [11, 12], nanobody-drug conjugates [13–15], imaging agents for positron emission tomography (PET) [16–18], fluorescent markers for immunoblot, immunofluorescence, or cytofluorometry, and live cell labeling to monitor immune cell interactions [19, 20]. As an illustration of this versatility, we conjugated two model VHHs, specific for mouse CD45 (VHH_{CD45}) or mouse class II MHC (VHH_{MHCII}) [12, 18, 21, 22], to three GGG-nucleophiles (GGGC-modified with dexamethasone, biotin or fluorophore AF647) with our optimized protocol. In small-scale reactions (30 μL), VHH_{CD45} and VHH_{MHCII} (100 μM) were conjugated with the GGG-nucleophiles (500 μM) using sortase 7M (10 μM) for 2 h at room temperature (Fig. 1). Sortase and any residual starting material were removed by depletion on Ni-NTA agarose beads exploiting the His-tag present on their C-terminus. The purified products were analyzed by gel electrophoresis, followed by in-gel fluorescence scanning and Western blot to confirm the presence of the fluorophore- and biotin-modified proteins. This afforded us

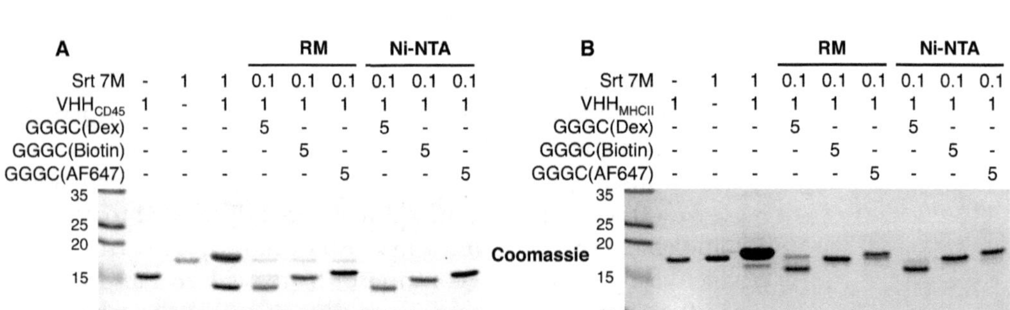

Fig. 1 VHH_{MHCII} and VHH_{CD45} conjugated with oligoglycine nucleophiles. (**A**) VHH_{CD45} and (**B**) VHH_{MHCII} were modified with GGG-nucleophiles (modified with dexamethasone (Dex), biotin or AF647) for two hours at room temperature. Sortase and residual starting material were removed by incubation on Ni-NTA beads. Proteins were separated by SDS-PAGE and visualized by Coomassie staining, in-gel fluorescence scanning (AF647), and western blotting (biotin)

six VHH-adducts with good conversion and high purity. To demonstrate the optimized preparatory workflow, we modified VHH_{MHCII} with dexamethasone at 1 mL /1.5 mg reaction scale under the same conditions (Fig. 2). After a 2-h reaction, sortase was removed by depletion on Ni-NTA beads. The resulting mixture was then purified by size-exclusion chromatography to remove the excess GGG-nucleophile. This yielded ~1.32 mg VHH_{MHCII}-Dex (88% isolated yield) in excellent purity (>95%). Together, these examples highlight the efficient workflow of the sortase reaction.

2 Materials

Prepare all solutions with ultrapure water (double deionized water with a resistance of 18.2 MΩ at 25°C). All reagents can be stored at room temperature unless specified otherwise.

2.1 Reagents and Buffers

1. Starting material: genetically engineered protein substrate that contains C-terminal LPETGGHHHHHH Sortag/ His-tag, purified by affinity chromatography over Ni-NTA agarose (>95% by SDS-PAGE analysis). Aliquot and store at −80 °C (*see* **Note 1**).
2. Sortase 7 M: expressed from pET30b in BL21 (DE3) bacteria in terrific broth at 30 °C for 16 h, purified by affinity chromatography over Ni-NTA agarose beads as described below (Fig. 3) (>95% by SDS-PAGE analysis) [23]. Plasmid is available from Addgene (51141). Aliquot and store at −80 °C.
3. GGG-nucleophile: 20–50 mM stock in DMSO or water for hydrophilic acceptors. These can be prepared by solid-phase peptide synthesis, or a range of GGG-peptides can be obtained commercially (*see* **Note 2**).
4. Sortase buffer (Tris-buffered saline (TBS)): 50 mM Tris-HCl (pH 7.4), 150 mM NaCl (*see* **Note 3**).

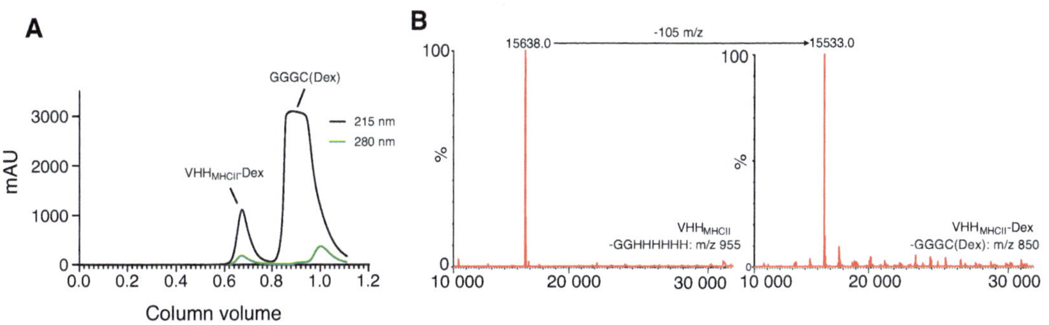

Fig. 2 Size-exclusion chromatography and liquid chromatography mass spectrometry (LCMS). (**A**) Size-exclusion chromatography affords complete removal of excess acceptor. (**B**) LCMS analysis of VHH_{MHCII}-Dex confirms the expected mass shift after sortase reaction

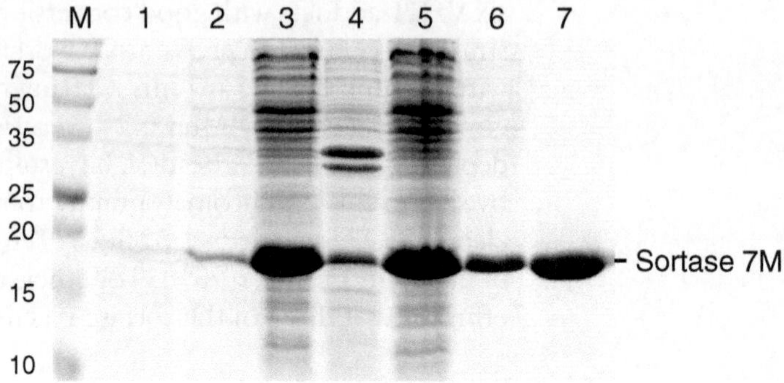

Fig. 3 Sortase 7 M expression and purification. Sortase 7 M was expressed in BL21 and purified with the described protocol yielding ~35 mg after purification from two liters TB culture. From left to right: (1) uninduced, (2) induced, (3) French press supernatant, (4) French press pellet, (5) column flow-through, (6) column wash, and (7) eluate

5. Lysis buffer: 50 mM Tris-HCl (pH 7.4), 150 mM NaCl, 0.1 mg/mL Roche EDTA-free Protease Inhibitor Cocktail (4693132001, Sigma-Aldrich), 0.1 mg/mL DNAse I (4536282001, Sigma-Aldrich).
6. Terrific Broth (TB): 47.6 g/L Terrific Broth powder, 0.4% (v/v) glycerol. Autoclave for 15 min at 121 °C. Store at 4 °C.
7. Luria-Bertani-Miller Broth (LB): 5 g/L yeast extract, 10 g/L tryptone powder, 10 g/L NaCl. Autoclave for 15 min at 121 °C. Store at 4 °C.
8. LB-agar kanamycin plates: 5 g/L yeast extract, 10 g/L tryptone powder, 10 g/L NaCl, 15 g/L agar. Autoclave for 15 min at 121 °C. Cool down to ~50 °C before adding kanamycin (50 μg/mL) and casting plates. Store plates at 4 °C.
9. LB-kanamycin: 50 μg/mL kanamycin in LB medium.
10. TB-kanamycin: 50 μg/mL kanamycin in TB medium.
11. Ni-NTA agarose beads (Qiagen, 30210).

2.2 SDS-PAGE Materials

1. Acrylamide/Bis solution (30% (w/v), 37.5:1).
2. 4× resolving buffer: 1.5 mM Tris-HCl (pH 8.8), 0.4% SDS.
3. 4× stacking buffer: 0.5 mM Tris-HCl (pH6.8), 0.4% SDS.
4. 10% (w/v) ammonium persulfate (APS) in water. Store at −20 °C.
5. N,N,N,N-tetramethyl-ethylenediamine (TMEDA or TEMED). Store at 4 °C (*see* **Note 4**).
6. 5× Sample buffer: 50% (v/v) glycerol, 250 mM Tris-HCl (pH 6.8), 10% (w/v) SDS, 10% (v/v) BME, 0.05% (w/v) bromophenol blue (*see* **Note 5**).

7. TGS running buffer: 25 mM Tris (3 g/L), 192 mM glycine (14.4 g/L), 0.1% SDS (1 g/L).
8. Gel stain: 1% (w/v) Coomassie® Brilliant Blue G-250, 10% AcOH (v/v), 40% MeOH (v/v) in water.
9. Gel de-stain: 10% (v/v) AcOH, 40% (v/v) MeOH in water.
10. Precision Plus Protein Standard Dual Color (Bio-Rad, 1610374).

2.3 Equipment

1. Orbital shaker incubator (Infors HT Multitron Triple or equivalent).
2. French press or Sonicator bacterial lysis machine.
3. Fritted column (7321011, Bio-Rad).
4. Size-exclusion FPLC (NGC chromatography system Bio-Rad or equivalent).
5. Gel imaging system (Chemidoc Bio-Rad or equivalent).
6. Amicon Ultra concentrators 10 kDa cut-off (Sigma-Aldrich, UFC801008). Cut-off range depends on the anticipated protein size.
7. SDS-PAGE equipment (Bio-Rad mini-PROTEAN).

3 Methods

All steps should be performed at room temperature unless specified otherwise.

3.1 Sortase Expression and Isolation

1. Isolate the pET30b Sortase 7M plasmid from DH5α *E. coli* and transform chemically competent BL21 (D3) *E. coli* by heat shock (45 s, 42 °C) and plate on LB-agar kanamycin plates.
2. Inoculate a starter culture with an individual colony of freshly transformed BL21 bacteria in 50 mL LB-kanamycin and incubate overnight (12–16 h) at 37 °C at 210 rpm to reach saturated growth (*see* **Note 6**).
3. Dilute 50 ml starter culture into 2 L TB-kanamycin and grow the culture at 37 °C at 210 rpm until the absorbance at 600 nm (OD600) reaches 0.4–0.6 (~3 h). Take an uninduced sample for SDS-PAGE analysis.
4. Induce expression by adding 1 mM (final concentration) isopropyl-β-D-thiogalactopyranoside (IPTG). Incubate overnight at 30°C at 210 rpm. Take an induced sample for SDS-PAGE analysis.
5. Spin down bacteria at 4000–6000 × *g* for 15 min at 4 °C (*see* **Note 7**).

6. Resuspend bacteria in ice-cold lysis buffer and perform French press lysis or Sonicator lysis (*see* **Note 8**).
7. Spin down lysate at 12,000 × *g* for 45 min at 4 °C (*see* **Note 9**). Sample supernatant and pellet for SDS-PAGE analysis.
8. Apply Ni-NTA agarose resin to a fritted column (1 mL resin per liter of bacterial culture) and wash with 20 column volumes (CV) of sortase buffer to remove storage solution.
9. Load the supernatant maintaining a flow rate of ~1 cm/min (*see* **Note 10**). Collect and save the flow-through for SDS-PAGE analysis.
10. Wash the column with 20 CV sortase buffer + 10 mM imidazole. Collect and save the washing step for SDS-PAGE analysis.
11. Elute Ni-NTA-bound protein with 2 CV sortase buffer + 250 mM imidazole (*see* **Note 11**).
12. Immediately follow up with PD10 gel filtration or size-exclusion chromatography (preferred) to remove the high concentration imidazole (*see* **Note 12**).
13. Wash and store Ni-NTA beads in 20% (v/v) EtOH in water for future use (*see* **Note 13**).
14. Analyze sortase purity and all processing steps by SDS-PAGE (Fig. 3).
15. Aliquot and store at sufficient concentration (>10 mg/mL) at −80 °C (*see* **Note 14**).

3.2 Sortase Reaction and Purification

1. Set up the following reaction mixture at the desired reaction scale (typically 1–2 mL): 100 μM protein substrate (1 eq.), 500 μM GGG-containing nucleophile (5 eq.), 10 μM sortase 7A (0.1 eq.) in sortase buffer (TBS pH 7.4) in a standard Eppendorf tube (*see* **Note 15**).
2. Gently agitate the reaction mixture for 2 h at room temperature, or until consumption of the starting material is complete, as observed per reaction optimization (*see* **Note 16**). Care should be taken not to excessively agitate the reaction mixture, as this could promote aggregation (*see* **Note 17**).
3. Prepare the following samples to serve as reference to identify reaction components on SDS-PAGE: (1) protein substrate starting material, (2) sortase alone, and (3) a hydrolysis sample (starting material + sortase incubated for 16 h at 37 °C).
4. Wash Ni-NTA agarose beads twice with TBS prior to use.
5. After 2 h, incubate the reaction mixture with Ni-NTA beads (0.5 mL slurry per 1 mL reaction mixture) for 10 min at room temperature, to capture the sortase and any residual starting material.

6. Remove the sortase-bound Ni-NTA agarose beads by gravity flow column chromatography or medium-speed centrifugation (5000× *g* for 1 min).
7. Wash the resin with an additional column bed volume to ensure all product is captured in the flow-through/supernatant (*see* **Note 18**).
8. Refine the product by size-exclusion chromatography (*see* **Note 19**). This affords the complete removal of the excess acceptor.
9. Pre-equilibrate a HiLoad Superdex S75 16/600 column with 1.1 column volume (CV) of TBS at a flow rate of 1 mL/min maintaining the column at 4 °C (*see* **Note 20**).
10. Inject the sortase-free product and elute with 1.1 CV maintaining the flow rate of 1 mL/min. Collect fractions of 1.3 mL.
11. Wash the column with an additional 1.1 CV buffer, 1.1 CV water and 1.1 CV 20% (v/v) EtOH in water (*see* **Note 21**).
12. On the basis of the UV trace collect the protein-containing fractions and analyze them by SDS-PAGE.
13. Combine the pure fractions and concentrate them to the desired concentration (e.g., 10 mg/mL) using an Amicon Ultra concentrator. Aliquot and store the product at −80 °C.
14. Analyze the purity of the final product by SDS-PAGE and LC-MS.

3.3 SDS-PAGE Analysis

1. Prepare the resolving gel by mixing 4.8 mL water, 5 mL resolving gel buffer, 10 mL acrylamide/bis solution, 200 μL APS and 20 μL TMEDA (*see* **Note 22**).
2. Cast the gel solution in a mini-PROTEAN cassette and cover the resolving gel with isopropanol. Allow for complete polymerization. Thoroughly remove all isopropanol.
3. Prepare the stacking gel by mixing 5.9 mL water, 2.5 mL stacking gel buffer, 1.5 mL acrylamide/bis solution, 100 μL APS, and 10 μL TMEDA. Add the stacking gel and insert a 10- or 15-slot comb.
4. Prepare samples with 5× sample buffer. Denature and reduce at 95 °C for 5 min. Spin down the sample to collect the condensation that might have formed in the lid.
5. Load gel with 2–5 μg protein per lane and run for 15 min at 80 V followed by 90 min at 100 V. Gel electrophoresis should be stopped when the bromophenol blue dye front reaches the bottom of the gel. This front will also contain the excess small molecule acceptor (especially evident for fluorescent GGG-nucleophiles) prior to removal by FPLC.

6. Gently open the cassette and place the gel in a container.
7. Perform in-gel fluorescence scanning to detect fluorophore-modified products if applicable, or directly continue with Coomassie staining.
8. Incubate the gel for 2–16 h with Coomassie stain (*see* **Note 23**), followed by destaining for 2–16 h until individual bands develop.
9. Run two identical gels to detect biotinylated proteins by Western blot.

4 Notes

1. If protein purity is insufficient after Ni-NTA purification, the substrate protein can be refined by size-exclusion or ion-exchange chromatography.
2. Sortase reactions typically tolerate up to 10% (v/v) DMSO, but lower percentages are preferred. The nucleophile should be free of any contaminants that might interfere with the sortase reaction, for instance polyglycine- or thiol-reactive contaminants (e.g., Michael acceptors). Preparatory HPLC should afford the peptide in sufficient purity (>95% by LCMS). If necessary, pH values of the assembled reaction mixtures should be estimated by spotting a small volume (2–4 μL) on pH paper.
3. The pH of Tris buffers is temperature-dependent and decreases ~0.03 pH for each 1 °C increase. Similarly, diluting 10× TBS (or PBS) to the final 1× concentration changes the pH considerably. We find that setting the pH of Tris buffers at the final usage conditions (1× TBS at correct temperature) offers the easiest solution.
4. TMEDA should be stored under protective atmosphere to prevent oxidation (e.g., argon gas). Storing at 4 °C also slows down oxidation. TMEDA will show yellow coloration upon oxidation, strongly colored TMEDA should not be used.
5. Sample buffer should be made without BME, aliquoted to 900 μL and stored at −20 °C. Thawed sample buffer might contain precipitated SDS, which is dissolved by warming the sample and vortexing. For non-reducing sample buffer, add 100 μL water, and store at room temperature. For reducing sample buffer, add fresh 100 μL β-mercaptoethanol (BME), and store at room temperature. BME should be stored under protective atmosphere to prevent oxidation (e.g., argon gas).

6. Freshly transformed bacteria typically express higher copy numbers of the plasmid.
7. Bacterial pellet can be stored at −80 °C almost indefinitely.
8. Lysis step should be strictly maintained at ice-cold (~4 °C) temperatures to preserve protein quality and enzymatic activity.
9. The bacterial lysate should be crystal-clear after high-speed centrifugation to facilitate efficient Ni-NTA affinity chromatography. If significant contamination is observed, subject the supernatant to a second round of ultracentrifugation.
10. To improve recovery, load supernatant twice. Alternatively, incubate Ni-NTA resin with the lysate for 2 h at 4 °C on an end-over-end rotator, or perform the separation on an NGC chromatography system equipped with a Ni-NTA affinity column.
11. Recalibrate the buffer pH after adding imidazole to TBS.
12. High imidazole concentration can lead to aggregation.
13. Ni-NTA agarose beads can be stripped, cleaned, and recharged with nickel sulfate to be reused up to ten times.
14. Concentration of the protein sample by membrane filtration (e.g., Amicon Ultra concentrators) affords good protein recovery.
15. Calcium(II)-dependent sortase variants (Srt A, Srt 2 M, Srt 3 M, and Srt 5 M) mandate the use of phosphate-free buffers to prevent the aggregation of calcium phosphate in the reaction mixture. Since sortase 7 M no longer requires calcium, PBS may be used instead of TBS.
16. Extended reaction times can lead to increased hydrolysis of the thioester intermediate, as the product can still partake in the sortase reaction (it contains an LPETGGGx motif at its C-terminus). This typically leads to lower product purity since the hydrolyzed sample can no longer be removed by affinity purification on Ni-NTA beads and is difficult to remove by size-exclusion or ion exchange chromatography. Reaction optimization of 20 μL reaction mixtures may be required to define the optimal conditions (e.g., time, concentration, GGG-nucleophile equivalents) for the particular substrate and oligoglycine nucleophile. We find that five equivalents of acceptor and 0.1 equivalent of sortase in a 12- to 16-h overnight reaction at 4 °C or 1- to 2-h incubation at room temperature offers the best results for most proteins.
17. Side-wise rolling prevents most aggregation observed for end-over-end rolling. Sortase 7 M offers the possibility to carry out reactions at 4 °C with limited loss of enzymatic

activity. However, lower temperatures sometimes promote aggregation of the reaction mixture.

18. The sortase-free product could be stored indefinitely at − 80 °C until further purification. However, we suggest storing the purified product.
19. Whilst other purification methods (e.g., PD10 gel filtration or size-exclusion membrane filtration) might offer time advantages, size-exclusion chromatography affords the best removal of small molecule contaminants (Fig. 2).
20. Size-exclusion chromatography also offers an excellent opportunity to perform buffer exchange, by equilibrating and eluting in the buffer of choice (e.g., PBS).
21. The flow rate should be reduced by half for 20% (v/v) ethanol/water mixture due to its high viscosity at 4 °C.
22. This makes four 0.75 mm gels or two 1.5 mm gels. 15% SDS-PAGE gels produce good separation in the 10–30 kD range.
23. Coomassie brilliant blue 250 is very economical. Colloidal Coomassie significantly speeds up Coomassie staining and de-staining, and removes the necessity of hazardous chemicals (MeOH and AcOH). SYPRO staining offers advantages in sensitivity but is not compatible with red/near-infrared fluorophores (e.g., AF647 or AF700).

References

1. Popp MW, Antos JM, Grotenbreg GM, Spooner E, Ploegh HL (2007) Sortagging: a versatile method for protein labeling. Nat Chem Biol 3:707–708. https://doi.org/10.1038/nchembio.2007.31
2. Hirakawa H, Ishikawa S, Nagamune T (2012) Design of Ca2+-independent *Staphylococcus aureus* sortase A mutants. Biotechnol Bioeng 109:2955–2961. https://doi.org/10.1002/bit.24585
3. Wu Q, Ploegh HL, Truttmann MC (2017) Hepta-mutant *Staphylococcus aureus* sortase A (SrtA7m) as a tool for in vivo protein labeling in *Caenorhabditis elegans*. ACS Chem Biol 12:664. https://doi.org/10.1021/acschembio.6b00998
4. Dorr BM, Ham HO, An C, Chaikof EL, Liu DR (2014) Reprogramming the specificity of sortase enzymes. Proc Natl Acad Sci U S A 111:13343–13348. https://doi.org/10.1073/pnas.1411179111
5. Le Gall CM, van der Schoot JMS, Ramos-Tomillero I, Khalily MP, van Dalen FJ, Wijfjes Z, Smeding L, van Dalen D, Cammarata A, Bonger KM, Figdor CG, Scheeren FA, Verdoes M (2021) Dual site-specific chemoenzymatic antibody fragment conjugation using CRISPR-based hybridoma engineering. Bioconjug Chem 32:301–310. https://doi.org/10.1021/acs.bioconjchem.0c00673
6. Huisgen R, Szeimies G, Möbius L (1967) 1.3-Dipolare Cycloadditionen, XXXII. Kinetik der Additionen organischer Azide an CC-Mehrfachbindungen. Chem Ber 100:2494–2507. https://doi.org/10.1002/cber.19671000806
7. Kolb HC, Finn MG, Sharpless KB (2001) Click chemistry: diverse chemical function from a few good reactions. Angew Chem Int Ed 40:2004–2021. https://doi.org/10.1002/1521-3773(20010601)40:11<2004::AID-ANIE2004>3.0.CO;2-5
8. Agard NJ, Prescher JA, Bertozzi CR (2005) A strain-promoted [3 + 2] azide–alkyne cycloaddition for covalent modification of biomolecules in living systems [J. Am. Chem. Soc. 2004, 126, 15046–15047]. J Am Chem Soc 127:11196–11196. https://doi.org/10.1021/ja059912x
9. Green NM (1975) Avidin. In: Anfinsen CB, Edsall JT, Richards FM (Eds.)Eds. Advances in protein chemistry. Academic Press, 85–133
10. Bruins JJ, Westphal AH, Albada B, Wagner K, Bartels L, Spits H, van Berkel WJH, van

Delft FL (2017) Inducible, site-specific protein labeling by tyrosine oxidation–strain-promoted (4 + 2) cycloaddition. Bioconjug Chem 28:1189–1193. https://doi.org/10.1021/acs.bioconjchem.7b00046

11. Pishesha N, Harmand TJ, Rothlauf PW, Praest P, Alexander RK, van den Doel R, Liebeskind MJ, Vakaki MA, McCaul N, Wijne C, Verhaar E, Pinney W, Heston H, Bloyet L-M, Pontelli MC, Ilagan MaXG, Jan Lebbink R, Buchser WJ, Wiertz EJHJ, Whelan SPJ, Ploegh HL (2021) A class II MHC-targeted vaccine elicits immunity against SARS-CoV-2 and its variants. Proc Natl Acad Sci 118:e2116147118. https://doi.org/10.1073/pnas.2116147118
12. Pishesha N, Harmand T, Smeding LY, Ma W, Ludwig LS, Janssen R, Islam A, Xie YJ, Fang T, McCaul N, Pinney W, Sugito HR, Rossotti MA, Gonzalez-Sapienza G, Ploegh HL (2021) Induction of antigen-specific tolerance by nanobody–antigen adducts that target class-II major histocompatibility complexes. Nat Biomed Eng 5:1389–1401. https://doi.org/10.1038/s41551-021-00738-5
13. Fang T, Duarte JN, Ling J, Li Z, Guzman JS, Ploegh HL (2016) Structurally defined α MHC-II nanobody–drug conjugates: a therapeutic and imaging system for B-cell lymphoma. Angew Chem Int Ed 55:2416–2420. https://doi.org/10.1002/anie.201509432
14. Pishesha N, Harmand T, Carpenet C, Liu X, Bhan A, Islam A, van den Doel R, Pinney W, Ploegh HL (2022) Targeted delivery of an anti-inflammatory corticosteroid to Ly6C/G-positive cells abates severity of influenza A symptoms. Proc Natl Acad Sci 119:e2211065119. https://doi.org/10.1073/pnas.2211065119
15. Liu X, Balligand T, Carpenet C, Ploegh HL (2023) An armed anti-immunoglobulin light chain nanobody protects mice against influenza A and B infections. Sci Immunol 8:eadg9459. https://doi.org/10.1126/sciimmunol.adg9459
16. Ingram JR, Dougan M, Rashidian M, Knoll M, Keliher EJ, Garrett S, Garforth S, Blomberg OS, Espinosa C, Bhan A, Almo SC, Weissleder R, Lodish H, Dougan SK, Ploegh HL (2017) PD-L1 is an activation-independent marker of brown adipocytes. Nat Commun 8: 647. https://doi.org/10.1038/s41467-017-00799-8
17. Rashidian M, LaFleur MW, Verschoor VL, Dongre A, Zhang Y, Nguyen TH, Kolifrath S, Aref AR, Lau CJ, Paweletz CP, Bu X, Freeman GJ, Barrasa MI, Weinberg RA, Sharpe AH, Ploegh HL (2019) Immuno-PET identifies the myeloid compartment as a key contributor to the outcome of the antitumor response under PD-1 blockade. Proc Natl Acad Sci U S A 116:16971–16980. https://doi.org/10.1073/pnas.1905005116
18. Salehi Farid A, Rowley JE, Allen HH, Kruger IG, Tavakolpour S, Neeley K, Cong M, Shahbazian H, Dorafshani N, Berrada A, MacDonagh AC, Padera RF, Brugarolas P, Packard AB, Rosenbaum MW, Divakaran S, Di Carli MF, Rashidian M (2025) CD45-PET is a robust, non-invasive tool for imaging inflammation. Nature 639(8053):214–224. https://doi.org/10.1038/s41586-024-08441-6
19. Pasqual G, Chudnovskiy A, Tas JMJ, Agudelo M, Schweitzer LD, Cui A, Hacohen N, Victora GD (2018) Monitoring T cell–dendritic cell interactions in vivo by intercellular enzymatic labelling. Nature 553:496–500. https://doi.org/10.1038/nature25442
20. Nakandakari-Higa S, Walker S, Canesso MCC, van der Heide V, Chudnovskiy A, Kim D-Y, Jacobsen JT, Parsa R, Bilanovic J, Parigi SM, Fiedorczuk K, Fuchs E, Bilate AM, Pasqual G, Mucida D, Kamphorst AO, Pritykin Y, Victora GD (2024) Universal recording of immune cell interactions in vivo. Nature 627:399–406. https://doi.org/10.1038/s41586-024-07134-4
21. Rossotti M, Tabares S, Alfaya L, Leizagoyen C, Moron G, González-Sapienza G (2015) Streamlined method for parallel identification of single domain antibodies to membrane receptors on whole cells. Biochim Biophys Acta 1850:1397–1404. https://doi.org/10.1016/j.bbagen.2015.03.009
22. Duarte JN, Cragnolini JJ, Swee LK, Bilate AM, Bader J, Ingram JR, Rashidfarrokhi A, Fang T, Schiepers A, Hanke L, Ploegh HL (2016) Generation of immunity against pathogens via single-domain antibody–antigen constructs. J Immunol 197:4838–4847. https://doi.org/10.4049/jimmunol.1600692
23. Guimaraes CP, Witte MD, Theile CS, Bozkurt G, Kundrat L, Blom AEM, Ploegh HL (2013) Site-specific C-terminal and internal loop labeling of proteins using sortase-mediated reactions. Nat Protoc 8:1787–1799. https://doi.org/10.1038/nprot.2013.101

Chapter 12

Attachment of 2-Pyridyltetrazine Affinity Bioorthogonal Chemistry (ABC) Tags to Proteins for Site-Selective Bioorthogonal Chemistry

Paramesh K. Ramaraj, Samuel L. Scinto and Joseph M. Fox

Abstract

Site-selective labeling of proteins with small molecules can be performed by using different chemical and enzymatic methods. However, many of these methods lead to incomplete conversion or side products resulting in a mixture where the desired product may be inseparable. Herein, we present a protocol for the site selective conjugation of 3-methyl-6-(2-pyridyl)tetrazine to proteins. 3-methyl-6-(2-pyridyl) tetrazine, in addition to serving a bioorthogonal reagent capable of reacting to strained alkenes such as *trans*-cyclooctene, also functions as an affinity tag due to its ability to bind to nickel immobilized on resin. This Affinity Bioorthogonal Chemistry (ABC) tag enables the separation of the conjugated product from the unreacted starting material and any side products that may form during the conjugation reaction. This protocol describes the synthesis of the required 3-methyl-6-(2-pyridyl)tetrazines for conjugation to proteins, and the subsequent conjugation and purification steps required for obtaining ABC-tag functionalized protein.

Keywords Tetrazine, Protein labeling, Protein purification, Inverse electron demand Diels-Alder, Bioorthogonal chemistry

1 Introduction

Site-selective modification of proteins is broadly used in many areas such as in vivo imaging [1], drug-release [2], proteomics [3] and material science [4]. Incorporation of bioorthogonal tags into proteins greatly increases the types of molecules that can be site-specifically conjugated. Various chemical and enzymatic methods have been developed for the site-specific modification of protein at the N-terminal [5–8], C-terminal [9–11], and at internal positions [12–14]. However, many of these methods result in low yields resulting in a mixture of reacted and unreacted protein products that are in most cases inseparable, especially when the protein modifications are small. While the bioorthogonal handle on the protein

Alexander Deiters (ed.), *Bioconjugation Methods*, Methods in Molecular Biology, vol. 3009,
https://doi.org/10.1007/978-1-0716-5112-4_12,

can be used to attach an affinity handle that would enable the separation of the conjugated from the unconjugated product, this would use up the bioorthogonal tag, disabling any further modification [15, 16].

Tetrazines are a very useful bioorthogonal handle, since they undergo inverse electron demand Diels-Alder (IEDDA) reactions with dienophiles such as norbornene [17], bicyclononye [18, 19] and *trans*-cyclooctene (TCO) [20–22]. In particular, reaction of tetrazines with TCOs proceeds with reaction rates as fast as k_2 10^4–10^6 $M^{-1}s^{-1}$ [23], which is the fastest bioorthogonal reaction known. Given their rapid kinetics, tetrazine ligation has found use in a variety of applications such as in vivo imaging [24, 25], cellular labeling [26–28], drug delivery [29, 30] and proteomics [31, 32]. In addition to their use as bioorthogonal tags, tetrazines also serve as ligands for coordination chemistry. Symmetric bidendate 3,6-di-(2-pyridyl)- tetrazine ligands can bind to metal centers such as Ni^{2+}, Cu^{2+}, Fe2+, and Ag^{+} in organic solvents [33, 34]. The coordination of 6-methyl-3-(2-pyridyl)tetrazine to iridium(III) center was used to develop luminogenic probes for live cell labeling [35]. This property of 2-pyridyltetrazines (PyTz) to coordinate metal centers was leveraged to develop Affinity Bioorthogonal Chemistry tags (ABC-tags) that can be conjugated to proteins and subsequently purified by immobilized metal-ion affinity chromatography (IMAC)-resins commonly used for protein purification (Fig. 1A) [36]. The labeled protein that is selectively purified can be subsequently used for tetrazine ligation to conjugate molecules of interest to the protein (Fig. 1B). These dual-purpose tags have been shown to work in conjunction with protein conjugation

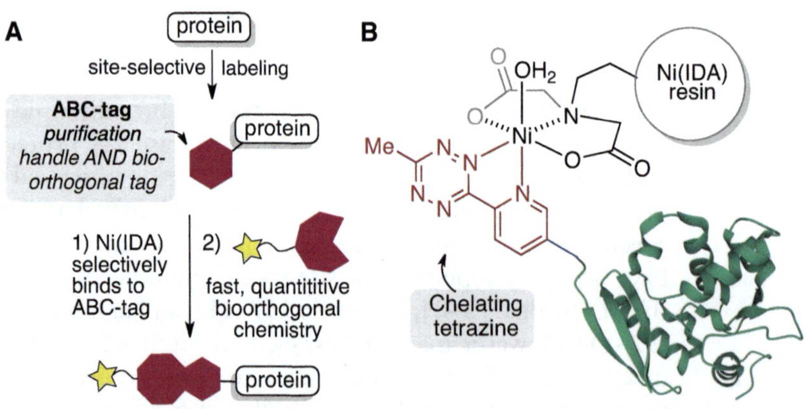

Fig. 1 (**A**) ABC-tags for Ni(IDA) purification of tagged proteins and subsequent bioorthogonal reaction with dienophile. (**B**) Pyridyl-tetrazine tagged proteins chelate to Ni(IDA) resin facilitating selective purification of tagged protein (PDB_ID 3KZY). Reproduced from Scinto S. L. et al. 2022, with permission from Wiley-VCH. © 2022 Wiley-VCH GmbH

methods at the N-terminus, C-terminus, and at internal positions. ABC-tags have been used to selectively purify various proteins differing in size and functionality such as sfGFP, SnapTag, 5F7 nanobody, Nanoluc luciferase, HaloTag, fibronectin type III domain 9-10, and Her2 affibody using C-terminal sortase ligation, N-terminal 2-cyanobenzothiazole ligation, and cysteine ligation. ABC-tags can also be used to perform on-resin coupling of proteins to synthesize and selectively purify homodimers and heterodimers. Proteins functionalized and purified using ABC-tags have been used to functionalize the surface of microfibers with active proteins [37].

Herein, we describe a protocol for the synthesis of two different 3-methyl-6-(2-pyridyl)tetrazines that can be used for cysteine ligation, and sortase enzyme-mediated ligation, respectively, subsequent method for the conjugation of these tags to proteins of interest and the IMAC protocol for the selective purification of the conjugated protein.

2 Materials

2.1 Synthesis of Pyridyl Tetrazine Compounds

1. 5-Amino-2-pyridinecarbonitrile.
2. Nickel(II) trifluoromethanesulfonate.
3. Anhydrous acetonitrile.
4. Anhydrous hydrazine.
5. Light mineral oil.
6. Ethyl acetate.
7. Sodium sulfate.
8. Dichloromethane.
9. Dess-Martin periodinane.
10. Silica gel.
11. Flash Chromatography column or chromatography system.
12. N,N-Diisopropylethylamine.
13. Bromoacteyl bromide.
14. 1 N hydrochloric acid.
15. Saturated sodium chloride (brine).
16. *N*-(*tert*-Butoxycarbonyl)glycine.
17. Isobutyl-chloroformate.
18. Saturated sodium bicarbonate.
19. Trifluoroacetic acid.
20. Inert gas (nitrogen used here).

2.2 Protein Expression and Purification

1. *E. coli* strain BL21(DE3).
2. LB media: Dissolve 6.35 g of LB media powder in 250 mL of deionized water and autoclave.
3. 50 mg/mL kanamycin: Dissolve 150 mg of kanamycin in 3 mL of deionized water. Store in aliquots at -20°C until use and avoid freeze thaw cycles.
4. Terrific Broth (TB): To prepare 1.8 L of TB media, dissolve 24 g of tryptone, 48 g of yeast extract and 8 mL of glycerol in deionized water and autoclave. Separately prepare 10× TB salts by dissolving 11.55 g of potassium phosphate monobasic (KH_2PO_4) and 62.76 g of potassium phosphate dibasic (K_2HPO4) in 500 mL of deionized water and autoclave.
5. 1 M 1-thio-β-D-galactopyranoside (IPTG) stock: Dissolve 596 mg of IPTG in 2.5 mL of deionized water. Store at –20°C until use and avoid freeze thaw cycles.
6. Phosphate buffer: For 1 L of 50 mM phosphate buffer, dissolve 10.1 g of sodium phosphate dibasic heptahydrate and 1.7 g of sodium phosphate monobasic monohydrate in 800 mL of deionized water. For 1 L of 25 mM phosphate buffer, dissolve 5.05 g of sodium phosphate dibasic heptahydrate and 0.85 g of sodium phosphate monobasic monohydrate in 800 mL of deionized water. For phosphate buffer containing 300 mM NaCl and 30 mM imidazole, add 17.5 g of NaCl and 2.04 g of imidazole. For phosphate buffer containing 300 mM NaCl and 400 mM imidazole, add 17.5 g of NaCl and 27.2 g of imidazole. For buffer containing 100 mM NaCl and 15 mM imidazole, add 5.84 g of NaCl and 1.02 g of imidazole. For phosphate buffer containing 100 mM NaCl, add 5.84 g of NaCl. Adjust to pH of 7.6 using concentrated hydrochloric acid or 10 M sodium hydroxide. Adjust volume to 1 L with deionized water.
7. Nickel chelated to iminodiacetic acid functionalized agarose resin (Ni-IDA resin).
8. 1,4-dithiothreitol (DTT).
9. Tobacco Etch Virus (TEV) protease: Can be purchased commercially or expressed and purified. Expression and purification of TEV protease are not described in this method.

2.3 Conjugation of ABC-Tags to Protein and Purification

1. Phosphate buffered saline (PBS).
2. Dimethyl sulfoxide (DMSO).
3. Sodium iodide.
4. 2-(*N*-morpholino)ethanesulfonic acid (MES) buffer: For 1 L of 25 mM MES buffer with 100 mM NaCl, dissolve 4.88 g of MES, and 5.84 g of NaCl in 800 mL of deionized water. For MES buffer containing 10 mM imidazole, add 0.68 g of imidazole.

Adjust to pH of 6.0 using concentrated hydrochloric acid or 10 M sodium hydroxide. Adjust volume to 1 L with deionized water.

5. Nap5 column (Cytiva).
6. 5 mL HiTrap Desalting column (GE Healthcare).

3 Methods

3.1 Synthesis of 6-(6-Methyl-1,2,4,5-Tetrazin-3-yl)Pyridin-3-Amine (Fig. 2)

Perform all the synthesis reactions in a properly ventilated fume hood.

1. Oven dry a 25 mL round bottom flask with an appropriately sized magnetic stir bar (*see* **Note 1**).
2. Cool the flask down to room temperature. To the flask add 456 mg of 5-amino-2-pyridinecarbonitrile (3.84 mmol) and 350 mg of nickel(II) trifluoromethanesulfonate (0.96 mmol).
3. Seal flask with a rubber septum. Secure septum by tying it with copper wire (*see* **Note 2**).
4. Pierce the rubber septum with a small needle and let it remain for venting.
5. Add 2 mL of anhydrous acetonitrile (39 mmol) and 3 mL of anhydrous hydrazine (95 mmol) to the flask via a needle.
6. Remove the venting needle.
7. Place a blast shield in front of the flask.
8. Heat the flask to 60°C using an oil bath and allow the reaction to proceed overnight.
9. The next day, cool reaction to room temperature and carefully vent the flask by piercing the septum with a needle (*see* **Note 3**).
10. Add 40 mL of water to the reaction mixture and transfer the contents to a separatory funnel. Add 20 mL of ethyl acetate to the separatory funnel. Cap the separatory funnel and shake vigorously by inverting the separatory funnel three to four times (*see* **Note 4**). Allow the organic and aqueous layer to settle. Collect the aqueous layer, which will be the bottom layer, in a 125 mL Erlenmeyer flask. Collect the organic layer in a separate 250 mL Erlenmeyer flask.
11. Add the aqueous layer back into the separatory funnel. Add 20 mL of ethyl acetate and repeat the extraction. Perform this step for a total of five times.
12. Dry the organic layer with sodium sulfate and concentrate down by rotary evaporation (*see* **Note 5**).

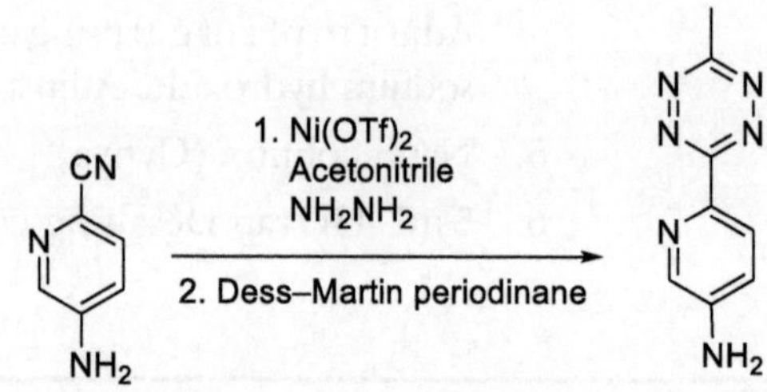

Fig. 2 Synthesis of 6-(6-Methyl-1,2,4,5-tetrazin-3-yl)pyridin-3-amine

13. Redissolve the mixture with 15 mL of dichloromethane and transfer to a 125 mL Erlenmeyer flask.
14. Oxidize the dihydrotetrazine that is formed using Dess-Martin periodinane (DMP). Add a little scoop of DMP to the solution and monitor the oxidation by thin-layer chromatography. Continue adding DMP till the dihydrotetrazine has been fully oxidized (*see* **Note 6**).
15. Dilute the solution with 30 mL of dichloromethane and extract three times with 10 mL brine.
16. Dry the organic layer with sodium sulfate. Filter by using a filter paper and concentrate by rotary evaporation.
17. Purify the compound by silica gel chromatography by using a gradient of 0%–100% ethyl acetate in hexanes.
18. The pure compound is obtained as a dark red solid with a yield of 14%. Successful isolation of product can be confirmed by NMR spectroscopy.

3.2 Synthesis of 2-Bromo-N-(6-(6-Methyl-1,2,4,5-Tetrazin-3-yl)Pyridin-3-yl)Acetamide (Fig. 3)

1. Oven dry a 5 mL round bottom flask with an appropriately sized stir bar.
2. Cool the flask down to room temperature. Add 20 mg of 6-(6-methyl-1,2,4,5-tetrazin-3-yl)pyridin-3-amine (0.106 mmol) to flask. Attach a nitrogen inlet adaptor to flask and fill flask with nitrogen.
3. Under nitrogen, add 1 mL of dichloromethane, followed by 37 μL of N,N-diisopropylethylamine.
4. Cool the reaction flask using an ice-bath.
5. Prepare a 1.15 M solution of bromoacteyl bromide in dichloromethane (*see* **Note 7**).
6. With the flask kept in the ice bath, add 138 μL of the bromoacetyl bromide solution to the reaction dropwise.
7. Allow the reaction to warm up to room temperature and stir at room temperature for 30 min.
8. Dilute the reaction with 8–10 mL dichloromethane and transfer to a 25 mL separatory funnel.

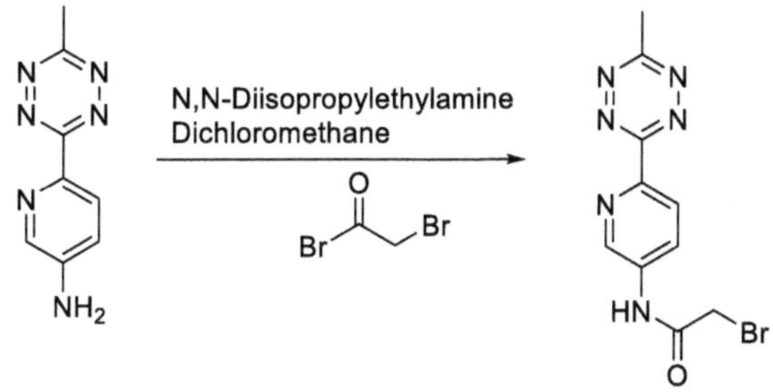

Fig. 3 Synthesis of 2-Bromo-N-(6-(6-methyl-1,2,4,5-tetrazin-3-yl)pyridin-3-yl) acetamide

9. Add 5 mL of 1 N HCl to the separatory funnel. Cap the separatory funnel and shake vigorously by inverting the flask, three to four times. Allow the organic and aqueous layer to settle. Collect the organic layer, which will be the bottom layer, in a 25 mL Erlenmeyer flask. Collect the aqueous layer in a separate 25 mL Erlenmeyer flask.
10. Repeat Step 9 one more time. Collect the aqueous and organic layers in their respective Erlenmeyer flasks.
11. Transfer the combined organic layer to the separatory funnel (*see* **Note 8**). Add 5 mL of brine solution to the flask. Cap the separatory funnel and shake vigorously by inverting the flask three to four times. Allow the organic and aqueous layer to settle. Collect the organic layer, which will be the bottom layer, in a 25 mL Erlenmeyer flask.
12. Dry the organic layer with sodium sulfate. Filter the solution using a filter paper and funnel into a 50 mL round bottom flask. Remove solvent by rotary evaporation.
13. Carry out flash chromatography of the crude using silica gel with a gradient of 0%–50% ethyl acetate in hexanes.
14. The product is obtained as a pink solid with a yield of 84%. Successful isolation of product can be confirmed by NMR spectroscopy.

3.3 Synthesis of Tert-Butyl (2-((6-(6-Methyl-1,2,4,5-Tetrazin-3-yl) Pyridin-3-yl)Amino)-2-Oxoethyl)Carbamate (Fig. 4)

1. Oven dry a 10 mL round bottom flask with an appropriately sized magnetic stir bar.
2. Cool down to room temperature. Add 372 mg of *N*-(*tert*-butoxycarbonyl)glycine (2.12 mmol) to the flask. Attach a nitrogen inlet adaptor to flask and fill flask with nitrogen.
3. Under nitrogen, add 3 mL of anhydrous THF followed by 512 μL of N-methyl-2-pyrrolidone to the flask.
4. Cool reaction to 0°C using an ice bath. While stirring, add 206 μL of isobutyl-chloroformate (1.59 mmol) to the flask dropwise. A white mixture will form almost instantaneously.

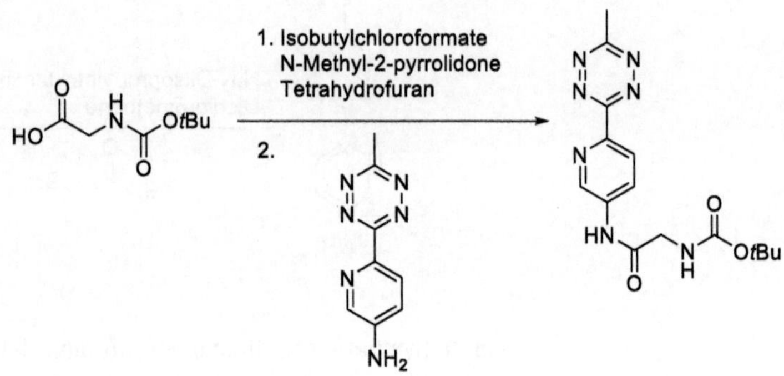

Fig. 4 Synthesis of tert-Butyl (2-((6-(6-methyl-1,2,4,5-tetrazin-3-yl)pyridin-3-yl)amino)-2-oxoethyl)carbamate

5. Allow the mixture to stir at 0°C for 30 min. Add 100 mg of 6-(6-methyl-1,2,4,5-tetrazin-3-yl)pyridin-3-amine(0.531 mmol) to the flask.
6. Remove the ice-bath and allow the reaction to stir at room temperature for 2 h.
7. Dilute the reaction with 8–10 mL of ethyl acetate and transfer to a separatory funnel. Add 5 mL of 1 N HCl the separatory. Cap the separatory funnel and shake vigorously by inverting the flask three to four times. Allow the organic and aqueous layer to settle. Collect the aqueous layer, which will be the bottom layer, in a 25 mL Erlenmeyer flask. Repeat the wash with 1 N HCl two more times.
8. With the organic layer still in the separatory funnel, add 5 mL of saturated sodium bicarbonate to the separatory funnel. Cap the separatory funnel and shake vigorously by inverting the flask three to four times. Allow the organic and aqueous layer to settle. Collect the aqueous layer in a 25 mL Erlenmeyer flask.
9. With the organic layer still in the separatory funnel, add 5 mL of brine to the separatory funnel. Cap the separatory funnel and shake vigorously by inverting the flask three to four times. Allow the organic and aqueous layer to settle. Collect the aqueous layer in the 25 mL Erlenmeyer flask.
10. Collect the organic layer in a separate 25 mL Erlenmeyer flask. Dry the organic layer with sodium sulfate. Filter the solution using a filter paper and funnel into a 50 mL round bottom flask. Remove solvent by rotary evaporation.
11. Carry out flash chromatography of the crude using silica gel with a gradient of 0%–100% ethyl acetate in hexanes.
12. The product is obtained as a pink solid with a yield of 70%. Successful isolation of product can be confirmed by NMR spectroscopy.

3.4 Synthesis of 2-Amino-N-(6-(6-Methyl-1,2,4,5-Tetrazin-3-yl)Pyridin-3-yl)Acetamide (Fig. 5)

1. Oven dry a 5 mL round bottom flask with an appropriately size magnetic stir bar.
2. Cool down to room temperature. Add 30 mg of tert-butyl (2-((6-(6-methyl-1,2,4,5-tetrazin-3-yl)pyridin-3-yl)amino)-2-oxoethyl)carbamate to the flask.
3. Add 1 mL of anhydrous dichloromethane to the flask followed by 1 mL of trifluoroacetic acid. Allow the reaction to stir at room temperature for 2 h.
4. Transfer the reaction mixture to a 20 mL scintillation vial. Evaporate reaction mixture to dryness by rotatory evaporation.
5. Redissolve the leftover mixture in 5 mL of dichloromethane and evaporate to dryness by rotatory evaporation. Repeat this step two more times (*see* **Note 9**).
6. The product is obtained as a red solid in quantitative yield and can be used without further purification.

3.5 DNA Sequences for Proteins

The protein used for conjugation reactions are:

1. His-TEV-sfGFP-SH—Superfolder GFP (sfGFP) with a C-terminal cysteine and an N-terminal His-Tag with a TEV cleavage sequence in a pET28a vector.
   ATGGGACATCATCATCATCATCACAGCAGCGG-
   CGAAAACCTGTATTTTCAGGGATCCGTTAGCAAAG-
   GTGAAGAACTGTTTACCGGCGTTGTGCCGATTCT-
   GGTGGAACTGGATGGTGATGTGAATGGCCATAAAT-
   TTAGCGTTCGTGGCGAAGGCGAAGGTGATGCGAC-
   CAACGGTAAACTGACCCTGAAATTTATTTGCAC-
   CACCGGTAAACTGCCGGTTCCGTGGCCGACCCT-
   GGTGACCACCCTGACCTATGGCGTTCAGTGCT-
   TTAGCCGCTATCCGGATCATATGAAACGCCATGAT-
   TTCTTTAAAAGCGCGATGCCGGAAGGCTATGTG-
   CAGGAACGTACCATTAGCTTCAAAGATGATGG-
   CACCTATAAAACCCGTGCGGAAGTTAAATTTGAA-
   GGCGATACCCTGGTGAACCGCATTGAACTGAAA-
   GGTATTGATTTTAAAGAAGATGGCAACATTCTGG-
   GTCATAAACTGGAATATAATTTCAACAGCCATAAT-

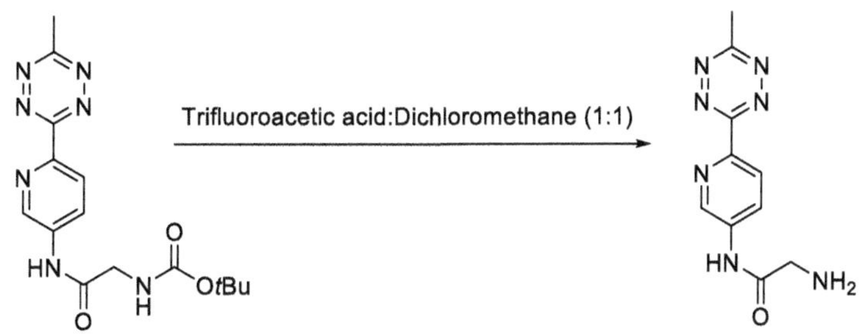

Fig. 5 Synthesis of 2-Amino-N-(6-(6-methyl-1,2,4,5-tetrazin-3-yl)pyridin-3-yl) acetamide

GTGTATATTACCGCCGATAAACAGAAAAATGG\
CATCAAAGCGAACTTTAAAATCCGTCACAACGTG-
GAAGATGGTAGCGTGCAGCTGGCGGATCATTATCAG-
CAGAATACCCCGATTGGTGATGGCCCGGTGCTGCTGC-
CGGATAATCATTATCTGAGCACCCAGAGCGTTCT-
GAGCAAAGATCCGAATGAAAAACGTGATCATATG-
GTGCTGCTGGAATTTGTTACCGCCGCGGGCAT-
TACCCACGGTATGGATGAACTGTATAAAGGCAGCTGC

2. His-TEV-HaloTag-LPETGG—HaloTag with a C-terminal LPETGG sequence and an N-terminal His-Tag with a TEV cleavage sequence in a pET28a vector

   ATGGGACATCATCATCATCATCACAGCAGCGG-
   CGAAAACCTGTATTTTCAGGGATCCGAAATCGG-
   TACTGGCTTTCCATTCGACCCCCATTATGTGGAA-
   GTCCTGGGCGAGCGCATGCACTACGTCGATGTTG-
   GTCCGCGCGATGGCACCCCTGTGCTGTTCCTG-
   CACGGTAACCCGACCTCCTCCTACGTGTGGCGCAA-
   CATCATCCCGCATGTTGCACCGACCCATCGCTG-
   CATTGCTCCAGACCTGATCGGTATGGGCAAATC-
   CGACAAACCAGACCTGGGTTATTTCTTCGACGAC-
   CACGTCCGCTTCATGGATGCCTTCATCGAAGCCCT-
   GGGTCTGGAAGAGGTCGTCCTGGTCATTCACGACT-
   GGGGCTCCGCTCTGGGTTTCCACTGGGCCAAG-
   CGCAATCCAGAGCGCGTCAAAGGTATTGCAT-
   TTATGGAGTTCATCCGCCCTATCCCGACCTGG-
   GACGAATGGCCAGAATTTGCCCGCGAGACCT-
   TCCAGGCCTTCCGCACCACCGACGTCGGCCG-
   CAAGCTGATCATCGATCAGAACGTTTTTATCGAG-
   GGTACGCTGCCGATGGGTGTCGTCCGCCCGCT-
   GACTGAAGTCGAGATGGACCATTACCGCGAGC-
   CGTTCCTGAATCCTGTTGACCGCGAGCCACT-
   GTGGCGCTTCCCAAACGAGCTGCCAATCGCCG-
   GTGAGCCAGCGAACATCGTCGCGCTGGTCGAA-
   GAATACATGGACTGGCTGCACCAGTCCCCT-
   GTCCCGAAGCTGCTGTTCTGGGGCACCCCAGG-
   CGTTCTGATCCCACCGGCCGAAGCCGCTCGCCT-
   GGCCAAAAGCCTGCCTAACTGCAAGGCTGTGGA-
   CATCGGCCCGGGTCTGAATCTGCTGCAAGAAGA-
   CAACCCGGACCTGATCGGCAGCGAGATCGCG-
   CGCTGGCTGTCTACTCTGGAGATTTCCGGTGGGAG-
   CGGTTCTCTGCCTGAGACAGGAGGT

SrtA7M-TEV-His – Hepta-mutant *Staphylococcus aureus* Sortase A with a C-terminal His-Tag and TEV cleavage sequence. TEV cleavage sequence ENLYFQ was inserted between SrtA7M and the C-terminal His-Tag using PCR mutagenesis. pet30b-7M SrtA was a gift from Hidde Ploegh (Addgene plasmid # 51141; http://n2t.net/addgene:51141; RRID:Addgene_51141).

3.6 Protein Expression

1. Transform plasmids into *E. coli* strain BL21(DE3) using heat-shock method (*see* **Note 10**).
2. In a 250 mL Erlenmeyer flask, add 50 mL of autoclaved LB media. Add 50 μL of 50 mg/mL kanamycin stock to reach a final concentration of 50 μg/mL (*see* **Note 11**).
3. Inoculate the media with a single colony from the transformed plate (*see* **Note 12**).
4. Grow the cells overnight at 37°C while shaking at 250 rpm.
5. For the expression of each protein, prepare 1.8 L of Terrific Broth (TB) and split into four 2 L Erlenmeyer flasks and autoclave the media. Also, autoclave 500 mL of 10× TB salts. Allow to cool to room temperature (*see* **Note 13**).
6. Under sterile conditions, add 50 mL of 10× TB salts to each 2 L flask containing TB media, along with 500 μL of 50 mg/mL kanamycin stock.
7. Add 5 mL of the overnight culture to each 2 L flask.
8. Grow cells at 37°C while shaking at 250 rpm. Periodically monitor the optical density (OD) at 600 nm. Once the OD_{600} reaches 0.8, induce protein expression by the addition of 500 μL of a 1 M solution of 1-thio-β-D-galactopyranoside (IPTG) to each flask (*see* **Note 14**).
9. Reduce the temperature of the incubator to 18°C and express the protein for 16–20 h.
10. Harvest cells by transferring the cells to centrifuge bottles and centrifuge at 3000 × *g* for 10 min. Pour off supernatant and proceed to protein purification. The pellets can also be stored at –80°C for processing later.

3.7 Protein Purification

1. Resuspend cell pellets in 60 mL of phosphate buffer containing 30 mM imidazole (50 mM phosphate, 300 mM NaCl, 30 mM imidazole, pH 7.6) and transfer to 50 mL conical tubes (*see* **Note 15**).
2. Cool resuspended cells in an ice-bath.
3. Lyse cells by using a French press (*see* **Notes 16 and 17**).
4. Place conical tubes in an ice bath and sonicate the cells at 50% amplitude for 5 s.
5. Transfer the lysate to appropriately sized centrifuge tubes and centrifuge at 16,000 × *g* for 20 min at 4°C.
6. Collect supernatant and repeat the centrifugation step to remove any remaining debris.
7. Filter all the supernatant through a 0.45 μM syringe filter.
8. Set up a gravity flow column containing 10 mL of Ni-IDA resin (*see* **Note 18**).

9. Equilibrate resin with 5 column volumes of wash buffer (50 mM phosphate, 300 mM NaCl, 30 mM imidazole, pH 7.6).
10. Load filtered supernatant onto resin and collect flowthrough (*see* **Note 19**).
11. Once all the solution has passed through the resin, wash the column with 15 column volumes of the wash buffer (50 mM phosphate, 100 mM NaCl, 30 mM imidazole, pH 7.6). Collect all wash fractions.
12. Elute the protein by adding five column volumes of elution buffer (50 mM phosphate, 300 mM NaCl, 400 mM imidazole, pH 7.6). Collect the elution in fractions (*see* **Note 20**).
13. Confirm the presence of purified protein by SDS-page or mass spectrometry.
14. Dialyze the protein in 2 L of phosphate buffer (50 mM, 100 mM NaCl, 1 mM 1,4-dithiothreitol [DTT] pH 7.6) for 3 hours at 4°C.
15. Exchange the buffer for fresh buffer (same as used in Step 14). Add 200 μL of a 70 μM solution of TEV protease to the protein in the dialysis bag. Allow dialysis and cleavage to proceed overnight at 4°C (*see* **Note 21**).
16. Perform two buffer exchanges in DTT-free phosphate buffer (50 mM, 100 mM NaCl, pH 7.6) for 2 h each (*see* **Note 22**).
17. Load the dialyzed protein solution on Ni-IDA packed column pre-equilibrated with phosphate buffer (50 mM, 100 mM NaCl, pH 7.6). Collect flowthrough.
18. Wash column with 3 column volumes of phosphate buffer containing 15 mM imidazole (50 mM phosphate, 100 mM NaCl, 15 mM imidazole pH 7.6) and collect protein (*see* **Note 23**).
19. Confirm successful cleavage and purity by SDS-PAGE or mass spectrometry.
20. Dialyze the combined protein fractions from flowthrough and the 15 mM imidazole wash, which contain pure protein, and dialyze in phosphate buffer (50 mM, 100 mM NaCl, pH 7.6) (*see* **Note 24**).
21. Concentrate the protein solution by diafiltration using Amicon Ultra-15 tubes.
22. Determine concentration of the protein using A280 (*see* **Note 25**).
23. Aliquot protein appropriately into 1.5 mL centrifuge tubes and store at –80°C until further use.

3.8 Conjugation of 2-Bromo-N-(6-(6-Methyl-1,2,4,5-Tetrazin-3-yl)Pyridin-3-yl)Acetamide to sfGFP-Cys

1. Prepare a gravity flow column containing 500 μL of Ni-IDA resin and equilibrate the column with five column volumes of MES buffer (25 mM, 100 mM NaCl, pH 6.0).
2. Prepare a 250 mM solution of DTT in deionized water (*see* **Note 26**).
3. Prepare 100 μM sfGFP-cys solution in 247.5 μL of phosphate buffer (25 mM, 100 mM NaCl, pH 7.6).
4. Reduce the sfGFP-Cys solution by adding 2.5 μL of the DTT stock for a final concentration of 2.5 mM and incubate at room temperature for 30 min.
5. Prepare a 100 mM stock of 2-bromo-N-(6-(6-methyl-1,2,4,5-tetrazin-3-yl)pyridin-3-yl)acetamide in anhydrous DMSO. Also, prepare a 10 M solution of sodium iodide (NaI) in water.
6. Equilibrate a Nap5 column with phosphate buffer (25 mM, 100 mM NaCl, pH 7.6).
7. Remove DTT from the reduced sfGFP by desalting into phosphate buffer (25 mM, 100 mM NaCl, pH 7.6) using the pre-equilibrated Nap5 column. The volume after desalting will be 500 μL (*see* **Notes 27 and 28**).
8. Add 2.5 μL of the tetrazine stock to sfGFP-Cys solution for a final concentration of 500 μM. Add 2.5 μL of the NaI stock to the reaction for a final concentration of 50 mM (*see* **Note 29**).
9. Incubate the reaction at 37°C in a water bath for 1 h. Take a small aliquot for mass spectrometry analysis to verify completion of reaction.
10. While the reaction proceeds, pre-equilibrate a 5 mL HiTrap Desalting column (GE Healthcare) with 25 mL of MES buffer (25 mM, 100 mM NaCl, pH 6.0) (*see* **Note 30**).
11. Upon completion of reaction, remove small molecules and buffer exchange into MES buffer (25 mM, 100 mM NaCl, pH 6.0) using the pre-equilibrated 5 mL Hitrap Desalting column using 1 mL of MES buffer to elute (*see* **Note 31**).
12. Concentrate the solution to 500 μL using an Amicon Ultra-0.5 mL centrifugal filter.
13. Load the protein solution to the pre-equilibrated Ni-IDA column and collect flowthrough.
14. Wash the column with five column volumes of MES buffer (25 mM, 100 mM NaCl, pH 6.0) and collect wash fractions.
15. Wash the column with five column volumes of MES buffer containing 10 mM imidazole (25 mM MES, 100 mM NaCl, 10 mM imidazole pH 6.0) and collect wash fractions (*see* **Note 32**).

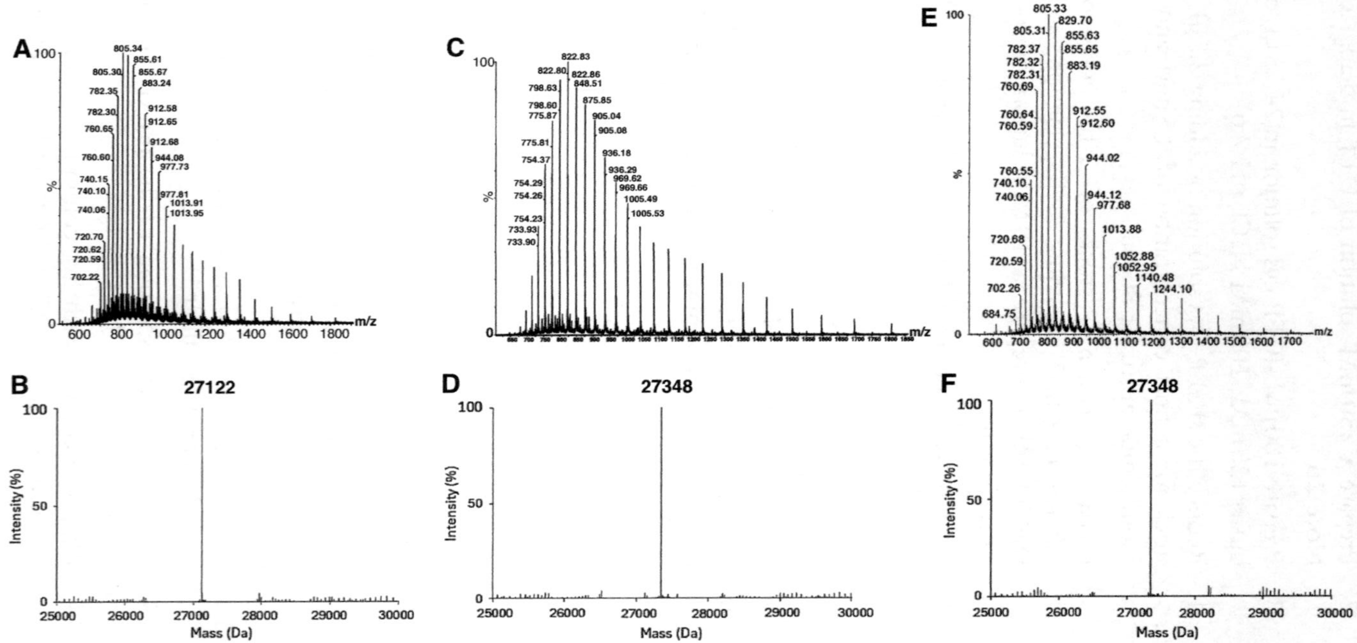

Fig. 6 **(A)** Mass spectrum of sfGFP-Cys. **(B)** Deconvoluted mass spectrum of sfGFP-PyTz (expected mass = 27,120, experimental mass = 27,122) **(C)** Mass spectrum of sfGFP-Cys + 2-bromo-N-(6-(6-methyl-1,2,4,5-tetrazin-3-yl)pyridin-3-yl)acetamide reaction mixture. **(D)** Deconvoluted mass spectrum of sfGFP-Cys + 2-bromo-N-(6-(6-methyl-1,2,4,5-tetrazin-3-yl)pyridin-3-yl)acetamide reaction mixture. **(E)** Mass spectrum of sfGFP-PyTz after purification using Ni-IDA resin. **(F)** Deconvoluted mass spectrum of sfGFP-PyTz after purification using Ni-IDA resin (expected mass = 27,348 Da, experimental mass = 27,348 Da)

16. Elute the protein conjugate with five column volumes of phosphate buffer containing 400 mM imidazole (25 mM phosphate, 100 mM NaCl, 400 mM imidazole, pH 7.6) and collect fractions.
17. Concentrate eluted fractions and buffer exchange the protein into phosphate buffer (25 mM, 100 mM NaCl, pH 7.6) using an Amicon Ultra-0.5 centrifugal filter (*see* **Note 33**).
18. Verify purity of the protein conjugate using mass spectrometry (Fig. 6).
19. Determine the concentration of the protein using the Bradford assay.

3.9 Conjugation of 2-Amino-N-(6-(6-Methyl-1,2,4,5-Tetrazin-3-yl)Pyridin-3-yl)Acetamide to HaloTag-LPETGG

1. Prepare a 100 mM stock of 2-amino-N-(6-(6-methyl-1,2,4,5-tetrazin-3-yl)pyridin-3-yl)acetamide in anhydrous DMSO.
2. Prepare a stock 500 μM solution of SrtA7M in phosphate buffer (25 mM, 100 mM NaCl, pH 7.6).
3. Prepare a 150 μM solution of HaloTag-LPETGG in 240 μL of phosphate buffer (25 mM, 100 mM NaCl, pH 7.6).
4. To this solution, add 5 μL of the 2-amino-N-(6-(6-methyl-1,2,4,5-tetrazin-3-yl)pyridin-3-yl)acetamide stock for a final concentration of 2 mM and 5 μL of SrtA7M for a final concentration of 10 μM.
5. Incubate at room temperature for 1 h and verify the formation of the conjugated product by mass spectrometry (Fig. 7 C–D) (see **Note 34**).
6. Follow Steps 10–17 of Sect. 3.8 for the purification of the tetrazine conjugated protein using Ni-IDA resin.
7. Verify purity of the protein conjugate using mass spectrometry (Fig. 7 E–F).
8. Determine the concentration of the protein using the Bradford assay.

4 Notes

1. Alternatively, reaction flasks can be flame dried before start of reaction.
2. Tie the copper wire around the septum 2–3 times with the end twisted multiple times till the wire is tightly wound.
3. The septum will have bulged due to the buildup of gases during the reaction. With a needle bent away from you, carefully pierce the septum to release pressure.

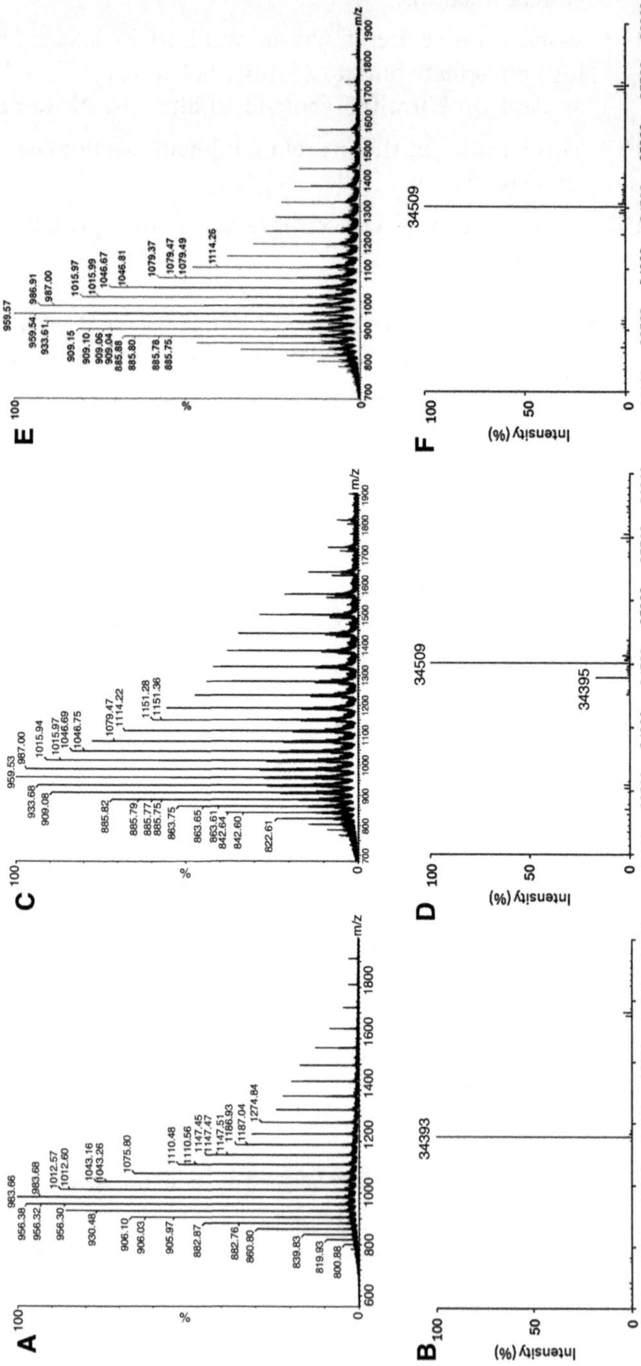

Fig. 7 (**A**) Mass spectrum of Halo-LPETGG. (**B**) Deconvoluted mass spectrum of Halo-LPETGG (expected mass = 34,394, experimental mass = 34,393) (**C**) Mass spectrum of Halo-LPETGG + 2-amino-N-(6-(6-methyl-1,2,4,5-tetrazin-3-yl)pyridin-3-yl)acetamide reaction mixture. (**D**) Deconvoluted mass spectrum of Halo-LPETGG + 2-amino-N-(6-(6-methyl-1,2,4,5-tetrazin-3-yl)pyridin-3-yl)acetamide reaction mixture. (**E**) Mass spectrum of Halo-PyTz after purification using Ni-IDA resin. (**F**) Deconvoluted mass spectrum of Halo-PyTz after purification using Ni-IDA resin (expected mass = 34,507 Da, experimental mass = 34,509 Da)

4. Make sure to vent the separatory funnel after shaking. By firmly grasping the separatory funnel with it sealed and while it is upside down, open the valve on the bottom of the flask. Close the valve before setting the separatory funnel right side up.
5. Add enough sodium sulfate till they are not clumped together, and some free-floating granules can be seen.
6. Do not add a large excess of DMP, since that makes the purification of the compound more tedious.
7. The bromoacetyl bromide solution will smoke when opened due to the release of HCl gas.
8. After transferring the combined organic layer, wash the Erlenmeyer flask with 10 mL of DCM and transfer to the separatory funnel to transfer any residual compound.
9. This step is to remove the residual trifluoroacetic acid that remains.
10. High Efficiency Transformation Protocol from New England Bio labs was followed. https://www.neb.com/en-us/protocols/0001/01/01/high-efficiency-transformation-protocol-c2987?srsltid=AfmBOoq5T86EffiDesLhnzsRLQs1JzKSpkmeCpWThQ-wH2bwSJ3B6s-6
11. When growing cells, ensure that the vessel size is at least four times the volume of the media being used for optimal growth of bacterial cells.
12. Alternatively, a colony from a plate streaked from a glycerol stock can be used.
13. For ease of workflow, media can be autoclaved the day before and stored with the top sealed.
14. Typically cells take between 2–4 hours to read OD600 of 0.8, depending on the amount of overnight culture added to the flasks.
15. Buffer composition needs to be compatible with protein properties. Keep the resuspended cells in an ice bath.
16. Cells can also be lysed by other methods such as chemical lysis or sonication.
17. From this point forward, maintain the lysate at 4°C using an ice bath and perform purification steps at 4°C as much as possible.
18. 10 mL of resin was used for 2 L of culture. Adjust amount of resin to the amount of cells lysed for purification.
19. Store flowthrough and all collected fractions in an ice bath.
20. BCA or Bradford assay can be used to determine the presence of protein in the different fractions and only the fractions that

contain detectable amounts of protein can be carried through in the remaining steps.

21. The efficiency of TEV cleavage will differ for different proteins. Complete cleavage may require a larger amount of TEV protease. Small-scale cleavage reactions can be set up to determine the optimal amount of TEV protease to use.
22. Presence of DTT in solution will reduce the nickel present in the resin and lead to His-tagged impurities as well free nickel present in the flowthrough.
23. Some amount of cleaved protein may still be nonspecifically bound to the column. Washing with buffer containing a very low amount of imidazole will help wash off the nonspecifically bound protein.
24. Buffer exchange can also be performed using a desalting column or by diafiltration.
25. Alternatively, concentration can also be determined using Bradford or BCA assay.
26. Always prepare the DTT stock fresh, right before use.
27. A 5 mL HiTrap Desalting column (GE Healthcare) can also be used for removing DTT. When performing the reaction at larger scales than specified here, it is recommended to use a HiTrap Desalting column or a similar column, since a Nap5 column will not be efficient enough to remove all the DTT from the solution.
28. Once the DTT is removed, sfGFP-Cys tends to dimerize by disulfide linkage rapidly. Perform the labeling step with 2-bromo-N-(6-(6-methyl-1,2,4,5-tetrazin-3-yl)pyridin-3-yl) acetamide quickly right after desalting.
29. Ensure that the final amount of DMSO does not exceed 5% of the total volume of the reaction. Excess DMSO in the solution will cause the protein to precipitate out of solution. In general, try to keep the volume of DMSO or any other organic solvent used as low as possible for labeling reactions.
30. The desalting column can be used with an FPLC system or a syringe pump.
31. Instead of using a desalting column, the excess small molecule can also be removed by ion exchange chromatography.
32. Some pyridyl tetrazine proteins may not bind as strongly as sfGFP-PyTz to Ni-IDA resin, and hence, the 10 mM imidazole wash step may wash away the bound protein. For such proteins, skip this step and directly proceed to elution step with 400 mM imidazole. It is advisable to save flowthrough and wash fractions and running an SDS-PAGE gel or using

mass spectrometry to determine loss of any of the pyridyl tetrazine conjugated protein product.

33. Alternatively, the protein can be buffer exchanged using a desalting column.

34. While in the case of HaloTag, the only other product seen in the mass spectrum is the unconjugated HaloTag-LPETGG, sortase mediated ligation reactions with other proteins can lead to the formation of other protein products. In these cases, the reaction can be optimized to maximize the formation of the desired conjugated product by varying the amount of the SrtA7M and 2-amino-N-(6-(6-methyl-1,2,4,5-tetrazin-3-yl)pyridin-3-yl)acetamide used in the reaction as well as the reaction time, if a higher reaction yield is desired.

References

1. Schneider AFL, Hackenberger CPR (2017) Fluorescent labelling in living cells. Curr Opin Biotechnol 48:61–68. https://doi.org/10.1016/j.copbio.2017.03.012
2. Agarwal P, Bertozzi CR (2015) Site-specific antibody–drug conjugates: the nexus of bioorthogonal chemistry, protein engineering, and drug development. Bioconjug Chem 26(2):176–192. https://doi.org/10.1021/bc5004982
3. Parker CG, Pratt MR (2020) Click chemistry in proteomic investigations. Cell 180(4):605–632. https://doi.org/10.1016/j.cell.2020.01.025
4. Spicer CD, Pashuck ET, Stevens MM (2018) Achieving controlled biomolecule–biomaterial conjugation. Chem Rev 118(16):7702–7743. https://doi.org/10.1021/acs.chemrev.8b00253
5. Theile CS et al (2013) Site-specific N-terminal labeling of proteins using sortase-mediated reactions. Nat Protoc 8(9):1800–1807. https://doi.org/10.1038/nprot.2013.102
6. Weeks AM, Wells JA (2018) Engineering peptide ligase specificity by proteomic identification of ligation sites. Nat Chem Biol 14(1):50–57. https://doi.org/10.1038/nchembio.2521
7. Nguyen GKT et al (2015) Site-specific N-terminal labeling of peptides and proteins using butelase 1 and thiodepsipeptide. Angew Chem Int Ed 54(52):15694–15698. https://doi.org/10.1002/anie.201506810
8. Kulkarni C, Kinzer-Ursem TL, Tirrell DA (2013) Selective functionalization of the protein N terminus with N-myristoyl transferase for bioconjugation in cell lysate. ChemBioChem 14(15):1958–1962. https://doi.org/10.1002/cbic.201300453
9. Schumacher D et al (2015) Versatile and efficient site-specific protein functionalization by tubulin tyrosine ligase. Angew Chem Int Ed 54(46):13787–13791. https://doi.org/10.1002/anie.201505456
10. Guimaraes CP et al (2013) Site-specific C-terminal and internal loop labeling of proteins using sortase-mediated reactions. Nat Protoc 8(9):1787–1799. https://doi.org/10.1038/nprot.2013.101
11. Rehm FBH et al (2019) Site-specific sequential protein labeling catalyzed by a single recombinant ligase. J Am Chem Soc 141(43):17388–17393. https://doi.org/10.1021/jacs.9b09166
12. Fernández-Suárez M et al (2007) Redirecting lipoic acid ligase for cell surface protein labeling with small-molecule probes. Nature Biotechnol 25(12):1483–1487. https://doi.org/10.1038/nbt1355
13. de Boer E et al (2003) Efficient biotinylation and single-step purification of tagged transcription factors in mammalian cells and transgenic mice. Proc Natl Acad Sci U S A 100(13):7480–7485. https://doi.org/10.1073/pnas.1332608100
14. Lopez Aguilar A et al (2017) Tools for studying glycans: recent advances in chemoenzymatic glycan labeling. ACS Chem Biol 12(3):611–621. https://doi.org/10.1021/acschembio.6b01089
15. Nessen MA et al (2009) Selective enrichment of azide-containing peptides from complex mixtures. J Proteome Res 8(7):3702–3711. https://doi.org/10.1021/pr900257z
16. Biedka S et al (2021) Reversible click chemistry tag for universal proteome sample preparation for top-down and bottom-up analysis. J Proteome Res 20(10):4787–4800. https://doi.org/10.1021/acs.jproteome.1c00443
17. Han H-S et al (2010) Development of a bioorthogonal and highly efficient

conjugation method for quantum dots using tetrazine–norbornene cycloaddition. J Am Chem Soc 132(23):7838–7839. https://doi.org/10.1021/ja101677r
18. Chen W et al (2012) Clicking 1,2,4,5-tetrazine and cyclooctynes with tunable reaction rates. Chem Commun 48(12):1736–1738. https://doi.org/10.1039/C2CC16716F
19. Lang K et al (2012) Genetic encoding of bicyclononynes and trans-cyclooctenes for site-specific protein labeling in vitro and in live mammalian cells via rapid fluorogenic Diels–Alder reactions. J Am Chem Soc 134(25):10317–10320. https://doi.org/10.1021/ja302832g
20. Blackman ML, Royzen M, Fox JM (2008) Tetrazine ligation: fast bioconjugation based on inverse-electron-demand Diels–Alder reactivity. J Am Chem Soc 130(41):13518–13519. https://doi.org/10.1021/ja8053805
21. Darko A et al (2014) Conformationally strained trans-cyclooctene with improved stability and excellent reactivity in tetrazine ligation. Chem Sci 5(10):3770–3776. https://doi.org/10.1039/C4SC01348D
22. Lambert WD et al (2017) Computationally guided discovery of a reactive, hydrophilic trans-5-oxocene dienophile for bioorthogonal labeling. Org Biomol Chem 15(31):6640–6644. https://doi.org/10.1039/C7OB01707C
23. Taylor MT, Blackman ML, Dmitrenko O, Fox JM (2011) Design and synthesis of highly reactive dienophiles for the tetrazine–trans-cyclooctene ligation. J Am Chem Soc 133(25):9646–9649. https://doi.org/10.1021/ja201844c
24. Rossin R et al (2010) In vivo chemistry for pretargeted tumor imaging in live mice. Angew Chem Int Ed 49(19):3375–3378. https://doi.org/10.1002/anie.200906294
25. Wang M et al (2019) Hydrophilic 18F-labeled trans-5-oxocene (oxoTCO) for efficient construction of PET agents with improved tumor-to-background ratios in neurotensin receptor (NTR) imaging. Chem Commun 55(17):2485–2488. https://doi.org/10.1039/C8CC09747J
26. Murrey HE et al (2015) Systematic evaluation of bioorthogonal reactions in live cells with clickable HaloTag ligands: implications for intracellular imaging. J Am Chem Soc 137(35):11461–11475. https://doi.org/10.1021/jacs.5b06847
27. Jemas A et al (2022) Catalytic activation of bioorthogonal chemistry with light (CABL) enables rapid, spatiotemporally controlled labeling and no-wash, subcellular 3D-patterning in live cells using long wavelength light. J Am Chem Soc 144(4):1647–1662. https://doi.org/10.1021/jacs.1c10390
28. Baalmann M et al (2019) Enzymatic and site-specific ligation of minimal-size tetrazines and triazines to proteins for bioconjugation and live-cell imaging. Bioconjug Chem 30(5):1405–1414. https://doi.org/10.1021/acs.bioconjchem.9b00157
29. Versteegen RM et al (2013) Click to release: instantaneous doxorubicin elimination upon tetrazine ligation. Angew Chem Int Ed 52(52):14112–14116. https://doi.org/10.1002/anie.201305969
30. van Onzen AHAM et al (2020) Bioorthogonal tetrazine carbamate cleavage by highly reactive trans-cyclooctene. J Am Chem Soc 142(25):10955–10963.https://doi.org/10.1021/jacs.0c00531
31. Kang K, Park J, Kim E (2017) Tetrazine ligation for chemical proteomics. Proteome Sci 15(1):15. https://doi.org/10.1186/s12953-017-0121-5
32. Qian L et al (2019) Live-cell imaging and profiling of c-Jun N-terminal kinases using covalent inhibitor-derived probes. Chem Commun 55(8):1092–1095. https://doi.org/10.1039/C8CC09558B
33. Stetsiuk O, Abhervé A, Avarvari N (2020) 1,2,4,5-Tetrazine based ligands and complexes. Dalton Trans 49(18):5759–5777. https://doi.org/10.1039/D0DT00827C
34. Kaim W (2002) The coordination chemistry of 1,2,4,5-tetrazines. Coord Chem Rev 230(1):127–139. https://doi.org/10.1016/S0010-8545(02)00044-9
35. Tang TS-M, Liu H-W, Lo KK-W (2017) Monochromophoric iridium(iii) pyridyl–tetrazine complexes as a unique design strategy for bioorthogonal probes with luminogenic behavior. Chem Commun 53(23):3299–3302. https://doi.org/10.1039/C7CC00427C
36. Scinto SL, Reagle TR, Fox JM (2022) Affinity bioorthogonal chemistry (ABC) tags for site-selective conjugation, on-resin protein-protein coupling, and purification of protein conjugates. Angew Chem Int Ed 61(45):e202207661. https://doi.org/10.1002/anie.202207661
37. Ramaraj PK et al (2024) Covalent attachment of functional proteins to microfiber surfaces via a general strategy for site-selective tetrazine ligation. ACS Appl Mater Interfaces 16(46):63195–63206. https://doi.org/10.1021/acsami.4c12609

Chapter 13

N-Acyl-*N*-Alkyl Sulfonamide Mediated Bioconjugation of His-Tag Fused Proteins

Vikram Thimaradka, Tomonori Tamura and Itaru Hamachi

Abstract

Chemical modification of proteins is an important tool for the development of protein-based therapeutics, a rapidly growing field in disease treatment. Currently, no universal approach exists for site-specifically introducing desired functionalities into proteins with minimal disruption. Here, we present a reactive peptide tag/probe pair system for site-specific covalent protein labeling, leveraging the interaction between a lysine-containing histidine tag (KH6 or H6K) and a binuclear nickel (II)-nitrilotriacetic acid ($BisNi^{2+}$-NTA) probe conjugated to a lysine-reactive *N*-acyl-*N*-alkyl sulfonamide (NASA) electrophilic group. This chapter describes the system's validation, preparation of a glutamate receptor δ2 (GluD2) targeting nanobody-biotin conjugate, and its functional assessment via live-cell imaging of GluD2. The method offers a versatile, efficient approach for protein labeling applicable to protein engineering and targeted therapeutics.

Keywords Covalent protein labeling, *N*-acyl-*N*-alkyl sulfonamide, Short peptide tag, His tag, Ni^{2+}-NTA

1 Introduction

Proteins are central to nearly all biological processes. Site-specific covalent protein labeling is an indispensable tool enabling precise control over protein functions, visualization, and development of therapeutic conjugates [1–3]. Traditional methods, such as usage of amine-reactive NHS esters, are widely employed to prepare protein conjugates. However, they often generate heterogeneous conjugates (mixtures), rendering them unsuitable for precision medicine. Cysteine-maleimide conjugation, another common approach, finds limited use when target proteins contain functionally critical cysteines. Enzymatic self-labeling tags, such as Halo-tag and SNAP-tag, present an alternative strategy [4, 5]. While highly effective, these tags are often large, and their size can interfere with the function of the target protein, a concern that is particularly relevant

Alexander Deiters (ed.), *Bioconjugation Methods*, Methods in Molecular Biology, vol. 3009,
https://doi.org/10.1007/978-1-0716-5112-4_13,

in the context of therapeutic conjugates, where preserving protein integrity is crucial. To address these challenges, we recently developed a lysine-reactive short-peptide/tag probe pair system as a universal approach for site-specific conjugation of protein of interest (POI) [6]. By incorporating a lysine residue into the oligo-histidine (His6) tag, we created a KH6/H6K tag which pairs with a Ni^{2+}-NTA chelator bearing a cleavable *N*-acyl-*N*-alkyl-sulfonamide (NASA) electrophile [7]. The noncovalent interaction between KH6/H6K tag and BisNi^{2+}-NTA probe drives the reaction between the tag's lysine residue and NASA reaction center, covalently incorporating the desired functional group into the tag-fused protein. The cleavable nature of the NASA electrophile renders the method traceless, as the Ni^{2+}-NTA chelator is released after the reaction and can be removed from the modified protein, allowing precise installation of the desired motif without leaving residual metals (Fig. 1). Given the widespread use of His6 tags in immobilized metal affinity chromatography (IMAC) for protein purification, incorporation of single additional lysine amino acid residue on the tag makes this method broadly applicable for preparing protein conjugates in research and therapeutics.

This chapter describes a protocol for KH6/H6K peptide modification, preparation of a nanobody-biotin conjugate and validation of its functionality by visualizing cell surface GluD2 in live cells. The BisNTA-NASA-Bt probe (Fig. 2), referred as probe **1**, can be synthesized as described previously [6] or obtained from our group upon request. This system can be adapted to conjugate various functional groups, such as fluorophores or drugs, making it versatile for diverse experimental needs.

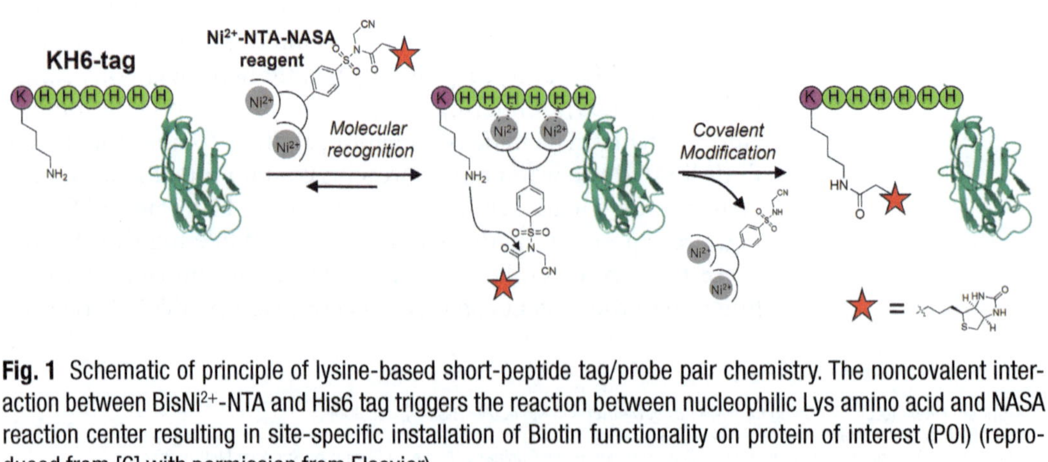

Fig. 1 Schematic of principle of lysine-based short-peptide tag/probe pair chemistry. The noncovalent interaction between BisNi^{2+}-NTA and His6 tag triggers the reaction between nucleophilic Lys amino acid and NASA reaction center resulting in site-specific installation of Biotin functionality on protein of interest (POI) (reproduced from [6] with permission from Elsevier)

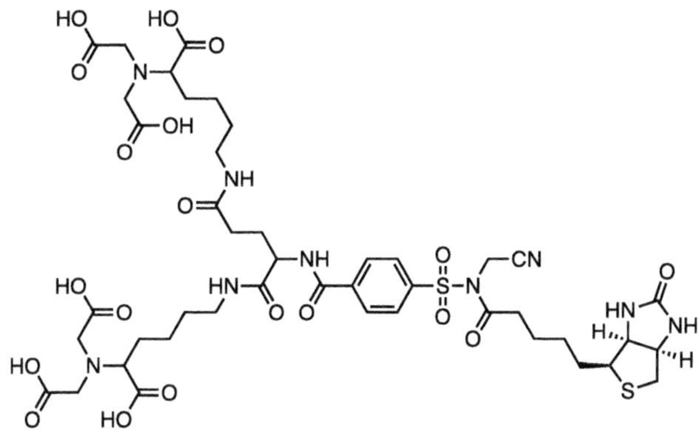

Fig. 2 Chemical structure of BisNTA-NASA-Bt (**1**) probe

2 Materials

2.1 General Biological Instruments

1. 1.5 mL plastic tubes
2. Heat block or heating incubator pre-set to 37°C.
3. HPLC instrument equipped with UV detector and C18 column (e.g., YMC-Pack ODS-A column 5 μm, 250 × 4.6 mm).
4. MALDI-TOF MS (e.g., Bruker Ultraflextreme).
5. 1 mL glass syringe.
6. Syringe filter (0.45 μm).
7. Glass bottom dish (35 mm).
8. 3.5 cm tissue culture dish.
9. Centrifugal filter units (e.g., MWCO 3 K).
10. Cell culture incubator maintained at 37°C, 5% CO_2.
11. Confocal microscopy setup.

2.2 Chemical or Biological Reagents

12. Ultrapure water.
13. Dimethyl sulfoxide (biological grade) (*see* **Note 1**).
14. Nickel Chloride solution (aq.) stock solution (100 mM).
15. BisNTA-NASA-Biotin (**1**) solution in DMSO (10 mM) (*see* **Note 1**).
16. Ac-KH_6AW peptide solution (1 μM, PBS) (*see* **Note 2**).
17. Purified recombinant protein to be conjugated (purified anti-GluD2 nanobody is used for demonstration).
18. Ethylenediaminetetraacetic acid disodium salt.
19. Phosphate-buffered saline (PBS 1×): 137 mM NaCl, 2.7 mM KCl, 1.5 mM KH_2PO_4, 8.1 mM Na_2HPO_4, pH 7.4.
20. Internal standard (e.g., benzene sulfonic acid).

21. α-Cyano-4-hydroxycinnamic acid (CHCA) matrix for MALDI-TOF (10 mg/mL in CH_3CN: H_2O 1:1, 0.1% TFA).
22. Acetonitrile (HPLC grade).
23. Trifluoracetic acid (TFA).
24. Size exclusion chromatography gel TOYOPEARL HW-40F.
25. Dulbecco's Modified Eagle Medium (DMEM), high glucose.
26. 0.25% Trypsin-EDTA solution.
27. Fetal bovine serum (FBS).
28. Penicillin/Streptomycin (Gibco).
29. pCAGGS(HA-GluD2) plasmid [8] (*see* **Note 3**).
30. Lipofectamine 2000 transfection reagent.
31. Poly-L-lysine.
32. 0.25% Trypsin-EDTA.
33. Secondary antibodies for immunostaining: Streptavidin-HiLyte647 (AnaSpec), antialexa488 (abcam) 5 μg/mL in DMEM-HEPES.
34. DMEM-HEPES.
35. Ice.

3 Methods

3.1 Peptide Labeling

The preparation of simple peptide conjugates enables the initial assessment of the labeling system's robustness, characterization of reaction kinetics, and optimization of reaction conditions. Peptides, due to their ease of synthesis at scale and straightforward handling, serve as versatile tools for testing and screening probes.

1. Just before the reaction, pipette 8 μL of $NiCl_2$ (100 mM, aq.) to 1.5 mL tube containing 92 μL of ultrapure water to make 8 mM stock solution. Pipette 4 μL of **1** to 1.5 mL tube containing 6 μL DMSO to make a 4 mM stock solution.
2. In a 1.5 mL tube mix **1** (4 mM) with $NiCl_2$ aq. (8 mM) in 1:1 volume. Mix well and incubate at room temperature for 5 min.
3. In a 1.5 mL tube pipette Ac-KH_6AW peptide solution (1 mL, 1 μM) containing benzene sulfonamide as internal standard (8 μM).
4. Add Ni(II) complexed solution of **1** (2 μL) to peptide solution (final concentration of **1** is 4 μM) (*see* **Note 4**).
5. At desired time point, filter the solution using 0.45 μm syringe filter to a fresh 1.5 mL tube. Inject the reaction mixture for HPLC analysis. The reaction progress can be monitored by measuring the absorbance at 220 nm/280 nm (Fig. 3).

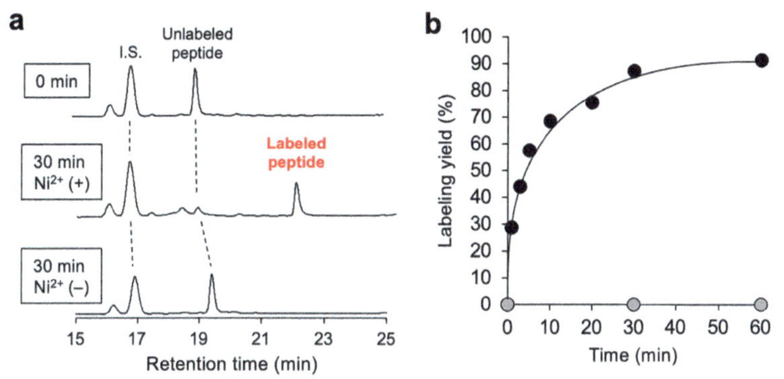

Fig. 3 HPLC (**a**) and time profile (**b**) of labeling Ac-KH6AW peptide (1 μM) by **1** (4 μM) in the presence and absence of Ni^{2+} (8 μM) (reproduced from [6] with permission from Elsevier)

Gradients; A (CH_3CN containing 0.1% TFA): B(H_2O containing 0.1% TFA) = 5: 95 (0 min) to 35: 65 (30 min) (*see* **Note 5**).

6. If MALDI-TOF analysis is desired, from the above reaction mixture, mix 1–2 μL of the labeled peptide fraction with equal volume of CHCA solution, spot on the MALDI-TOF MS plate and allow it to dry at room temperature or in a vacuum chamber. Analyze the sample by MALDI-TOF MS/MS (Fig. 4) (*see* **Note 6**).

3.2 GluD2 Targeting Nanobody Conjugate Preparation and Application in Live Cell Imaging

Chemically functionalized nanobodies, valued for their high specificity, thermal stability, and small size (~14 kDa), are widely used in fundamental research and medical application. NASA chemistry enables the production of modified nanobodies with precise control over the tethering site and number of synthetic probes, overcoming challenges posed by multiple lysines and two conserved cysteines in the nanobody scaffold. This method facilitates site-specific modification by introducing a single lysine adjacent to His6 tag. As an example, we demonstrate site-specific conjugation of biotin to GluD2 targeting nanobody (Nb-H6K).

7. In a 1.5 mL tube, mix **1** (4 mM, DMSO) with $NiCl_2$ (8 mM, aq.) in 1:1 volume. Incubate at room temperature for 5 min.
8. Pipette 1.5 μL of the Ni(II) complexed **1** to the Nb-H6K (1 μM, 1 mL in PBS) in a 1.5 mL tube (final concentration 1–3 μM) and incubate at room temperature for 1 h (*see* **Notes 7 and 8**).
9. The biotin modification can be monitored using MALDI-TOF MS (Fig. 5a, b) (*see* **Note 9**).
10. Excess small molecule can be removed using gel filtration (e.g., TOYOPEARL HW-40F) according to manufacturer's instructions.

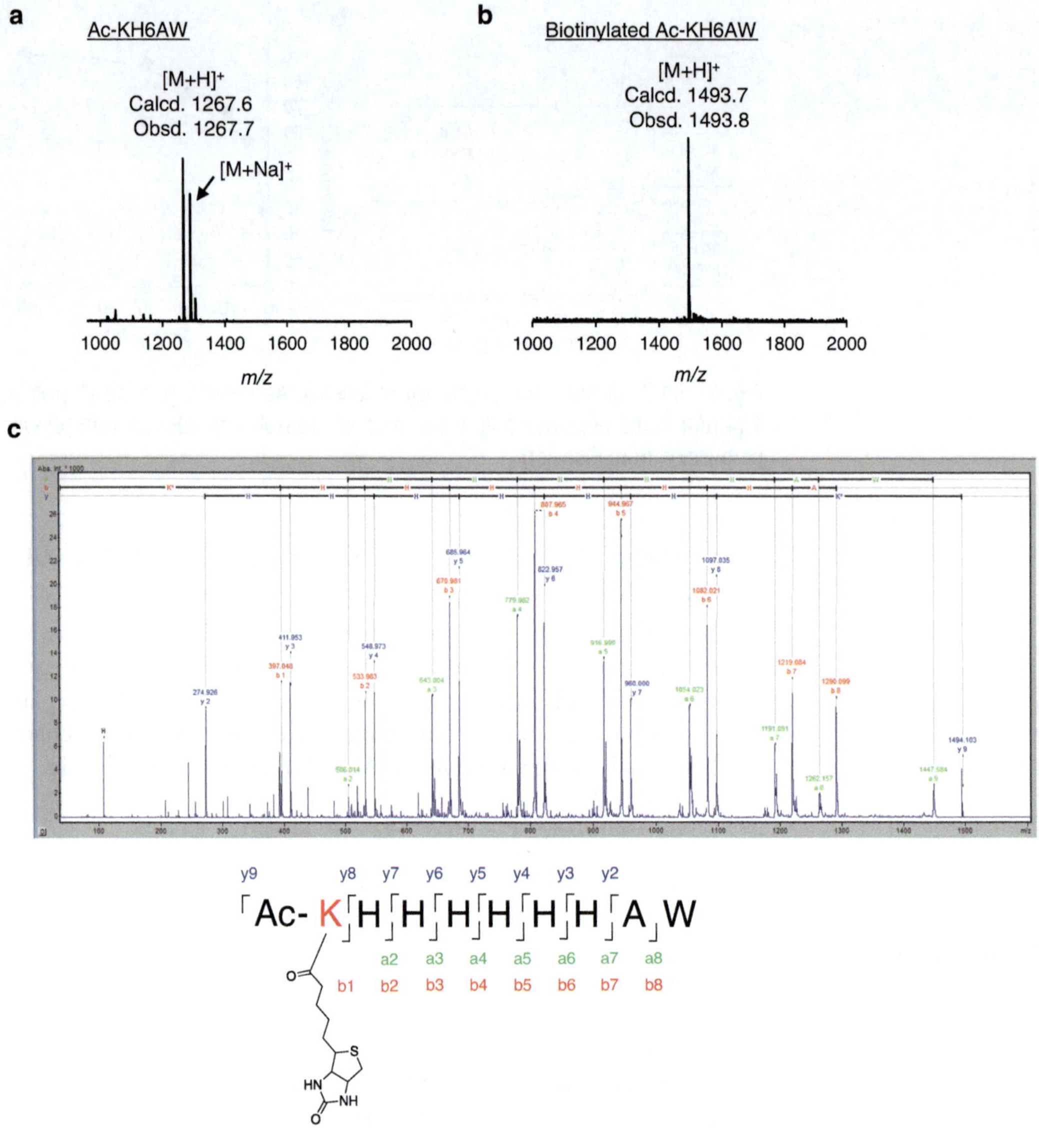

Fig. 4 Mass spectral analysis of modification of Ac-KH6AW peptide using probe **1**. Mass of peptide before (**a**) and after (**b**) covalent modification. Reaction condition: 1 μM peptide, 4 μM **1**, 8 μM $NiCl_2$, 37°C, PBS, pH 7.4, 1 h. (**c**) MALDI-TOF/TOF MS analysis of the peptide labeled with **1** showing covalent attachment of biotin functional group at the e-amino group of lysine residue in Ac-KH6AW peptide (reproduced from [6] with permission from Elsevier)

11. Analyze each fraction for modified nanobody conjugate by MALDI-TOF MS. Combine the fractions together and concentrate using centrifugal filter unit (*see* **Note 9**).
12. Plate 2 × 10^5 HEK293T cells in 3.5 cm cell culture dish.
13. Incubate in culture medium (DMEM, High glucose) supplemented with 10% fetal bovine serum (FBS), penicillin (100 units/ml) and streptomycin (100 mg/mL), for 24 h at 37°C under 5% CO_2.

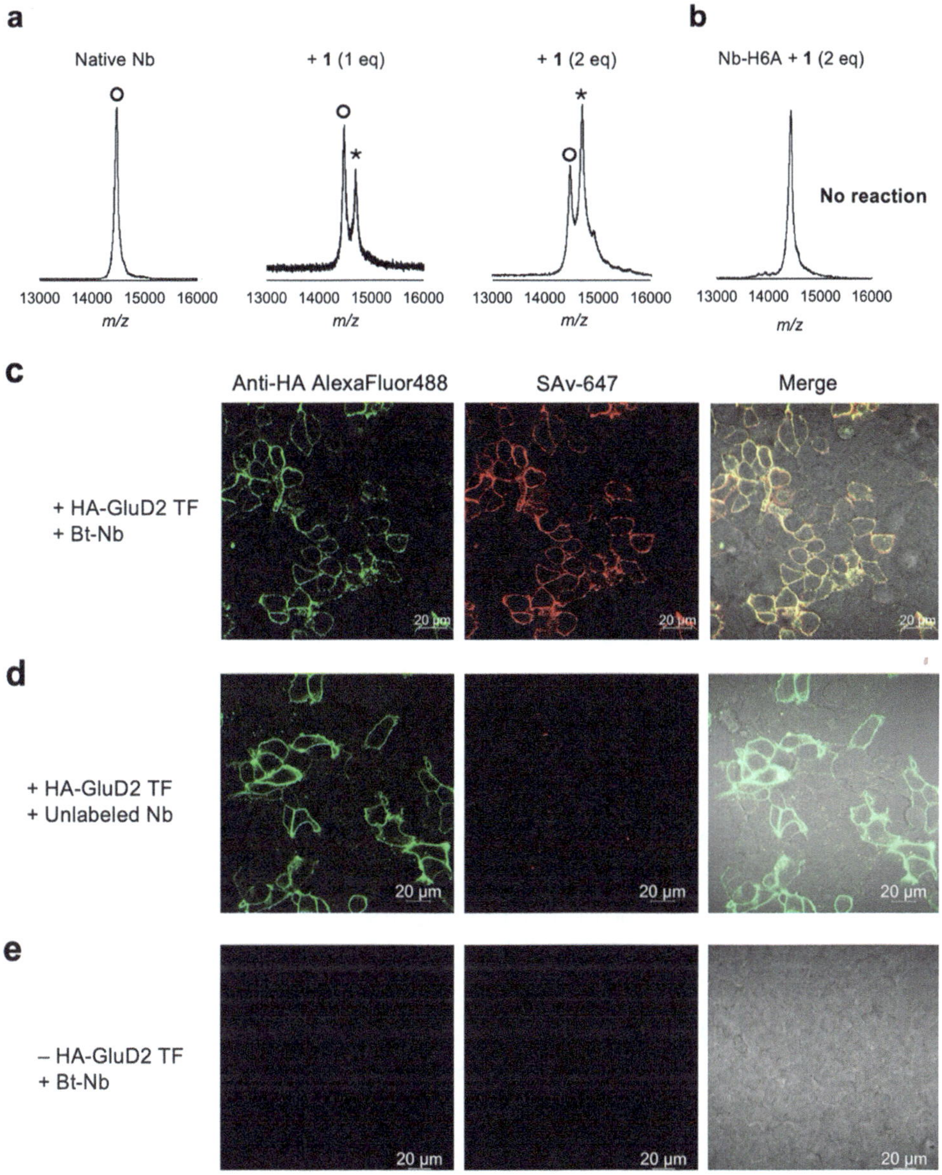

Fig. 5 MALDI-TOF MS analysis of conjugation of biotin functional group to anti-GluD2 nanobody tagged with H6K (**a**) or H6A (**b**) at its C-terminus. Reaction condition: 1 μM nanobody, 1–2 μM **1**, 2–4 μM $NiCl_2$, 37°C, PBS, pH 7.4, 1 h. ○, native nanobody (obsd *m/z*: 14476); *, biotinylated nanobody (obsd *m/z*: 14,702). Confocal imaging of HEK293T cells transiently expressing with HA-tagged GluD2. (**c**) Cell surface GluD2 was visualized by treatment of an anti-HA antibody conjugated with AlexaFluor488 (green) and the biotinylated nanobody (Bt-Nb) followed by SAv-647 (red). (**d**) Using non-biotinylated nanobody or (**e**) nontransfected cells, SAv-647 signals were not observed. TF, transfection. Excitation/Detection wavelength: 488 nm/410–617 nm for anti-HA AlexaFlur488, 640 nm/650–700 nm for SAv-647 (reproduced from [6] with permission from Elsevier)

14. Transfect the cells with 2 μg of pCAGGS(HA-GluD2) plasmid using Lipofectamine 2000 according to the manufacturer's instructions. Incubate at 37°C under 5% CO_2.
15. Coat the glass-bottomed dish with 100 μL poly-L-lysine and incubate for 5 min at room temperature.
16. Wash the dish with room temperature 200 μL PBS 3 times (*see* **Note 10**).
17. After 24–30 h post-transfection, carefully wash the transfected cells using PBS (1 mL). Add 400 μL trypsin-EDTA solution on top of cells. Incubate the cells at room temperature for about 2 min (if cells are difficult to detach, cells can also be incubated at 37°C or user can make use of commercially available cell scrapers).
18. Add 1 mL fresh medium and slowly pipette the cells to avoid any clumps. Count the cells using cell counter, place desired number of cells at the center of poly-L-lysine coated glass bottom dish. Add 1 mL fresh medium around the center and allow cells to attach to the bottom. Incubate at 37°C under 5% CO_2.
19. After overnight incubation, carefully remove the medium.
20. Place the dish on ice. Carefully wash the cells using PBS and add 100–200 μL of 1 μM nanobody-biotin conjugate to the center of the dish (*see* **Note 8**).
21. Continue incubating on ice for 3–5 min.
22. Remove the medium. Wash the cells one time using 100–200 μL DMEM-HEPES.
23. Add 100–200 μL of antibody fluorophore conjugate solution to the cells. Continue incubating on ice for 3 min.
24. Wash the cells on ice for two times using 100–200 μL DMEM-HEPES. Mount on microscope for confocal imaging (Fig. 5c-e).

4 Notes

1. DMSO should be free of moisture, as residual water can readily lead to the hydrolysis of the NASA warhead during long-term storage of the probe stock solution. Since moisture accumulation is inevitable during repeated freeze-thaw cycles, it is advisable to initially prepare small aliquots of the dry compound and prepare fresh DMSO stock solutions at regular intervals to minimize this risk. In the event of probe decomposition or nonspecific reactions become an issue, it might be worth testing the more stable ArNASA warhead developed by our group [9].

2. Ac-KH6AW was synthesized using standard Fmoc solid-phase peptide synthesis [10]. It can easily be purchased through custom peptide synthesis.
3. pCAGGS (HA-GluD2) plasmid was prepared by insertion of a cDNA encoding the amino acid sequence of mouse GluD2 (residues 20–1007) fused at the N-terminus with the sequence [MVLLVSLAAVANANTGYPYDVPDYAGGHHHHHHKGT] into the pCAGGS vector.
4. The reaction can be stopped at any time point by adding strong chelating agent such as EDTA. The reaction progress can be monitored by HPLC, and modified product can be characterized by MALDI-TOF MS.
5. The HPLC gradient should be optimized, as it can vary depending on factors such as the instrument used, column type, and other conditions.
6. The raw MS data files were analyzed by FlexAnalysis 3.4 (Bruker) to create peak lists based on the recorded spectra. The mass peak assigned as the biotinylated peptide was selected and subjected to MS/MS analysis with the LIFT mode with a precursor mass tolerance and a fragment ion mass tolerance of 0.5 Da. The MS/MS spectra were analyzed by BioTools and Sequence editor (Bruker).
7. Labeling conditions should be optimized based on the protein of interest. When two equivalents of Probe **1** were used with the GluD2 nanobody, we observed a minor population of the double-modified nanobody. However, we consider this an artifact, as no off-target labeling was observed when the lysine residue in H6K was mutated to alanine (H6A).
8. The anti-GluD2 nanobody was kindly provided by Prof. A. Radu Aricescu. The H6K-fused Nanobody was prepared according to a previous report [11].
9. After adding the nanobody, all subsequent steps should be performed quickly on ice. We observed internalization of GluD2 at higher temperatures or with extended incubation times. This effect may vary depending on the specific protein of interest.
10. One should be careful while adding reagent/washing steps since HEK93T cells easily detach from the glass-bottomed dish.

References

1. Krall N, da Cruz FP, Boutureira O, Bernardes GJL (2016) Site-selective protein-modification chemistry for basic biology and drug development. Nature Chem 8:103–113. https://doi.org/10.1038/nchem.2393
2. Horn JM, Obermeyer AC (2021) Genetic and covalent protein modification strategies to facilitate intracellular delivery. Biomacromolecules 22:4883–4904. https://doi.org/10.1021/acs.biomac.1c00745

3. Chalker JM, Bernardes GJL, Davis BG (2011) A “Tag-and-Modify” approach to site-selective protein modification. Acc Chem Res 44:730–741. https://doi.org/10.1021/ar200056q
4. Los GV, Encell LP, McDougall MG et al (2008) HaloTag: a novel protein labeling technology for cell imaging and protein analysis. ACS Chem Biol 3:373–382
5. Gronemeyer T, Chidley C, Juillerat A et al (2006) Directed evolution of O 6-alkylguanine-DNA alkyltransferase for applications in protein labeling. Protein Eng Des Sel 19:309–316
6. Thimaradka V, Oh JH, Heroven C et al (2021) Site-specific covalent labeling of His-tag fused proteins with N-acyl-N-alkyl sulfonamide reagent. Bioorg Med Chem 30:115947
7. Tamura T, Ueda T, Goto T et al (2018) Rapid labelling and covalent inhibition of intracellular native proteins using ligand-directed N-acyl-N-alkyl sulfonamide. Nat Commun 9: :1870. https://doi.org/10.1038/s41467-018-04343-0
8. Kiyonaka S, Kubota R, Michibata Y et al (2016) Allosteric activation of membrane-bound glutamate receptors using coordination chemistry within living cells. Nat Chem 8:958–967
9. Kawano M, Murakawa S, Higashiguchi K et al (2023) Lysine-reactive N-Acyl-N-aryl sulfonamide warheads: improved reaction properties and application in the covalent inhibition of an ibrutinib-resistant BTK mutant. J Am Chem Soc 145:26202–26212. https://doi.org/10.1021/jacs.3c08740
10. Fields GB, Noble RL (1990) Solid phase peptide synthesis utilizing 9-fluorenylmethoxycarbonyl amino acids. Int J Pept Protein Res 35:161–214. https://doi.org/10.1111/j.1399-3011.1990.tb00939.x
11. Pardon E, Laeremans T, Triest S et al (2014) A general protocol for the generation of Nanobodies for structural biology. Nat Protoc 9:674–693

Check for updates

Chapter 14

Chemoselective Histidine Bioconjugation Using Thiophosphorodichloridate Reagents

Xinyuan Li, Fung Kit Tang and Shang Jia

Abstract

Chemoselective bioconjugation techniques are essential tools for precise modification and functional analysis of proteins. Here, we describe a detailed protocol for histidine-specific labeling using thiophosphorodichloridate reagents, forming stable thiophosphoramidate linkages. The procedure includes synthesis of the labeling reagents, protein modification, purification by ion-exchange chromatography, and subsequent copper-catalyzed azide-alkyne cycloaddition (CuAAC) for further conjugation.

Keywords Histidine labeling, Chemoselective bioconjugation, Phosphohistidine, Phosphoramidate, Protein analysis, CuAAC

1 Introduction

Chemoselective bioconjugation serves as a powerful chemical approach for the investigation and expansion of protein functions [1–4]. While traditional chemoselective bioconjugation methods primarily target thiol groups on cysteine [5, 6] and amino groups on lysine [7]. Recent developments have significantly broadened this toolbox. Over the past two decades, innovative methods have been developed to chemoselectively label specific amino acids residues on native proteins, including traditionally targeted serine and cysteine residues via nucleophilic addition or transition-metal-mediated labeling [8, 9], as well as more recently methionine [10, 11], tryptophan [12, 13], glutamate [14], and tyrosine [15]. These advancements in the bioconjugation toolbox provides versatile methods for generating engineered proteins, enabling the development of new probes or pharmaceuticals [16–19]. Additionally, they facilitate the identification of novel drug/protein binding sites through profiling techniques and the study of the specific amino acid residues in protein functions [20–24]. Thus, the continuous development

Alexander Deiters (ed.), *Bioconjugation Methods*, Methods in Molecular Biology, vol. 3009,
https://doi.org/10.1007/978-1-0716-5112-4_14,

of site-selective bioconjugation methods is reshaping our ability to engineer and study protein function precisely.

Among these targeted amino acids, the selective modification of histidine has attracted much interest. On one hand, the low natural abundance of histidine in the proteome (~2.2%) inherently enhances labeling specificity when targeting this residue, enabling the creation of bioconjugates with more defined modification stoichiometries [25–27]. On the other hand, histidine is an essential amino acid residue that requires further analysis with the help of profiling methods. The imidazole side-chain of histidine features pKa around six with two nitrogen atoms, enabling it to simultaneously accept a proton on one side and donate another on the other, effectively acting as a proton shuttle in biological systems [28]. The metal binding capability of imidazole also makes histidine a frequently utilized residue in metalloproteins [29]. These features enable histidine to appear in over 20% of enzyme active sites, making it the most prevalent catalytic residue [27, 28].

Against this backdrop, a series of bioconjugation methods for histidine have been reported [30–33], enabling studies such as profiling of histidine reactivity and metal binding activity. Among these methods, labeling with electrophiles avoids the potential side reactions on reductive residues such as cysteine and methionine, but faces the competition from stronger nucleophiles such as lysine and cysteine that may require preblocking with certain methods [32].

This chapter details the method for histidine modification on native proteins using phosphorus(V) electrophiles, which is the first example of chemoselective labeling of single histidine residues in nondenaturing environments [30, 34]. This method uses thiophosphorodichloridate reagent to generate thiophosphoramidate linkage at the histidine residue, which is an analog of naturally occurring phosphohistidine. The labeling procedure generally comprises of two steps, where the first step introduces a bioconjugation handle onto the labeled histidine residue, and the second step introduces additional building blocks using this handle (Fig. 1). Since its development, this method has enabled the modification on spike proteins for vaccine development [35], identification of metal-binding site of metalloproteins [34], and immobilization of proteins for atomic force microscopy [36]. We expect this method to continue to facilitate research and engineering in situations requiring selective histidine bioconjugation.

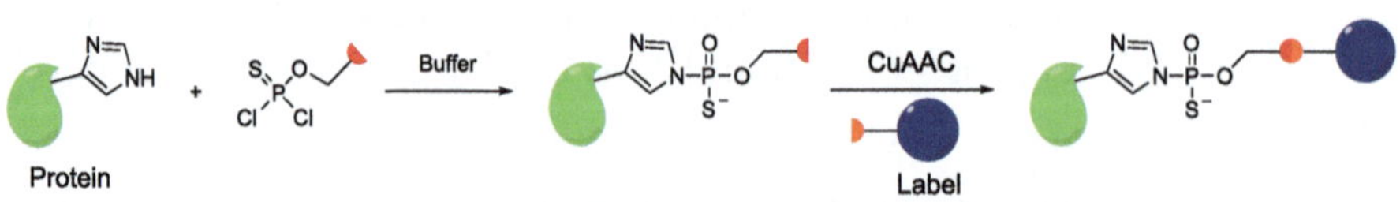

Fig. 1 Labeling of histidine residues on a protein via thiophosphorodichloridate tagging followed by click chemistry

2 Materials

2.1 Reagents for the Synthesis of Thiophosphorodichloridate

All reagents and solvents for organic synthesis can be purchased from commercial suppliers.

1. Thiophosphoryl chloride.
2. Alcohol: propargyl alcohol, 2-azidoethanol (synthesized [37] or commercial), or other (*see* **Note 1**).
3. Base: Potassium carbonate (K_2CO_3) or butyllithium (*n*-BuLi) (*see* **Note 1**).
4. Solvents: Dichloromethane (CH_2Cl_2), acetone, and tetrahydrofuran (THF).
5. Ethyl acetate and hexanes for column chromatography.

2.2 Reagents for Protein Labeling

1. HEPES buffer: 25 mM HEPES, pH 8.5. Weight 6.0 g HEPES and transfer to a 1 L media bottle or Erlenmeyer flask. Add deionized water to a volume of 900 mL. Mix and adjust pH to 8.5 by using 5 M NaOH (10 g NaOH in 50 mL deionized water). Make up to 1 L with deionized water. Store at room temperature.
2. Protein substrate solution: 20 μM RNase A in 25 mM HEPES, pH 8.5. Weigh 1.0 mg RNase A. Transfer to a 5 mL conical tube. Add 25 mM HEPES buffer, pH 8.5, to a final volume of 3655 μL, and mix until fully dissolved. Store at −20°C.
3. Stock solution of thiophosphorodichloridate reagent: 100 mM stock solution in acetonitrile. Weight 18.9 mg of propargyl thiophosphorodichloridate reagent (synthesized in Sect. 3.1) into a 1.5 mL microcentrifuge tube. Add a volume of 1 mL of dry acetonitrile and dissolve the liquid. Store at −20°C over Drierite.

2.3 Reagents for Protein Separation

1. Labeled protein (e.g., RNase A) product (from Sect. 3.2.1).
2. Elution Buffer 1: 25 mM Sodium succinate, pH 4.2. Weight 3.4 g sodium succinate dibasic hexahydrate and transfer to a 1 L Erlenmeyer flask. Add deionized water to a volume of 400 mL. Mix and adjust pH to 4.2 by using 1 M HCl (82.6 mL 37% HCl in 917.4 mL deionized water). Make up to 500 mL with deionized water and filter through 0.4 μm membrane into a media bottle. Store at room temperature.
3. Elution Buffer 2: 25 mM Sodium succinate with 1000 mM NaCl, pH 4.2. Weight 3.4 g sodium succinate dibasic hexahydrate and 29 g sodium chloride. Transfer everything to a 1 L glass beaker or Erlenmeyer flask. Add deionized water to a volume of 400 mL. Mix and adjust pH to 4.2 by using 1 M HCl

(82.6 mL 37% HCl in 917.4 mL deionized water). Make up to 500 mL with deionized water and filter through 0.4 μm membrane into a media bottle. Store at room temperature.

4. Desired buffer for buffer exchange.

2.4 Reagents for Click Reaction

1. Acetone.
2. 1 M HEPES buffer, pH 7.5 and 1 × PBS buffer: Both can be purchased directly from vendors. For PBS buffer, 10 × PBS buffer can be purchased as well. To make 1 × PBS buffer, add 100 mL 10 × PBS buffer in a 1 L glass beaker or Erlenmeyer flask. Add deionized water to a volume of 1 L and mix well. Store under room temperature.
3. PBS buffer with 0.1% SDS: 1× PBS buffer with 0.1% SDS, pH 7.4. Weight 20 mg sodium dodecyl sulfate (SDS). Transfer to a 50 mL conical tube containing 50 mL of store-bought 1× PBS buffer. Mix until fully dissolved. Store at room temperature.
4. Aminoguanidine Hydrochloride Stock Solution: 100 mM aminoguanidine hydrochloride stock solution. Weigh 11.16 mg aminoguanidine hydrochloride. Transfer to a 1.5 mL microcentrifuge tube. Add 1 mL deionized water and mix until fully dissolved. Make the solution before the experiment.
5. Copper (II) Sulfate Stock Solution: 10 mM $CuSO_4$. Weigh 2.5 mg copper(II) sulfate pentahydrate ($CuSO_4{\cdot}5\,H_2O$). Transfer to a 1.5 mL microcentrifuge tube. Add 10 mL deionized water and mix until fully dissolved. Store at room temperature.
6. THPTA or BTTAA Stock Solution: 50 mM THPTA or BTTAA. Weigh the appropriate amount of THPTA or BTTAA based on molecular weight. For THPTA, weigh 28.7 mg; for BTTAA, weigh 34.1 mg. Transfer to a 1.5 mL microcentrifuge tube. Add 1 mL deionized water and mix until fully dissolved. Store at –20°C.
7. Cy3-N_3 Stock Solution: 10 mM Cy3-N_3 in DMSO. Weigh 8.06 mg Cy3-azide (Cy3-N_3, MW = 806.2 g/mol). Transfer to a 1.5 mL amber microcentrifuge tube or a normal 1.5 mL microcentrifuge tube wrapped with aluminum foil. Add 1 mL anhydrous DMSO and mix until fully dissolved. Store at –20°C protected from light.
8. Sodium Ascorbate Stock Solution: 500 mM sodium ascorbate. Weigh 99.05 mg sodium ascorbate. Transfer to a 1.5 mL microcentrifuge tube. Add 1 mL deionized water and mix until fully dissolved. Make the solution before the experiment.
9. EDTA Solution: 50 mM EDTA. Weigh 18.6 mg disodium EDTA. Transfer to a 1.5 mL microcentrifuge tube. Add 1 mL

deionized water and mix until fully dissolved. Store at room temperature or 4°C.

2.5 Reagents for Digested Protein Analysis

1. Labeled protein (e.g., RNase A) product from Sect. 3.2.1.
2. Acetone.
3. HEPES Buffer 1: 25 mM HEPES, pH 8.0. Weigh 1.5 g HEPES free acid. Transfer to a 1 L glass beaker or Erlenmeyer flask. Add deionized water to a volume of 800 mL and mix until mostly dissolved. Adjust pH to 8.0 using 1 M NaOH (40 g NaOH in 1 L deionized water). Make up to 1 L with deionized water. Store at room temperature.
4. HEPES Buffer 2: 25 mM HEPES with 6 M urea, pH 8.0. Weigh 1.5 g HEPES free acid and 360.4 g urea. Transfer to a 1 L glass beaker or Erlenmeyer flask. Add deionized water to a volume of 800 mL and mix until fully dissolved. Adjust pH to 8.0 using 1 M NaOH (40 g NaOH in 1 L deionized water). Make up to 1 L with deionized water. Make this solution before the experiment.
5. TCEP Solution: 500 mM TCEP. Weigh 143 mg tris(2-carboxyethyl)phosphine hydrochloride (TCEP·HCl). Transfer to a 1.5 mL microcentrifuge tube. Add 1.0 mL deionized water and mix until fully dissolved. Make the solution before the experiment.
6. Iodoacetamide Solution: 1 M iodoacetamide. Weigh 185 mg iodoacetamide. Transfer to a 1.5 mL microcentrifuge tube wrapped with aluminum foil. Add 1.0 mL deionized water mix until fully dissolved. Prepare fresh immediately before use and protect from light.
7. Sequencing grade modified trypsin (Promega).

2.6 Instrumentation

1. NMR spectrometer (e.g., Bruker AVB-400 or AVQ-400, for chemical characterization).
2. LC-MS (e.g., Synapt G2-Si mass spectrometer coupled with an Acquity M-class LC system, for analysis of the degree of modification).
3. LC-MS/MS spectrometer with ETD fragmentation (e.g., LTQ-Orbitrap-XL mass spectrometer coupled with an UltiMate3000 RSLCnano LC system with a reversed-phase analytical column, for analysis of labeling site).
4. HPLC or FPLC with ion exchange column (e.g., HiTrap SP HP cation exchange chromatography column).
5. Fluorescence Gel Imaging System.

2.7 Software

1. Proteomics analysis software (e.g., Trans Proteomic Pipeline [38], free and open-sourced, for analysis of labeling site).
2. Mass spectra deconvolution software (e.g., UniDec [39], free and open-sourced, for analysis of mass of modified protein).
3. NMR analysis software (e.g., TopSpin).

3 Methods

3.1 Synthesis of Thiophosphorodichloridate Reagents

This part of the protocol provides details of synthesizing thiophosphorodichloridate reagents (Fig. 2). This protocol uses propargyl alcohol and 2-azidoethanol as examples, but other alcohols can be utilized to suit the specific needs. The following are examples of using propargyl alcohol and 2-azidoethanol. For other alcohols, the synthesis should be performed similarly (*see* **Note 1**).

3.1.1 Synthesis of Propargyl Thiophosphoryl Chloridate (1)

1. In a 50 mL round-bottom flask, add 10 mL of CH_2Cl_2 containing 1 mL of thiophosphoryl chloride ($PSCl_3$, 9.8 mmol).
2. Add 0.57 mL of propargyl alcohol (9.8 mmol) and 1.36 g potassium carbonate (K_2CO_3, 9.8 mmol) plus a magnetic stir bar into the flask.
3. Stir the solution on a magnetic stirrer overnight.
4. Filter the mixed solution through a Buchner funnel. Discard the solid and use a rotary evaporator to remove CH_2Cl_2 under a decreased pressure.
5. Purify the crude product by column chromatography using a 50:1 hexanes/ethyl acetate eluent (*see* **Note 2**). The final product **1** is a colorless oil with a pungent smell (*see* **Note 3**).

The structure and purity of product **1** can be characterized by ^{1}H NMR, ^{13}C NMR and ^{31}P NMR (*see* **Note 4**) and optionally HRMS with APCI source. ^{1}H NMR (400 MHz, $CDCl_3$) δ 4.94 (dd, J = 15.9, 2.5 Hz, 2 H), 2.72 (t, J = 2.5 Hz, 1 H). ^{13}C NMR (101 MHz, $CDCl_3$) δ 78.37, 75.48 (d, J = 11.1 Hz), 58.52 (d, J = 7.4 Hz). ^{31}P NMR (162 MHz, $CDCl_3$) δ 59.68. HRMS ($APCI^+$) m/z calcd 188.9092, found 188.9115 for $C_3H_4Cl_2OPS^+$ ($M+H^+$).

3.1.2 Synthesis of 2-Azidoethyl Phosphorodichloridothioate (2)

1. Dissolve 2-azidoethanol (0.43 g, 4.9 mmol) in dry THF (15 mL) under N_2.
2. Cool the solution using a dry ice/acetone bath.

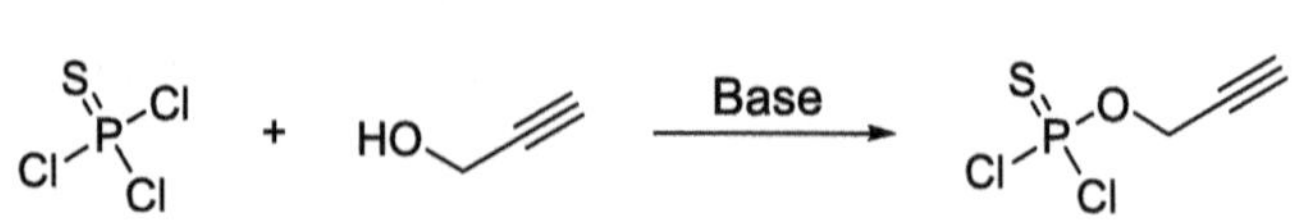

Fig. 2 Synthesis of thiophosphorodichloridate reagents

3. Add *n*-butyllithium (2.0 mL, 2.5 M in hexanes, 5 mmol) dropwise to the cooled solution.
4. Stir the resulting mixture at room temperature for 20 min to form the lithium salt.
5. In a separate flask, add dry THF (15 mL) and $PSCl_3$ (1.0 mL, 4.9 mmol) under N_2.
6. Cool this second flask in a dry ice/acetone bath.
7. Add the lithium salt solution dropwise into the cooled $PSCl_3$ solution.
8. Stir the combined reaction mixture at room temperature for 1 h.
9. Concentrate the reaction mixture under reduced pressure.
10. Dilute the residue with CH_2Cl_2 and filter to remove lithium chloride.
11. Purify the filtrate by column chromatography using a 50:1 hexanes/ethyl acetate eluent (*see* **Note 2**).
12. Collect product **2** as a light-yellow oil.

The structure and purity of product **2** can be characterized by ^{1}H NMR, ^{13}C NMR, ^{31}P NMR and optionally HRMS with APCI source. ^{1}H NMR (400 MHz, CDCl3) δ 4.45 (dt, J = 11.0, 5.1 Hz, 2 H), 3.63 (t, J = 4.4 Hz, 2 H). ^{13}C NMR (101 MHz, $CDCl_3$) δ 69.57 (d, J = 9.9 Hz), 50.11 (d, J = 10.6 Hz). ^{31}P NMR (162 MHz, $CDCl_3$) δ 59.05. HRMS ($APCI^+$) m/z calcd 191.9202, found 191.9219 for $C_2H_5Cl_2NOPS^+$ ($M-N_2+H^+$).

3.2 Protein Labeling with Thiophosphorodichloridate 1

The following protocol details the labeling of protein substrate (RNase A) with **1** (Fig. 3), with further conjugation on its alkyne handle. Steps can be adapted when labeling other proteins with other custom thiophosphorodichloridate reagents.

3.2.1 Labeling of Protein with Thiophosphorodichloridate 1

1. To a 1.5 mL microcentrifuge tube, add 2 μL stock solution of thiophosphorodichloridate reagent **1** (*see* **Note 5**).
2. Add 100 μL of 20 μM RNase A in 25 mM HEPES solution (*see* **Note 6**) into the tube containing the stock solution, followed by immediate pipetting or vortexing.
3. Incubate the reaction at room temperature for 1 h.

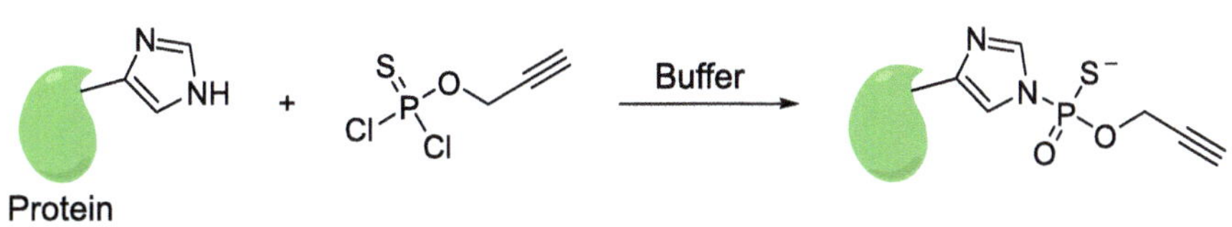

Fig. 3 Protein labeling with thiophosphorodichloridate (**1**)

4. Cleanup the protein by buffer exchange with ultracentrifugation or purify by ion exchange chromatography (*see* Sect. 3.2.2).
5. Confirm the labeling by protein mass spectrometry (*see* Sect. 3.3.1 and 3.3.2).

3.2.2 Separation of Proteins by Degree of Modification

Because the modification turns a slightly basic residue into a strong acid, such charge difference can be utilized for their separation by ion exchange chromatography. The following are steps separating protein substrate (RNase A) with different degrees of modification on a strong cation exchange column. For separation of other proteins, anion exchange columns can be used as well with appropriate buffers.

1. Buffer-exchange the **1**-labeled protein substrate (RNase A) reaction mixture into 25 mM sodium succinate buffer, pH 4.2.
2. Load the diluted sample onto a cation exchange chromatography column (1 mL).
3. Elute the protein using a linear gradient from 25 mM sodium succinate, pH 4.2 to 25 mM sodium succinate with 1 M NaCl, pH 4.2 at a flow rate of 1.0 mL/min over 40 min.
4. Collect the protein-containing fractions.
5. Concentrate the collected fractions.
6. Buffer-exchange the concentrated protein into the desired buffer for subsequent click reaction or LC/MS analysis.

The following is an example of chromatogram (Fig. 4) during the ion exchange separation.

3.2.3 Further Modification of 1-Labeled Protein with Copper-Catalyzed Cycloaddition (CuAAC) Reaction

We followed the protocol recommended by Finn et al. [40] and used Cy3-N_3 as the label for demonstration.

1. The protein labeled by **1** should be separated or buffered-exchanged into appropriate buffer to remove excess unreacted

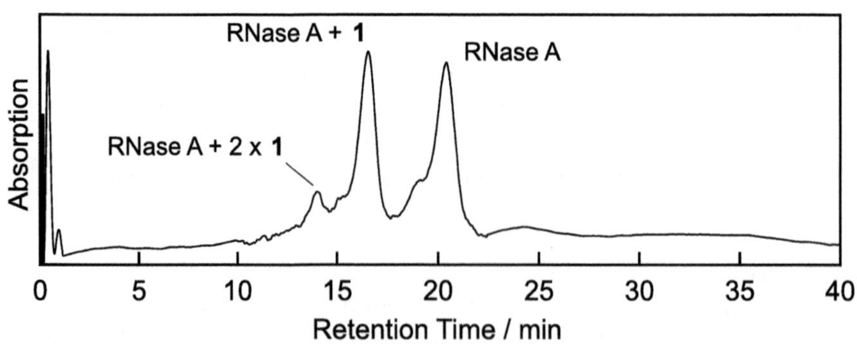

Fig. 4 Separation of protein (RNase A) labelled with **1** by ion exchange chromatography. Chromatogram at 280 nm shows the separation of the reaction mixture between the native RNase A, single **1**-labeled RNase A, and double **1**-labeled RNase A

reagent. Alternatively, for applications that tolerate denatured protein, the mixture can be precipitated with acetone and redissolved in PBS containing 0.1% SDS. The scale of the reaction can be alternated.

2. Add the following reagents to the protein solution from their 100× stock solution to a final concentration as follows:
   (a) 1 mM aminoguanidine hydrochloride
   (b) 100 μM $CuSO_4$
   (c) 500 μM THPTA or BTTAA
   (d) 100 μM azide parter (e.g., Cy3-N_3)
   (e) 5 mM sodium ascorbate (add last among these five reagents)
3. Mix thoroughly and incubate the reaction in the dark at room temperature for 1 h.
4. Quench the reaction with a final concentration of 500 μM EDTA by adding its 100× stock solution.
5. The mixture can be buffer-exchanged or desalted for further analysis or application.

3.3 Analysis of Labeled Protein

The labeled protein can be characterized by mass spectrometry, which is detailed below. For certain modifications, such as fluorescent labeling, the degree of labeling can also be qualitatively analyzed by SDS-PAGE followed by in-gel fluorescence.

3.3.1 Analysis of Degree of Modification by Mass Spectrometry

The labeling stoichiometry of proteins can be determined by intact protein mass spectrometry. The protein sample can be analyzed either by liquid chromatography/mass spectrometry (LC/MS) or by direct infusion following buffer exchange into an MS-compatible solution. The acquired mass spectra can be deconvoluted using software tools such as UniDec to convert the multiply charged ion series into a neutral mass spectrum. To confirm the presence of modifications, the experimental mass can be compared against the theoretical mass calculated by summing the unmodified protein mass with the expected mass shift from the modification(s). The relative yield of singly, doubly, or higher-modified species can be quantified by dividing the intensity of each modified peak by the summed intensity of all protein peaks in the deconvoluted spectrum.

The following example is the analysis of protein substrate (RNase A) labeled by **1** (Fig. 5). The unmodified protein substrate (RNase A) has a molecular weight of 13,683 Da. The singly labeled protein will have a mass shift of 13,816 Da, which is a 133 Da increase from the labeling agent, and doubly labeled protein has a molecular weight of 13,950 Da and so on.

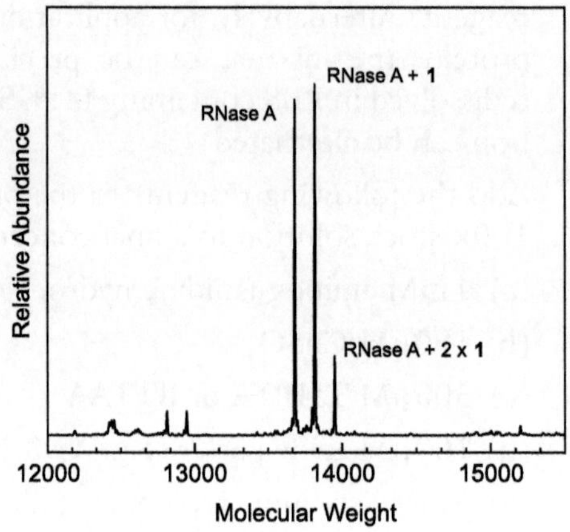

Fig. 5 Analysis of protein (RNase A) labeled with **1** by intact protein mass spectrometry. After deconvolution, the unmodified protein (RNase A) reports a molecular weight of 13,683 Da, while the single **1**-labeled and double **1**-labeled protein substrates show mass increase of 133 Da per labeling, corresponding to the covalent attachment of the labeling reagent

The following example is the analysis of CuAAC reaction between singly-modified protein substrate (RNase A) with Cy3 azide or desthiobiotin azide (Fig. 6), all showing expected mass that confirms the labeling.

3.3.2 Digested Protein Analysis

The site of modification can be determined by LC-MS/MS analysis of the digested protein.

1. 20 μL of resuspended RNase A (from Sect. 3.2.2) is labeled with 2 mM of **1**. Hydrolyzed byproducts of **1** are removed by

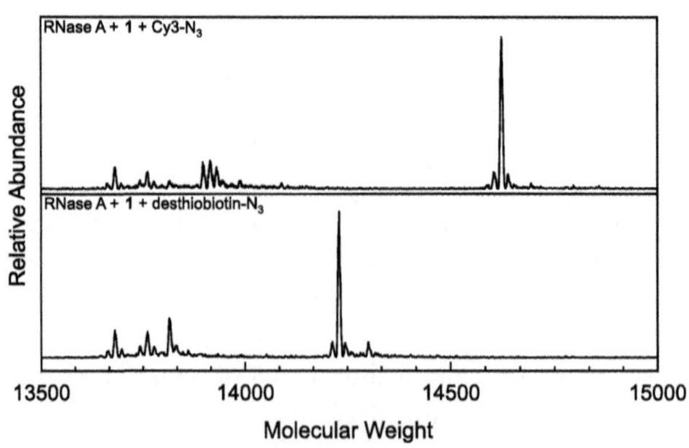

Fig. 6 Analysis of protein (RNase A) labeled with **1** and CuAAC by intact protein mass spectrometry. Products of single **1**-labeled RNase A with desthiobiotin-N_3 and Cy3-N_3 report mass of 14,231 Da and 14,623 Da, where the correct mass increase confirms successful CuAAC modification

1:4 acetone precipitation: Add 80 μL of cold acetone into the microcentrifuge tube containing the RNase A solution and freeze the mixture at −20°C for 30 min. Centrifuge the mixture at 17,000 × *g*, 4°C for 10 min.

2. The resulting pellet is dissolved in 20 μL of 25 mM HEPES buffer (pH 8.0) containing 6 M urea.
3. The sample is then reduced by adding 0.2 μL 100× stock solution of TCEP to a final concentration of 5 mM TCEP for 20 min and alkylated by adding 0.2 μL 100× stock solution of iodoacetamide to a final concentration of 10 mM iodoacetamide for 40 min at 37°C in the dark.
4. The protein is precipitated again using acetone, resuspended in 25 mM HEPES buffer (pH 8.0), and digested overnight with trypsin (1:20 dilution) at 37°C.
5. The samples are loaded into a reverse-phase LC-MS/MS system with an electrospray ionization (ESI) source, operated in the positive ion mode with ETD fragmentation for secondary MS analysis (*see* **Note 7**).
6. Proteomics software (such as Trans-Proteomic Pipline) can be used to search for the modification on the spectra data acquired. Correct mass shift of labeling needs to be set to run the search, and bundled tools can be used to analyze the modified peptides and the site of modification on each peptide. Note that the intensity of parent mass should be taken into consideration when analyzing the major site of modification, because the high sensitivity of mass spectrometry will reveal side reactions on other residues even if with low abundance.

4 Notes

1. Primary and secondary alcohols without competing nucleophilic sites can be incorporated into thiophosphorodichloridate reagents. To facilitate bioconjugation, such alcohol can carry bioconjugation handles such as terminal alkynes, azides or maleimides, for subsequent modification with corresponding conjugation moiety after histidine is labeled. It can also carry simple moieties such as PEG chains. Our first report focuses on the use of propargyl alcohol as the starting material. While propargyl is effective on model proteins, it experiences instability during copper-assisted click reactions on small molecule model reactions, likely due to metal-induced hydrolysis of the propargyl group [41]. To address this issue, terminal alkynes with longer methylene spacers are recommended to enhance the stability and effectiveness of bioconjugation reactions

[34]. The choice of base should be determined by the acidity of the alcohol. For primary alcohols with adjacent electron-ic-withdrawing groups, such as propargyl alcohol, and phenol derivatives, use of K_2CO_3 is sufficient as detailed in Sect. 3.1.1; otherwise, stronger base such as *n*-BuLi should be used to promote the reactivity as detailed in Sect. 3.1.2.

2. Unlike phosphorodichloridate, thiophosphorodichloridate is stable on TLC plate and silica columns during the course of synthesis owing to the reduced electronegativity from sulfur. The product can be analyzed via Thin-Layer-Chromatography (TLC). The product is generally visualized under 254 nm light and can also be verified by $KMnO_4$ stain.
3. Due to potential toxicity, all chemical procedures should be performed in a well-ventilated fume hood to minimize exposure to hazardous vapors and ensure safe handling of volatile or reactive compounds.
4. Despite the lack of added new protons and carbons comparing to the starting alcohol, the ^{1}H NMR signal of the α-proton should exhibit a large coupling constant (~15 Hz), and ^{13}C NMR signal of the α- and β-carbon should exhibit large coupling constants as well (~10 Hz). This is characteristic of nuclei positioned close to phosphorus. The ^{31}P NMR should show a single peak in the typical region for phosphorochloridate compounds (around 59 ppm), confirming the presence of the phosphorus center. These spectral features support the structure and connectivity of the synthesized compound.
5. To ensure efficient labeling, compound **1** is used at a relatively high concentration (2 mM), significantly exceeding the concentration of protein (20 μM). This is necessary because **1** is prone towards rapid hydrolysis in aqueous environments, producing hydrochloric acid as a byproduct. Specifically, the hydrolysis of 1 mM **1** can generate approximately 3 mM of acid, which can quench its activity and affect reaction efficiency.
6. We recommend using slightly basic buffer (pH 8.5) with higher concentration of buffering agent (25 mM HEPES). The labeling reaction is favored in slightly basic conditions. Moreover, the hydrolysis of the excess amount of thiophosphorodichloridate affords hydrochloric acid and thiophosphoric acid, which will acidify the buffer with insufficient buffering capacity.
7. The LC/MS-MS system must have ETD mode to detect the modification site of compound **1**. This is because a significant neutral loss of the phosphohistidine modification will occur during CID or HCD fragmentation, similar to the phosphorylated peptides [42]. Therefore, ETD mode is necessary to observe the peptide fragments containing modification with **1** in secondary MS.

Acknowledgments

We thank Rutgers, The State University of New Jersey, for supporting this work.

References

1. Stephanopoulos N, Francis MB, Stephanopoulos N, Francis MB (2011) Choosing an effective protein bioconjugation strategy. Nat Chem Biol 7:876–884
2. Boutureira O, Bernardes GJL (2015) Advances in chemical protein modification. Chem Rev 115:2174–2195
3. Koniev O, Wagner A, Koniev O, Wagner A (2015) Developments and recent advancements in the field of endogenous amino acid selective bond forming reactions for bioconjugation. Chem Soc Rev 44:5495–5551
4. Chauhan P, Ragendu V, Kumar M, Molla R, Mishra SD, Basa S, Rai V, Chauhan P, Ragendu V, Kumar M, Molla R, Mishra SD, Basa S, Rai V (2024) Chemical technology principles for selective bioconjugation of proteins and antibodies. Chem Soc Rev 53:380–449
5. Chalker JM, Bernardes GJL, Lin YA, Davis BG (2009) Chemical modification of proteins at cysteine: opportunities in chemistry and biology. Chem Asian J 4:630–640
6. Gunnoo SB, Madder A (2016) Chemical protein modification through cysteine. ChemBioChem 17:529–553
7. Matos MJ, Oliveira BL, Martínez-Sáez N, Guerreiro A, Cal PMSD, Bertoldo J, Maneiro M, Perkins E, Howard J, Deery MJ, Chalker JM, Corzana F, Jiménez-Osés G, Bernardes GJL (2018) Chemo- and regioselective lysine modification on native proteins. J Am Chem Soc 140:4004–4017
8. Liu Y, Patricelli MP, Cravatt BF (1999) Activity-based protein profiling: the serine hydrolases. PNAS 96:14694–14699
9. Vinogradova EV, Zhang C, Spokoyny AM, Pentelute BL, Buchwald SL (2015) Organometallic palladium reagents for cysteine bioconjugation. Nature 526:687–691
10. Taylor MT, Nelson JE, Suero MG, Gaunt MJ (2018) A protein functionalization platform based on selective reactions at methionine residues. Nature 562:563–568
11. Zang J, Chen Y, Zhu W, Lin S (2020) Chemoselective methionine bioconjugation on a polypeptide, protein, and proteome. Biochemistry 59:132–138
12. Tower SJ, Hetcher WJ, Myers TE, Kuehl NJ, Taylor MT (2020) Selective modification of tryptophan residues in peptides and proteins using a biomimetic electron transfer process. J Am Chem Soc 142:9112–9118
13. Xie X, Moon PJ, Crossley SWM, Bischoff AJ, He D, Li G, Dao N, Gonzalez-Valero A, Reeves AG, McKenna JM, Elledge SK, Wells JA, Toste FD, Chang CJ (2024) Oxidative cyclization reagents reveal tryptophan cation–π interactions. Nature 627:680–687
14. Jennings ML, Anderson MP (1987) Chemical modification and labeling of glutamate residues at the stilbenedisulfonate site of human red blood cell band 3 protein. J Biol Chem 262:1691–1697
15. Alvarez Dorta D, Deniaud D, Mével M, Gouin SG (2020) Tyrosine conjugation methods for protein labelling. Chem Eur J 26:14257–14269
16. Agarwal P, Bertozzi CR (2015) Site-specific antibody–drug conjugates: the nexus of bioorthogonal chemistry, protein engineering, and drug development. Bioconjug Chem 26:176–192
17. McCombs JR, Owen SC (2015) Antibody drug conjugates: design and selection of linker, payload and conjugation chemistry. AAPS J 17:339–351
18. Schumacher D, Hackenberger CPR, Leonhardt H, Helma J (2016) Current status: site-specific antibody drug conjugates. J Clin Immunol 36:100–107
19. Tsuchikama K, An Z (2016) Antibody-drug conjugates: recent advances in conjugation and linker chemistries. Protein Cell 9:33–46
20. Nomura DK, Dix MM, Cravatt BF (2010) Activity-based protein profiling for biochemical pathway discovery in cancer. Nat Rev Cancer 10:630–638
21. Niphakis MJ, Cravatt BF (2014) Enzyme inhibitor discovery by activity-based protein profiling. Annu Rev Biochem 83:341–377
22. Roberts AM, Ward CC, Nomura DK (2017) Activity-based protein profiling for mapping and pharmacologically interrogating proteome-wide ligandable hotspots. Curr Opin Biotechnol 43:25–33
23. Gavade A, Nagraj AK, Patel R, Pais R, Dhanure P, Scheele J, Seiz W, Patil J (2024) Understanding the specific implications of amino acids in the antibody development. Protein J 43:405–424

24. Niphakis MJ, Cravatt BF (2024) Ligand discovery by activity-based protein profiling. Cell Chem Biol 31:1636–1651
25. Dyer KF (1971) The quiet revolution: a new synthesis of biological knowledge. J Biol Educ 5:15–24
26. Tekaia F, Yeramian E, Dujon B (2002) Amino acid composition of genomes, lifestyles of organisms, and evolutionary trends: a global picture with correspondence analysis. Gene 297:51–60
27. Holliday GL, Almonacid DE, Bartlett GJ, O'Boyle NM, Torrance JW, Murray-Rust P, Mitchell JBO, Thornton JM (2006) MACiE (Mechanism, Annotation and Classification in Enzymes): novel tools for searching catalytic mechanisms. Nucleic Acids Res 35:D515–D520
28. GutteridgeA,ThorntonJM(2005)Understanding nature's catalytic toolkit. Trends Biochem Sci 30:622–629
29. Dokmanic I, Sikic M, Tomic S (2008) Metals in proteins: correlation between the metal-ion type, coordination number and the amino-acid residues involved in the coordination. Acta Cryst D 64:257–263
30. Jia S, He D, Chang CJ (2019) Bioinspired thiophosphorodichloridate reagents for chemoselective histidine bioconjugation. J Am Chem Soc 141:7294–7301
31. Wan C, Wang Y, Lian C, Chang Q, An Y, Chen J, Sun J, Hou Z, Yang D, Guo X, Yin F, Wang R, Li Z (2022) Histidine-specific bioconjugation via visible-light-promoted thioacetal activation. Chem Sci 13:8289–8296
32. Li J, Zhou J, Xu H, Tian K, Zhu H, Chen Y, Huang Y, Wang G, Gong Z, Qin H, Ye M (2023) ACR-based probe for the quantitative profiling of histidine reactivity in the human proteome. J Am Chem Soc 145:5252–5260
33. Zhai Y, Zhang X, Chen Z, Yan D, Zhu L, Zhang Z, Wang X, Tian K, Huang Y, Yang X, Sun W, Wang D, Tsai Y-H, Luo T, Li G (2024) Global profiling of functional histidines in live cells using small-molecule photosensitizer and chemical probe relay labelling. Nat Chem 16:1546–1557
34. Xiao T, He D, Liu D, Jia S, Chen Q, Silverman D, Maitra N, Huang AY, Nguyen TT, Rao G, Tillage R, Deng K, Weinshenker D, Britt RD, Kelly MJS, Dan Y, Chang CJ (2025) RAD23B acquires a copper metalloadaptor function in amphibian-to-reptile evolution to increase metabolism and regulate genomic integrity. Mol Cell 85:3443–3459
35. van der Velden YU, Grobben M, Caniels TG, Burger JA, Poniman M, Oomen M, Rijnstra ESV, Tejjani K, Guerra D, Kempers R, Stegmann T, van Gils MJ, Sanders RW, van der Velden YU, Grobben M, Caniels TG, Burger JA, Poniman M, Oomen M, Rijnstra ESV, Tejjani K, Guerra D, Kempers R, Stegmann T, van Gils MJ, Sanders RW (2022) A SARS-CoV-2 Wuhan spike virosome vaccine induces superior neutralization breadth compared to one using the Beta spike. Sci Rep 12:3884
36. Lei H, Zhang J, Li Y, Wang X, Qin M, Wang W, Cao Y (2022) Histidine-specific bioconjugation for single-molecule force spectroscopy. ACS Nano. J 16:15440–15449
37. Hounsou C, Margathe J-F, Oueslati N, Belhocine A, Dupuis E, Thomas C, Mann A, Ilien B, Rognan D, Trinquet E, Hibert M, Pin J-P, Bonnet D, Durroux T (2014) Time-resolved FRET binding assay to investigate hetero-oligomer binding properties: proof of concept with dopamine D1/D3 heterodimer. ACS Chem Biol 10:466–474
38. Deutsch EW, Mendoza L, Shteynberg DD, Hoopmann MR, Sun Z, Eng JK, Moritz RL (2023) Trans-proteomic pipeline: robust mass spectrometry-based proteomics data analysis suite. J Proteome Res 22:615–624
39. Marty MT, Baldwin AJ, Marklund EG, Hochberg GKA, Benesch JLP, Robinson CV (2015) Bayesian deconvolution of mass and ion mobility spectra: from binary interactions to polydisperse ensembles. Anal Chem 87:4370–4376
40. Hong V, Presolski SI, Ma C, Finn MG (2009) Analysis and optimization of copper-catalyzed azide–alkyne cycloaddition for bioconjugation. Angew Chem Int Ed 48:9879–9883
41. Sabatino V, Unnikrishnan VB, Bernardes GJL (2022) Transition metal mediated bioorthogonal release. Chem Catal 2:39–51
42. Boersema PJ, Mohammed S, Heck AJR (2009) Phosphopeptide fragmentation and analysis by mass spectrometry. J Mass Spectrom 44:861–878

Chapter 15

Reactive DNA Probes for Cross-Linking with Specific Amino Acids of Protein

Denise-Liu, Gracias Leone, Ambra Spampinato and Michal Hocek

Abstract

This chapter serves as a practical guide to designing and synthesizing reactive DNA probes for site-selective covalent cross-linking with proteins. By using enzymatic incorporation of modified nucleotides, researchers can create probes that selectively target amino acid residues such as cysteine, lysine, arginine, and tryptophan. The approach exploits the proximity effect for precise, reagent-free cross-linking under mild conditions, making it a powerful tool for mapping protein–DNA interactions and enabling stable bioconjugation for applications in chemical biology, diagnostics, and therapeutics.

Keywords DNA–protein crosslinks, Bioconjugation, Nucleotides, Oligonucleotides, Proteins

1 Introduction

Protein–DNA interactions play a crucial role in various biological processes [1], especially in transcription regulation [2]. Gaining insights into these interactions is the key to explain the mechanisms behind gene regulation, chromatin structure dynamics, and DNA repair pathways.

One of the most effective ways to investigate protein–DNA interactions is through covalent cross-linking [3], which stabilizes these complexes and enables comprehensive biochemical and structural analysis. Beyond their role in molecular biology research, covalent protein–DNA conjugates have also found applications in diverse fields, including nucleic acid therapeutics [4], chemical biology [5], nanotechnology [6], and bioanalysis [7]. These conjugates are widely explored for targeted drug delivery [8], biosensing [9, 10–12], and diagnostic applications [11, 12]. The potential applications of protein–DNA conjugates highlight the need for highly efficient and selective cross-linking strategies [13].

Traditional approaches to elucidate protein–DNA interaction often rely on nonspecific approaches such as photochemical

Alexander Deiters (ed.), *Bioconjugation Methods*, Methods in Molecular Biology, vol. 3009,
https://doi.org/10.1007/978-1-0716-5112-4_15,

cross-linking methods [14] that generate reactive radicals or carbenes in DNA. These reactive species randomly activate C–H bonds, which non-selectively react with nearby amino acids in proteins, leading to covalent conjugates that are not selective and are complicate the interpretation of the original protein structure.

In contrast, site-specific cross-linking methods allow to target specific amino acid residues within proteins. This high precision of the method is achieved through an attachment or incorporation of reactive functional groups that selectively react with certain amino acids under biocompatible conditions. This specificity is highly desirable for both studying protein–DNA interactions and for applications in bioconjugation, where specific and reproducible modifications are necessary. Additionally, reactive DNA modifications are frequently employed for postsynthetic labeling further expanding their utility in chemical biology and bioanalytical applications (Fig. 1).

A major limitation in developing site-specific cross-linking methods is the incompatibility of many reactive functional groups with traditional chemical phosphoramidite DNA synthesis. The instability of highly reactive groups often leads to degradation or undesired side reactions during oligonucleotide synthesis.

To overcome this problem, the enzymatic synthesis has emerged as a convenient alternative for incorporating reactive modifications into DNA [15, 16]. This approach relies on enzymatic DNA synthesis using 5-modified 2′-deoxyribonucleoside triphosphates (dNTPs) that carry reactive functional groups, which can be attached through cross-coupling reactions [17] or CuAAC click chemistry [18]. In our group, we have successfully developed the synthesis of modified dNTPs bearing vinylsulfonamide [19], acrylsulfonamide [20], chloroacetamide [21], squaramate [22], 1,3-diketones [23], glyoxal [24], or ABNOH [25], groups and their use for the polymerase synthesis of reactive DNA probes. Each of these modifications is designed to target a specific nucleophilic side amino acids chain.

Vinylsulfonamide and chloroacetamide-modified DNA can react with cysteine through Michael addition or nucleophilic substitution, respectively (Fig. 2a, b). Notably, chloroacetamide can alkylate cysteine but also modify histidine residues, expanding its utility for protein cross-linking. Squaramate-linked DNA has been introduced for the

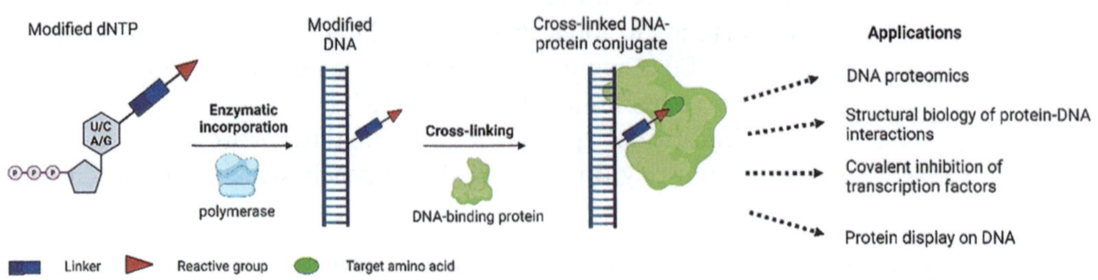

Fig. 1 General approach introduced in this chapter for site-specific cross-linking DNA and applications

formation of stable covalent conjugates with lysine-containing peptides and proteins without the need for external reagents (Fig. 2c). Arginine is an important nucleophilic amino acid commonly found in DNA-binding proteins. It has a guanidinium functional group, which presents challenges for selective modification. However, specific cross-linking strategies utilizing 1,3-diketones or glyoxal-functionalized DNA have been developed to covalently interact with arginine residues, through formation of pyrimidine or imizadolone ring, respectively (Fig. 2d, e).

A particular recent attention has been paid to targeting non-nucleophilic amino acid, i.e., Trp. The most useful reported reagent was the 9-azabicyclo[3.3.1]nonan-3-one (ABNO) 9-oxoammonium cation that is typically generated in situ by oxidation of the corresponding *N*-oxyl radical (ABNO) or hydroxylamine (ABNOH) by $NaNO_2$ and O_2, which reacts further with tryptophan. Therefore, ABNOH-linked DNA has been developed for cross-linking and bioconjugations with Trp-containing peptides and proteins (Fig. 2f).

A key driving force in the efficient covalent cross-linking reactions of the binding proteins with reactive DNA is the proximity effect. Only in case of proper recognition of the DNA sequence by the protein and when the reactive group is positioned near the target amino acid, the cross-link is formed with high conversion.

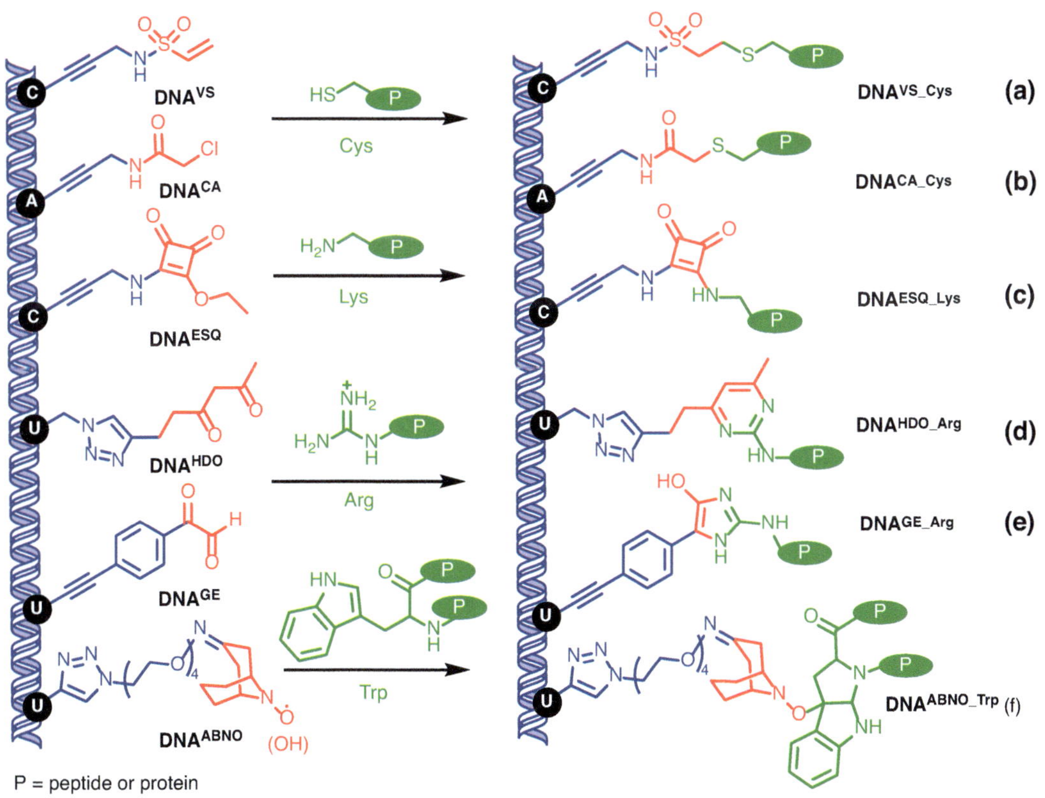

Fig. 2 Cross-linking groups and their conjugates

This precise spatial requirement minimizes off-target reactions and could be used for mapping of protein–DNA interaction. In contrast, cross-linking with peptides containing the same amino acid (but lack the affinity and proximity effect) needs a very high excess of the peptide and gives low conversions.

The presented advanced bioconjugation strategies may have diverse practical applications in chemical biology, biosensing, and targeted delivery of nucleic acid therapeutics, especially for their stability and specificity. These bioconjugation strategies enable the formation of stable, covalent protein–DNA conjugates in a reagent-free manner. This approach combines the precision of enzymatic synthesis and the proximity effect to enhance reactivity, overcoming the limitations of traditional methods. As a result, it provides a powerful tool for studying protein–DNA interactions and has a potential for the development of covalent inhibitors of transcription factors or for protein display on DNA.

2 Materials

2.1 Materials: Chemical Synthesis

Reagents: Chemistry

- **PO(OMe)$_3$**: the solvent needs to be distilled from a commercially available source [>95%–99% purity] (*see* **Note 1**) and stored under argon at 4°C and under molecular sieves.
- **POCl$_3$**: the reagent needs to be distilled from a commercially available source [with > 95%–99% purity] and stored at 4°C under argon (*see* **Note 2**).
- **Bis(tributylammonium) pyrophosphate (Bu$_3$NH)$_2$H$_2$P$_2$O$_7$**: Na$_4$P$_2$O$_7$ *10 H$_2$O; Dowex 50W; Pyridine, H$_2$O, HCl (*see* **Note 3**).
- **Triethylamine.**
- **CO$_2$.**
- **Diethyl squarate (ESQ)** is commercially available.
- ***p*-Bromo glyoxal (GE)** is commercially available.
- **Miscellaneous:** all chemicals (Et$_3$N, 2-chloroethanesulfonyl chloride, propargylamine, Chloroacetyl chloride, acetylacetone, BuLi, propargyl bromide, NaH, PO(OMe)$_3$, POCl$_3$, (NHBu$_3$)$_2$H$_2$P$_2$O$_7$, diethyl squarate, *p*-bromo glyoxal, CuI, Pd(OAc)$_2$, TPPTS, *N*-(propargyl)vinylsulfonamide, CuI, Pd(OAc)$_2$, N-(propargyl)chloroacetamide, CuSO$_4$(II)*5H$_2$O sodium ascorbate, oct-7-yne-2,4-dione (HDO), keto-ABNO, TBTA) purchased are reagent grade, Milli Q water is used in all the preparations. **Nucleotides**: 5-iodo-deoxyadenine triphosphate (**dA^{I}TP**) 5-propargylamino-2′-deoxycitidine triphosphate (**dC^{NH2}TP**), 5-azido-2′-deoxythymidine triphosphate

(**dT^{N3}TP**), 5-phenylethynylglyoxal-2′-deoxyuridine triphosphate (**dU^{GE}TP**), 5-ethynyl-2′-deoxyuridine triphosphate (**dU^{E}TP**), 5-iodo-2′-deoxycitidine triphosphate (**dC^{I}TP**).

Equipment: Chemistry

- For **PO(OMe)$_3$**: distillation apparatus, three bottle-neck adaptor, thermometer, oil bath, distillation flask, three receiving flasks with activated molecular sieves in it;
- For **POCl$_3$**: distillation apparatus, thermometer, oil bath, distillation flask, receiving flasks;
- For **Bis(tributylammonium) pyrophosphate (Bu$_3$NH)$_2$H$_2$P$_2$O$_7$**: 5 L glass column (diameter ca. 10 cm), 1–2 L flasks, 5 L glass bottle;
- For **TEAB 2 M**: Carbon dioxide pressure bottle with low-pressure gas regulator model SGT500/ 40; pH-meter, 3 L cylindrical glass pressure reactor; large (2/3 of the diameter of the glass reactor) magnetic stir bar (*see* **Note 4**).

Miscellaneous:

Silica gel 40–63 μm is used as stationary phase in flash-column chromatography.

TLC Silica gel 60 F254 is used to monitor the reactions;

HPFC (High-performance flash chromatography);

Biotage SP1 apparatus on DEAE SEPHADEX A-25 sodium form columns;

ISCO Combiflash Rf system on RediSep Rf Gold Silica Gel Disposable columns;

HPLC (Waters modular HPLC system);

HPLC Columns: Luna X-Bridge-prep (RP 18.5 μm OBD 19 × 150 mm), Kinetex (RP 18.5 μm EVO C18 100 LC-Column 250 × 21.2 mm).

2.2 Materials Biochemistry

Reagents: Biochemistry

Gels for Electrophoresis:

- 12.5% resolving PAGE: acrylamide/methylenebisacrylamide 19:1 (40%, 19 mL), 2× TBE (30 mL), H_2O (15 mL), TEMED (24 μL), 10% APS (480 μL).

- 20% resolving PAGE: acrylamide/methylenebisacrylamide 19:1 (40%, 30 mL), 5× TBE (30 mL), TEMED (24 μL), 10% APS (480 μL).
- 5% stacking SDS-PAGE: 30% acrylamide (37.5:1 mono:bis) (500 μL), 1 M TRIS (HCl) pH 6.8 (375 μL), H_2O (2100 μL), 20% SDS (14.2 μL), 10% APS (15 μL), TEMED (3 μL).
- 12% resolving SDS-PAGE: 30% acrylamide (37.5:1 mono:bis) (2414.5 μL), 1 M TRIS (HCl) pH 8.8 (2240 μL), H_2O (725.2 mL), 20% SDS (32.8 μL), 10% APS (20 μL), TEMED (4 μL).
- 20% resolving SDS-PAGE: 30 % acrylamide (37.5:1 mono:bis) (3018 μL), 1 M TRIS (HCl) pH 6.8 (2240 μL), H_2O (2100 mL), 20% SDS (32.8 μL), 10% APS (20 μL), TEMED (4 μL).

Conditions for PAGE runs: Preheating: 35 mA, 50°C, 30 min. Run: 35–42 mA, 50°C, 1-3 h.

Conditions for SDS-PAGE runs: 120–200 V, 25°C, 1 h.

Synthetic oligonucleotides: unmodified and modified primers and templates are purchased (Generi Biotech).

Natural nucleoside triphosphates: dATP, dGTP, dTTP, and dCTP are purchased (Thermo Scientific).

Cross-linking proteins: Full length p53 proteins (p53fl, residues 1–393), p53fl_R248W and p53fl_R249S need to be expressed from pT7-7p53fl vectors and purified without tag according to protocol described previously [13]; BSA and histone human recombinant proteins (H2A, H2B, H3.1, and H4) are purchased (NEB); *bs*GntR (amino acid residues 2–243) is cloned into pET151/D-TOPO expression plasmid and the recombinant protein is overexpressed in *E. coli* BL21 (DE3) [7].

DNA Polymerases: KOD XL DNA polymerase and correspondent polymerase reaction buffer are purchased (Merck Millipore).

Buffers:

- General buffers: binding buffer (3 × 200 μL; 10 mM Tris, 1 mM EDTA, 100 mM NaCl, pH 7.5), 1× TBE (1× solution containing 89 mM Tris, 89 mM boric acid, 2 mM EDTA, pH ~8.3), washing buffer (10 mM Tris, 1 mM EDTA, 500 mM NaCl, pH 7.5).
- For cross-linking: TrisHCl buffer (0.05 M, pH 10), TEAA buffer (0.3 M, pH 8.3), borate buffer (0.5 M, pH 9), phosphate buffer

NaH_2PO_4/Na_2HPO_4 (4.5 mM, pH 7.4), $NaHCO_3$ buffer (0.5 M, pH 10), HEPES/$KHCO_3$ buffer (0.05 M, pH 8.5), VP buffer (50 mM Tris, 0.1% Triton-X100, pH 7.6; 2 μL).

Loading Buffers:

- PAGE stop solution: 95% [v/v] formamide, 0.5 mM EDTA, 0.025% [w/v] bromophenol blue, 0.025% [w/v] xylene cyanol, 0.025% SDS);
- TBE (Tris/Borate/EDTA): 54 g of Tris base, 27.5 g of boronic acid, 20 mL of 0.5 M EDTA. This buffer is used to run the PAGE gels.
- VPS loading buffer: 0.05 M TRIS (pH 6.8), 17% glycerol, 16 mM mercaptoethanol, 3.5% SDS.

Equipment: Biochemistry

Equipment for gel run/analysis: Typhoon FLA 9500 (GE Healthcare); PowerPac™ HV Power supply (Bio-Rad, 850 VA, 50/60 Hz, 8 A Type T); Electrophoresis glass plates (40 × 20 × 20 × 0.5 cm), Bio Rad SDS-PAGE apparatus.

Equipment for purification of modified oligonucleotides: QIAquick Nucleotide Removal Kit QIAGEN (Merk life Sciences); Streptavidin magnetic particles were purchased (Merk life Sciences).

Equipment for conjugate mass analysis: bioZen Intact C4 column (Phenomenex), conjugates are separated by gradient of acetonitrile in water (both mobile phases modified by 0.1% formic acid). The separation is carried out by LC system (I-class, Waters) coupled to Mass Spectrometer (Synapt G2, Waters) to acquire m/z by positive electrospray ionization. Raw mass spectra of chromatographic peaks containing conjugate are combined, subtracted, and deconvoluted by MaxEnt1 algorithm.

Miscellaneous: Every protein gel has as a reference the Thermo Scientific PageRuler Prestained Protein Ladder, which is a mixture of 10 blue-, orange-, and green-stained proteins (10 to 180 kDa). All buffers are prepared with milliQ water. Other chemicals used are of analytical grade.

3 Methods: Chemistry

All glassware is oven-dried before every use. All reactions are performed under argon atmosphere (also specified in the text) beforehand flushing several times the septum-sealed vial or flask with argon/vacuum by inserting a needle through septum (avoiding its opening to the air). *p*-Anisaldehyde staining is used to reveal the spots after drying the TLC plate with heat-gun (preparation in **Note 4**).

3.1 Bis(tributyl-ammonium) Pyrophosphate $(Bu_3NH)_2H_2P_2O_7$

The reagent needs to be prepared in advance according to the method described in the next section starting from commercially available $Na_4P_2O_7$ *10 H_2O (*see* **Note 3**). This procedure takes one week and needs proper equipment (*see* **Note 5**).

Procedure: The cation exchange is performed using the Dowex 50W-X8, mesh 50–100 anionic resin. The resin (1.5 kg) is usually regenerated after each use (see below regeneration procedure) and stored in H_2O at 25°C in a plastic bottle. After pouring carefully the resin in 5L glass column follow the procedure:

1. Wash the regenerated resin with pyridine (3 L, 20% solution in H_2O), at the end of the washing, leave the resin in the solution over the weekend directly on the column.
2. Dissolve $Na_4P_2O_7$ *10 H_2O (44.6 g, 100 mmol) in 0.5 L of 20% pyridine and heat it to 50–55°C (it will not be completely dissolved); Decant the dissolved fraction on the column (do not cool it down) and start collecting. Dissolve the remaining solid in another 0.5 L of 20% pyridine, again at 50–55°C and pour it onto the column when the first part is almost drained. After the second portion is drained, wash it with 20 % pyridine (3.0 L). Collect the eluate into a 5 L flask, which is cooled to 0°C with ice (use bucket, shake it sometimes). Elute it from the column slowly (no pressure applied).
3. Pour Bu_3N (60 mL, 250 mmol) into the collection flask, shake it vigorously and then put it at 4°C for 16–18 h to allow the cation exchange.
4. Divide the entire solution in 1 L aliquots and evaporate until it is almost dry with a consistency of glue (smells quite a bit, it takes approx. 1 h, better to have rotavators under fume hood or open windows).
5. Transfer the evaporated fractions into one flask (wash the flasks with 2 × H_2O and 2 × MeCN) and again evaporate it until it is almost dry. Then add approx. 10 mL of DMF and again evaporate until almost dry (to remove the remaining mixture of H_2O/pyridine). Now it is ready for crystallization.
6. To crystallize the pyrophosphate, dilute the solution with acetone (approx. 50 mL, depending on the amount, as low volume as possible) and then add Et_2O slowly until the solution starts to be cloudy (approx. 15 mL). The ratio of acetone/Et_2O is around 2 to 1. Then put the flask into a freezer and wait at least 2 days. Collect and dry the crystals in a high

vacuum for 30 min. Store the crystals at −20 °C under argon. It is advisable to repeat the crystallization at least once more.

7. It is good to measure the purity (amount of tributylammonium salt) of prepared pyrophosphate by elemental analysis.
8. Afterward, regenerate the resin according to the following protocol:

Regeneration of the resin (Dowex 50 W):

1. Wash the resin with H_2O (2 L, to get rid of most of the pyridine/water);
2. Wash the resin with NaOH (2.5–3 L, 2 M solution in H_2O, or until pH >13). At the end of the washing, leave it 16–18 h (covered);
3. Wash the resin with H_2O (3 L, or until pH = 7);
4. Regenerate the resin with HCl (2.5–3 L, 2 M solution in H_2O, or until pH <1);
5. Wash the resin with H_2O (2–3 L, or until pH = 7).

Activation of dry resin prior to first use (Dowex 50 W):

1. HCl (2.5–3 L, 2 M solution in H_2O, or until pH <1);
2. H_2O (2–3 L, or until pH = 7).

3.2 TEAB 2 M

For the preparation, follow exactly the published procedure reported below [26].

Procedure:

1. Check a 3 L cylindrical glass pressure reactor for cracks and then fill with triethylamine (556 mL, 4 mol), deionized water (1480 mL), and a large magnetic stir bar (2–3″).
2. After sealing, place the reactor in an ice-water bath for cooling and stir rapidly (>1000 rpm). The headspace is briefly flushed with CO_2, then pressurized to 20–25 psi. After 45–60 min, gas flow slows and a clear to slightly cloudy 2 M TEAB solution formed (internal temperature: 20 ± 5°C).
3. Upon venting the pressure, measure the pH which should be at 7.6 ± 0.2, and the solution mass increase by 172–178 g (98%–101% of expected CO_2 uptake, $n = 6$).
4. Store the resulting 2 M TEAB stock can be stored at 4°C in a sealed serum bottle for up to two days and diluted as needed for chromatography. This procedure takes 1–3 h and needs proper equipment.

3.3 Reactive Moieties

Below is reported the protocol to synthesize the reactive moieties, for more info, check the cited source (Fig. 3).

3.3.1 N-(propargyl) vinylsulfonamide (VS) [19, 20]

1. Cool down (with ice until 0°C) dry DCM (8 mL) in an appropriate glass flask with septa, under argon atmosphere;
2. Add dropwise Et_3N (0.75 mL, 5 mmol) and then add dropwise 2-chloroethanesulfonyl chloride (0.50 mL, 0.55 mmol);
3. Add dropwise propargylamine (0.32 mL, 5 mmol) and stir the resulting yellow solution for 5 min keeping the flask in the ice;
4. Add 5 mL of brine to quench the reaction;
5. In a separatory funnel wash the organic layer (DCM) with 10 mL of brine two times;
6. Collect the organic layer, dry it over $MgSO_4$ and evaporate under reduced pressure;
7. Purify the product by flash column chromatography (hexane/ethyl acetate, 1:1);
8. Collect all fractions and evaporate under reduced pressure; the product appears as an orange oil (yield: 41%).

3.3.2 N-(propargyl) chloroacetamide (CA) [21]

1. Cool down (in an ice bath and a few spoons of NaCl in order to reach −5°C) 30 mL of dry THF and add chloroacetyl chloride (5.97mL ,75 mmol);
2. Mix propargylamine (3.47 mL, 50.6 mmol) and dry Et_3N (8.36 mL, 60 mmol) in 50 mL of THF, then stir for 2 h at −5°C;
3. Add 40 mL of 0.1 HCl and transfer in a separation funnel;
4. Extract x three times with 25 mL $CHCl_3$, collect and dry the organic phase over Mg_2SO_4;
5. Evaporate under reduced pressure until dryness and purify the resulting brown solid using dry column vacuum chromatography (DCVC) with 3 cm diameter column using a gradient 0–50% EtOAc in Hexane, after evaporation the product appears as a white solid (yield: 60%).

3.3.3 Oct-7-yne-2,4-dione (HDO) [23]

1. Mix NaH (65.9 mg, 2.74 mmol), acetylacetone (250 mg, 2.69 mmol), in 2.5 mL of dry THF in an appropriate dried flask under argon atmosphere and stir the solution for 10 min;

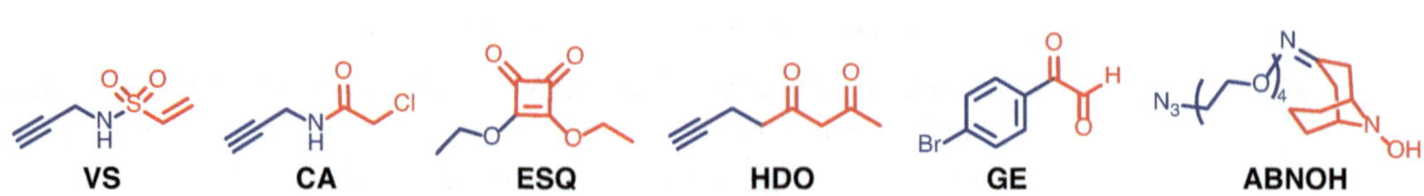

Fig. 3 Scheme of the reactive moieties. (VS = N-(propargyl)vinylsulfonamide; CA= N-(propargyl)chloroacetamide; ESQ = Diethyl squarate; HDO = Oct-7-yne-2,4-dione; GE= pBromo glyoxal; ABNOH = ABNOH-PEG4-N3)

2. After that, cool the reaction to −78°C (using a mixture of dry ice and acetone as bath) and slowly add dropwise BuLi (1.56 mL, 1.6 M in THF) and stir the solution for 30 min;
3. Add propargyl bromide (306 μL, 2.75 mmol), leave the reaction to warm up to 25°C and stir for 3 h;
4. Quench the solution with 10 mL of NH_4Cl at 0°C, dilute with EtOAc 10 mL and wash three times with H_2O and brine in a separation funnel. Collect the organic layer and dry it under $MgSO_4$;
5. Evaporate the solvent and purify the product by flash-chromatography using linear gradient DCM/MeOH (0–20%) which will appear as a yellow oil (yield: 34%).

3.3.4 ABNOH-PEG$_4$-N$_3$ (ABNOH) [25]

1. Dissolve ATD (82 mg, 0.216 mmol) in 11.2 mL of dry EtOH (>99% purity, 11.2 mL) in an appropriate dried septum sealed flask, by flushing it with vacuum (using a needle through the septum);
2. Leave the reaction mixture stirring for 26 h at room temperature;
3. Filter the resulting white precipitate and then evaporate the filtrate solution under reduced pressure;
4. Add to the evaporated residue hexane/$CHCl_3$ (1:1), filter off again the white precipitate and evaporate the filtrate under reduced pressure;
5. Repeat the previous point (4) twice till the formation of a colorless oil;
6. Dissolve the colorless oil in degassed methanol (11.3 mL), while flushing argon into the flask. Ensure that the air exits the flask through a needle puncture in the cap;
7. Stir the reaction mixture placing the flask in oil silicone bath and heat up reaching 50°C;
8. Add keto-ABNOH (38.43 mg, 248 mmol) and AcOH (26 μL, 0.455 mmol) to the reaction mixture and leave it stirring for 16 h at 50°C;
9. Evaporate the reaction solution under reduced pressure and purify the residue by flash column chromatography using CH_2Cl_2/EtOAc (3:2) to EtOAc/MeOH (5:1) as eluent. The yield of **ABNOH-PEG$_3$-N$_3$**, obtained as colorless oil, is 26% over two steps.

3.4 General Protocol for Triphosphorylation of Modified Nucleosides

The procedure for the triphosphorylation is common to each starting nucleoside (ca. 200 mg scale) (Fig. 4).

1. In a septum sealed vial dry the nucleoside powder (0.700 mmol) under vacuum at 25°C (mix A), by the use of a needle through the septum, for 16–18 h (*see* **Note 7**);

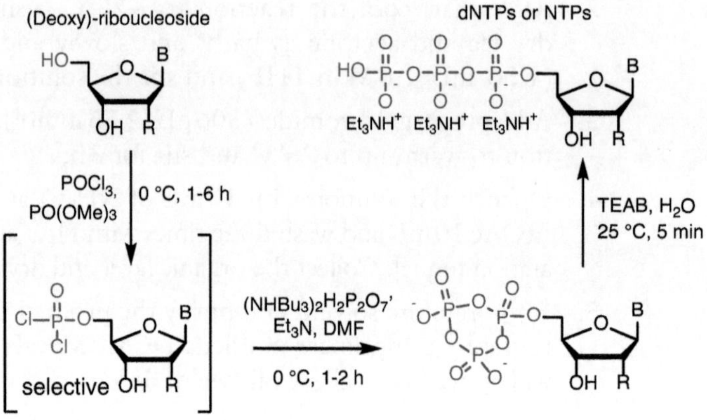

R = H or OH

B =natural nucleobase (A,G,C,T)
or modified nucleobase (A^X,G^X,C^X,T^X)

Fig. 4 Synthetic scheme of triphosphorylation procedure

2. Cool the vial on ice (0°C) and add freshly distilled $PO(OMe)_3$ (1.5–1.7 mL, keeping the concentration of nucleoside around 0.4–0.5 M depending on its solubility);
3. Add $POCl_3$ (1.07 mmol, 100 μL) with a Hamilton syringe (*see* **Note 8**);
4. Stir the reaction for 1–6 h at 0°C and check via TLC its progression (*see* **Note 9**);
5. In a separate sealed vial (mix B) dissolve $(NHBu_3)_2H_2P_2O_7$ (2.6 g, 4.9 mmol) and Bu_3N (4.3 mmol, 600 μL) in 6 mL of dry DMF under argon atmosphere;
6. After cooling the mix B to 0°C add it dropwise to mix A. Stir the resulting mixture for 2 h at 0°C;
7. Quench the reaction by slow addition of 1 mL TEAB 2 M while stirring for 5 min (*see* **Note 10**);
8. **Purification 1**: The reaction needs a pre-purification by H_2O/TEAB 2 M (0%–60%) buffer as eluent, using Biotage SP1 apparatus on DEAE SEPHADEX A-25 (*see* **Note 11**);
9. **Purification 2**: Concentrate the crude product and remove excess TEAB by evaporation and co-evaporation by H_2O. The triphosphate fraction from Purification 1 was purified on RP-HPLC with use of linear gradient of 0.1 M TEAB to 0.1 M TEAB/MeOH (1:1) to MeOH as eluent (*see* **Note 12**);
10. Evaporate, under reduced pressure, the fraction and co-evaporate with water several times to remove TEAB as much as possible by freeze-drying;

11. Check the purity of triphosphate by ^{1}H-NMR, ^{31}P-NMR, and MS (*see* **Note 13**);
12. The isolated yield of triphosphate is usually around 30%–40% (*see* **Note 14**).

3.5 Synthesis of Modified Triphosphates for Cross-Linking Reactions

3.5.1 Synthesis of dC^{VS}TP [19, 20]

dC^{I}TP was synthesized according to general protocol for triphosphorylation of modified nucleosides, starting from **dC^{I}** (see Fig. 4).

1. In a septum sealed vial dissolve in previously degassed water/acetonitrile (2:1, 2.5 mL), **dC^{I}TP** (30 mg, 0.044 mmol), N-(propargyl)vinylsulfonamide (9.6 mg, 0.088 mmol), CuI (1.3 mg, 6.6 μmol, 15 mol%), Pd(OAc)$_2$ (1.0 mg, 4.4 μmol, 10 mol%), TPPTS (5.0 mg, 8.8 μmol, 20 mol%) and flush the content of the vial with argon (using a needle through the septum);
2. Add DIPEA (77 μL, 0.441 mmol, 10 equiv.) dropwise and stir for 45 min at 75°C;
3. Evaporate the reaction under reduced pressure and redissolve the residue in milliQ water (5 mL) and purify by C18 reverse phase HPLC using water-methanol (5% to 100%) containing 0.1 M TEAB buffer as eluent;
4. Co-distill with water several times and convert it to sodium salt (Dowex 50WX8 in Na^{+} cycle) followed by freeze-drying from water gave the desired **dC^{VS}TP** as pale yellow solid (isolated yield: 23%).

3.5.2 Synthesis of dA^{CA}TP [21]

dA^{I}TP is synthesized according to general protocol for triphosphorylation of modified nucleosides, starting from **dA^{I}** (see Sect. 3.4 , Scheme 4).

1. In a septum sealed vial dissolve in previously degassed water/acetonitrile (2:1, 2.5 mL) **dA^{I}TP** (30 mg, 0.049 mmol), N-(propargyl)chloroacetamide (7.73 mg, 0.058 mmol), CuI (0.931 mg, 4.9 μmol), Pd(OAc)$_2$ (0.55 mg, 2.45 μmol), TPPTS (2.78 mg, 4.9 μmol) and flush the content of the vial with argon (using a needle through the septum);
2. Add dropwise DIPEA (85 μL, 0.49 mmol);
3. Evaporate the reaction under reduced pressure, redissolve in milliQ water (5 mL) and purify by C18 reverse phase HPLC using water-methanol (5–100%) containing 0.1 M TEAB buffer as eluent;

4. Co-distill with water several times and convert to sodium salt (Dowex 50WX8 in Na^+ cycle) followed by freeze-drying from water gave the desired **dA^{CA}TP** as pale yellow solid (yield: 32 %);

3.5.3 Synthesis of dC^{ESQ} [22]

For preparation of **dC^{NH2}** see [27].

1. Dissolve **dC^{NH2}** (0.2 g, 0.713 mmol) in 4 mL of ethanol in a proper round-bottom flask, sealed with a septum, while stirring;
2. Add dropwise diethyl squarate (0.211 mL, 1.43 mmol) to the stirred solution and leave it for 1 h at 25°C;
3. Purify the crude mixture by flash column chromatography using DCM/MeOH (8:2) as eluent;
4. Collect all fractions of the desired product **dC^{ESQ}** and evaporate them under reduced pressure to obtain the final product **dC^{ESQ}** as white powder Rf = 0.9 (Fig. 5);
5. Check the purity of **dC^{ESQ}** by ^{1}H-NMR, ^{31}P-NMR, and MS (yield: 80%).

3.5.4 Synthesis of dC^{ESQ}TP [22]

dC^{ESQ}TP is synthesized according to general protocol for triphosphorylation of modified nucleosides, starting from **dC^{ESQ}** (see Fig. 4). In this case, the timing for the second step, after addition of $(NHBu_3)_2H_2P_2O_7$, is 45 min instead of 2 h (see point 6 in Sect. 3.4).

3.5.5 Synthesis of dT^{HDO}TP [23]

dT^{N3}TP is synthesized according to general protocol for triphosphorylation of modified nucleosides, starting from **dT^{N3}** (see Fig. 4);

1. In the first septum sealed vial, dissolve $CuSO_4$(II) *5H_2O (2.75 mg, 20 %mol) in 500 μL of H_2O/tBuOH (1:1) and degass for 10 min (mix A) flushing the content of the vial with argon (using a needle through the septum);
2. In a second septum sealed vial, dissolve **dT^{N3}** (28.8 mg, 0.05 mmol) in 500 μL of H_2O/tBuOH (1:1) in a microwave vial and degas, with argon, for 10 min (mix B) flushing the content of the vial with argon (using a needle through the septum);
3. In a third septum sealed vial, dissolve oct-7-yne-2,4-dione (HDO) (31 mg, 0.022 mmol) in 100 μL of DMSO and degas for 10 min (mix C) flushing the content of the vial with argon (using a needle through the septum);
4. In a fourth septum sealed vial, dissolve sodium ascorbate (43.6 mg, 0.22 mmol) in 100 μL of H_2O/tBuOH (1:1) (mix D) and flush the content of the vial with argon (using a needle through the septum);
5. Then, add mix B, mix C, and mix D to mix A (first septum sealed vial) and degas the resulting mixture for 10 min flushing again the content of the vial with argon (using a needle through the septum);

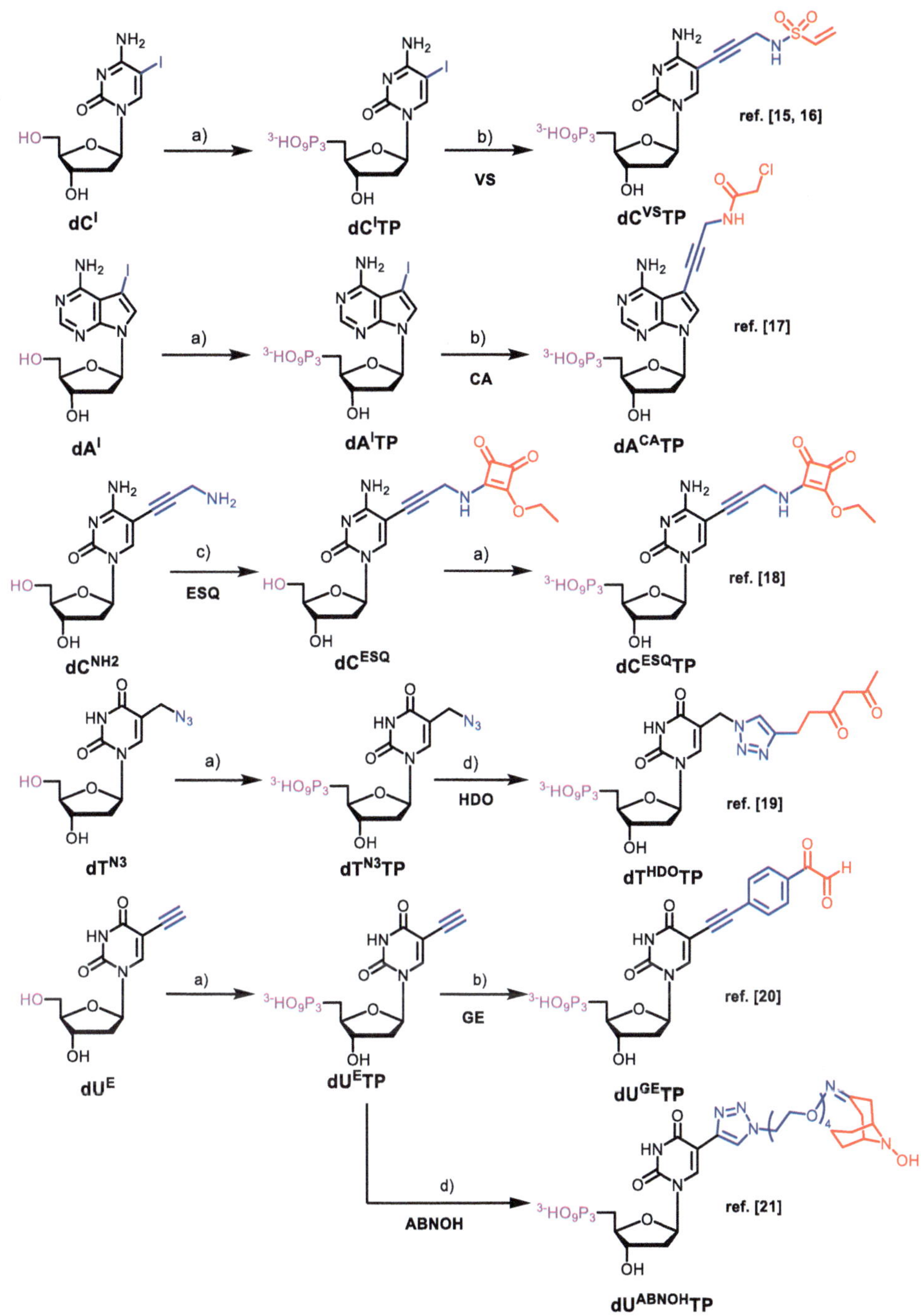

Fig. 5 Synthetic schemes of modified dNTPs. Conditions: (a) Triphosphorylation (for general conditions see Sect. 3.4); (b) Sonogashira coupling ($Pd(OAc)_2$, CuI, TPPTS, DIPEA, Sects. 3.5.1-2-6 –2-6); (c) Addition-elimination reaction (Sect. 3.5.3); (d) CuAAC reaction ($CuSO4*5H_2O$, sodium ascorbate, TBTA, Sects. 3.5.5)

6. Stir the reaction for 18 h at 25°C;
7. To quench the reaction, add 1 mL saturated solution of EDTA and stir for 5 min (*see* **Note 15**);
8. Purify the product on RP-HPLC with use of linear gradient of 0.1 M TEAB (Triethylammonium bicarbonate in H_2O) to 0.1 M TEAB in H_2O/MeOH (1:1) to MeOH as eluent. The triphosphate should elute between 5% and 20% B;
9. Collect all fractions containing the desired product. After several co-distillations with water followed by freeze drying from water you have the product **dT^{HDO} TP** as red powder Rf = 0.80 (IPAV) (Fig. 5);
10. Check the purity **dT^{HDO}TP** of ^{1}H-NMR, ^{31}P-NMR, and MS (yield: 17.8 %).

3.5.6 Synthesis of dU^{GE}TP [24]

dU^{GE}TP is synthesized according to general protocol for triphosphorylation of modified nucleosides, starting from **dU^E** (see Fig. 4);

1. In the first septum sealed vial, dissolve CuI (0.87 mg, 8 %mol), $Pd(OAc)_2$ (0.9 mg, 7% mol), TPPTS (3.6 mg, 11% mol), and p-bromo glyoxal (GE) (48.3 mg, 0.23 mmol) in 400 μL of H_2O/MeCN/DMSO (1:1:0.1), previously degassed (mix A) by flushing the content of the vial with argon (using a needle through the septum);
2. In a second septum sealed vial, dissolve **dU^{E}TP** (40 mg, 0.057 mmol) in 600 μL of H_2O/MeCN/DMSO (1:1:0.1), previously degassed by flushing the content of the vial with argon (using a needle through the septum) (mix B);
3. In a third septum sealed vial, add mix B to mix A, then add DIPEA (200 μL, 1.14 mmol) and stir the reaction at 25°C for 2 h and flush again the content of the vial with argon (using a needle through the septum);
4. To quench the reaction, add 1 mL sat. solution of EDTA and stir for 5 min;
5. Purify the product on RP-HPLC (using Kinetex and subsequently X-Bridge prep column) with use of linear gradient of 0.1 M TEAB (Triethylammonium bicarbonate in H_2O) to 0.1 M TEAB in H_2O/MeOH (1:1) to MeOH as eluent;
6. Collect all fractions of the desired product. Several co-distillations with water followed by freeze drying from water gives the product **dU^{GE}TP** as red powder Rf = 0.1 (IPAV) (Fig. 5);
7. Check the purity of **dU^{GE}TP** by ^{1}H-NMR, ^{31}P-NMR, and MS (yield: 20%).

3.5.7 Synthesis of dU^{ABNOH}TP [25]

1. Prepare 5-ethynyl-2′-deoxyuridine-5′-*O*-triphosphate **dU^{E}TP** according to general procedure for triphoshorylation of

modified nucleoside, starting from 5-ethynyl-2′-deoxyuridine, **dU^E^** (see Fig. 4);

2. In the first septum sealed vial, dissolve **dU^E^TP** (7 mg, 143 μ mol) in 0.15 mL DMSO, previously degassed by flushing the content of the vial with argon (using a needle through the septum) (mix A);
3. In a second septum sealed vial, dissolve ABNOH-PEG_4-N_3 (11 mg, 29.63 μmol) in 0.15 mL DMSO, which was previously prepared and degassed by flushing the content of the vial with argon (using a needle through the septum) (mix B) (see Sect. 3.3.4);
4. In a third septum sealed vial dissolve $CuSO_4*5H_2O$ (1.9 mg, 11.9 μmol), sodium ascorbate (3.7 mg, 18.67 μmol), and TBTA (4 mg, 7.53 μmol) in 0.25 mL DMSO, which was previously degassed by flushing the content of the vial with argon (using a needle through the septum) (mix C);
5. Add dropwise, with a syringe, mix A and B to the third septum sealed vial containing mix C and leave stirring vigorously for 2 h at 25°C;
6. Quench the reaction with 0.5 mL saturated solution of EDTA;
7. Purify the crude mixture by RP-HPLC X-Bridge prep column (see Sect. 2.1. , miscellaneous) with use of linear gradient of 0.1 M TEAB (Triethylammonium bicarbonate in H_2O) to 0.1 M TEAB in H_2O/MeOH (1:1) to MeOH as eluent;
8. Evaporate all fractions of the desired product under reduced pressure and several co-distillations with water followed by freeze drying from water give the product **$dU^{ABNOH}TP$** as white powder Rf = 0.30 (IPAV) (Fig. 5);
9. The dried powder may be stored in the sealed tube at −20°C protected from humidity;
10. Check the purity of **$dU^{ABNOH}TP$** by ^{1}H-NMR, ^{31}P-NMR, and MS (yield: 60%).

3.6 Methods: Biochemistry

The modified triphosphate will be incorporated in a dsDNA by enzymatic incorporation using primer extension reaction (PEX) (Fig. 6).

3.6.1 Enzymatical Incorporation of Modified Triphosphates (PEX Reaction)

Reagents: 5’-6-FAM labeled primer, template (**ON1Y, Y**: C or T or A or G), KOD XL DNA polymerase, KOD XL buffer (10×), 2’-deoxynucleosides triphosphates (dCTP, dGTP, dATP, dTTP) and modified nucleotide (**dN^XTP**). **X**: modification (***see* Note 16**).

Analytical scale (the final volume of the reaction is 20 μL)

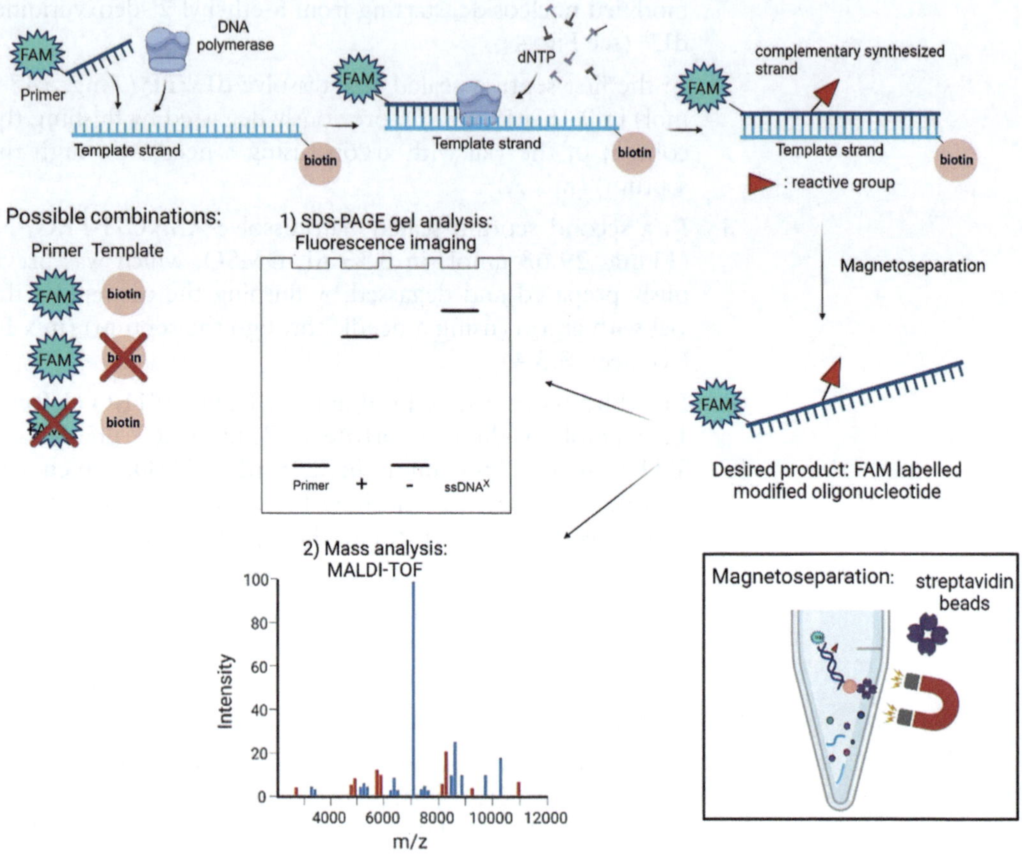

Fig. 6 Schematic representation of the work-flow for the enzymatic synthesis of modified oligonucleotides and their characterization by SDS-PAGE gel and mass analyses.

1. Prepare all stock solutions, in the appropriate concentration, of dNTPs (4 mM), template **ON1Y** (3 μM) and 5'-6-FAM labeled primer (3 μM);
2. Prepare the master mixture (MM) containing dNTP (4 mM, 1.5 μL), template **ON1Y** (3 μM, 4.5 μL), 5'-6-FAM labeled or not labeled primer (3 μM, 3 μL), KOD XL DNA polymerase (0.25 U/ 0.6 μL), KOD XL buffer (10X, 6.3 μL) and then add 44.1 μL of water to bring to total volume of 60 μL (3×). The MM is used for positive control, negative control and for the reaction mixture;
3. For positive control, mix 19 μL of MM and 1 μL of natural dNTP (4 mM) in the first Eppendorf tube. For negative control mix 19 μL of MM and 1 μL of additional water in the second Eppendorf tube. For the reaction mixture, mix 19 μL of MM and 1 μL of modified **dN^{X}TP** in the third Eppendorf tube (Fig. 7);

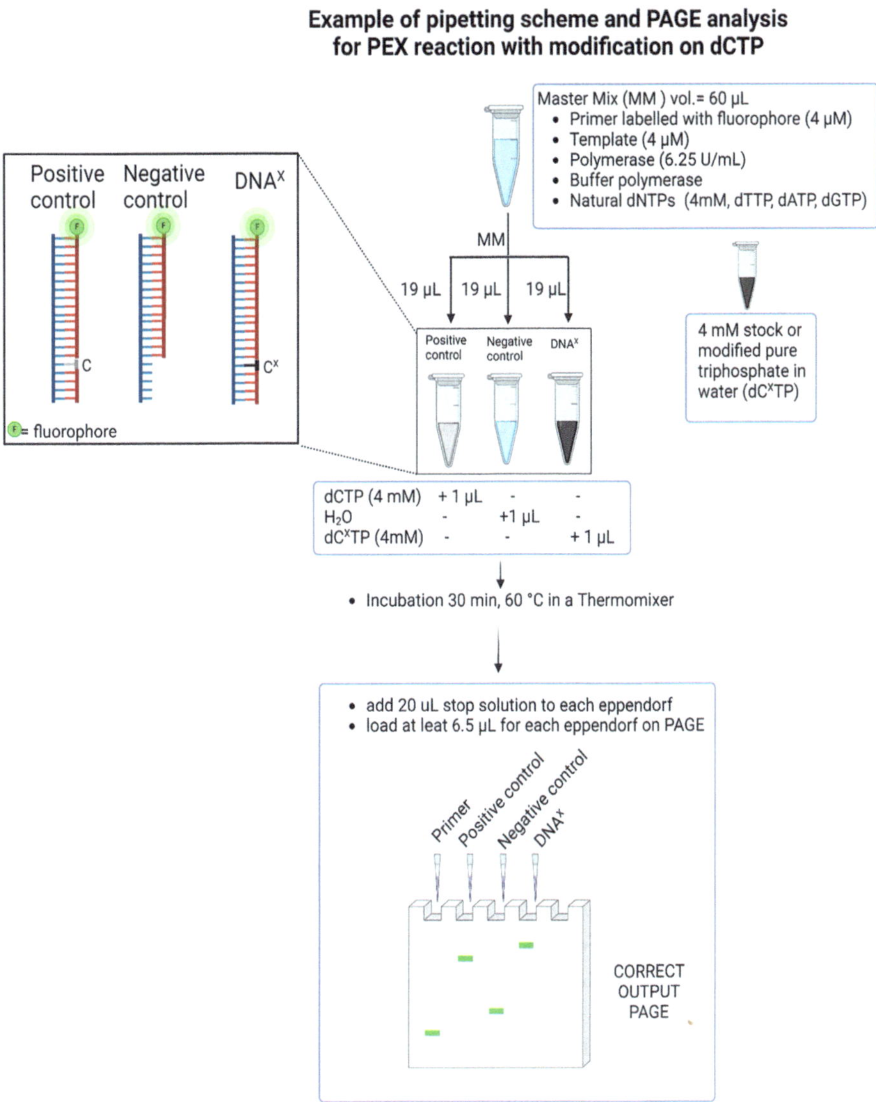

Fig. 7 Pipetting scheme for analytical scale PEX reaction (*see* **Note 17 and 18**)

4. Place these three Eppendorf tubes in the thermomixer and leave them in incubation for 30 min at 60°C;
5. Stop the reaction placing them in ice bath to cool down to 4°C;
6. Then, add in each Eppendorf tube the PAGE stop solution (see Sect. 2.2) as same as the volume of reaction mixture (20 μL) and heat in the thermomixer for 3 min at 95°C (to denature the double strand of DNA of each sample) (*see* **Note 19**);
7. Prepare denaturing 12.5% PAGE gel for electrophoresis and preheat it (see Sect. 2.2);

8. Load 6.5 μL of each sample in separated wells and run the gel (see Sect. 2.2) for gel run condition), then visualize it using fluorescence imaging (*see* **Note 25**).

Semipreparative scale (the final volume of the reaction is 50 μL)

1. Prepare all stock solutions, in the appropriate concentration, of natural dNTPs (4 mM), template **ON1Y** (50 μM) and 5'-6-FAM labeled or not labeled primer (50 μM);
2. Prepare the master mixture (MM) containing dCTP (4 mM, 7.5 μL), template **ON1Y** (50 μM, 22.5 μL), 5'-6-FAM labeled primer (50 μM, 15 μL), KOD XL DNA polymerase (0.25 U/7.5 μL), KOD XL buffer (10×, 31.5 μL) and then add 66 μL of water to bring to total volume of 150 μL (3×). Also in this case, the MM (*see* **Note 20**) is used for positive control, negative control and the reaction mixture, as described in analytical scale;
3. For positive control mix 48.5 μL of MM and 1.5 μL of natural dNTP (4 mM) in the first Eppendorf tube;
4. For negative control mix 48.5 μL of MM and 1.5 μL of extra water in the second Eppendorf tube;
5. For reaction mixture mix 48.5 μL of MM and 1.5 μL of modified **dN^{X}TP** (4 mM) in the third Eppendorf tube;
6. Place these three Eppendorf tubes in the thermomixer and leave them in incubation for 30 min at 60°C;
7. Stop the reaction placing them in ice bath to cool down to 4°C;
8. Then proceed with magneto-separation purification procedure as described below which will be analyzed by MALDI-TOF mass spectrometry or QIAGEN kit purification (*see* **Note 22**) in order to purify the modified strand (Fig. 2).

3.7 DB Streptavidin Magneto: Separation Procedure

Streptavidin particles (Roche; 50 μL) are washed with a binding buffer (3 × 200 μL; 10 mM Tris, 1 mM EDTA, 100 mM NaCl, pH 7.5). Then, the PEX solution (50 μL) is mixed with a binding buffer (200 μL) and incubated for 30 min at 15°C and 1400 rpm. The magnetic beads are collected on a magnet (DynaMagTM-2, Invitrogen), and washed with wash buffer (3 × 300 μL; 10 mM Tris, 1 mM EDTA, 500 mM NaCl, pH 7.5) and water (4 × 300 μL). Then water (50 μL) is added and the sample is denatured for 2 min at 900 rpm and 75°C. The beads were collected on a magnet and the solution was transferred into a clean vial. The product is evaporated to dryness, then dissolved in water and analyzed by MALDI-TOF mass spectrometry.

3.8 Cross-Linking with Peptides and Proteins (Fig. 8)

3.8.1 DNA^{VS} with Cysteine-Containing Peptides [19, 20]

Gel Analysis

Reagents: 30 mer **DNA^{VS}**, 30 mer **DNA^{Nat}**, L-cysteine, glutathione, Cys-containing undecapeptide (TAKSVGGACPA-OH).

1. Prepare 30 mer **DNA^{VS}** and 30 mer **DNA^{Nat}** (*see* **Note 18**) are prepared by PEX as described above (Sect. 3.6.1) using a ^{31}P-labeled primer;
2. In a final volume of 30 μL mix **DNA^{VS}** (4.3 μM) and L-cysteine, glutathione or Cys-containing undecapeptide in crescent concentrations (2.2 mM; 8.8 mM; 22 mM) in TEAA buffer (0.3 M, pH 8.3). Follow the same procedure for **DNA^{Nat}**;
3. Incubate the reactions in a thermomixer at 25°C for 18 h. After incubation add to each Eppendorf directly 30 μL of PAGE stop solution;
4. Incubate the reactions at 65°C for 10 min for denaturation and separate them by 20% PAGE under denaturing conditions (see Sect. 2.2);
5. Visualize the product by phosphoimaging.

Mass Analysis:

1. Prepare 30 mer **DNA^{VS}** by PEX in semi-preparative scale as described above (Sect. 3.6.1) using a 5'-biotinylated template;

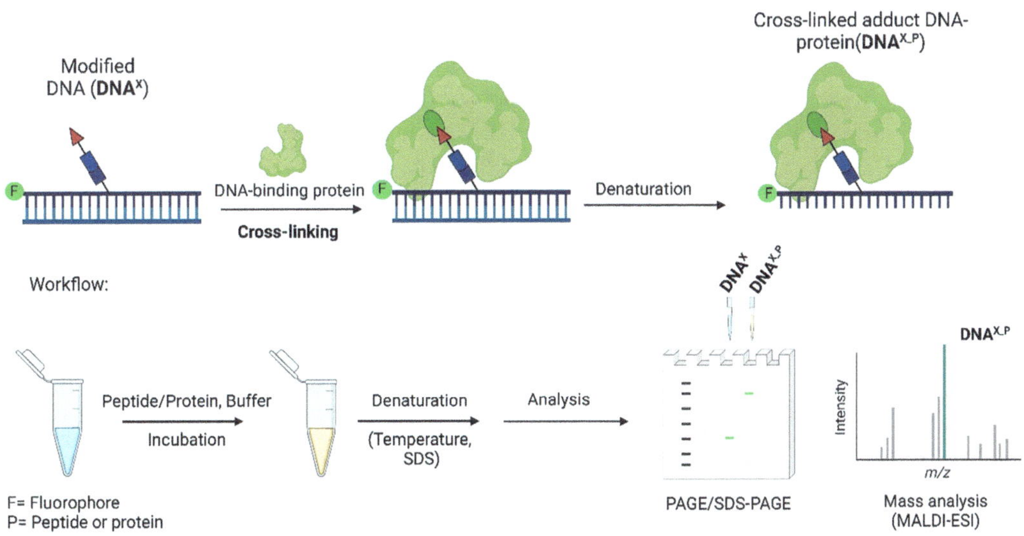

Fig. 8 General approach for analysis of DNA-protein cross-linking

2. In a final volume of 50 μL mix **DNA**^{VS} (40 μM) and L-cysteine, glutathione or Cys-containing undecapeptide (50 mM) in TEAA buffer (0.3 M, pH 8.3) (*see* **Note 21**);
3. Incubate the reactions in a thermomixer at 25°C for 18 h at 300 rpm;
4. Purify the reaction by magneto-separation as described above (Sect. 3.7) and analyze it by MALDI-TOF mass spectrometry (*see* **Note 26**) (Fig. 6).

3.8.2 DNA^{CA} with Cysteine-Containing Peptides [21]

Reagents: 30 mer **DNA**^{CA}, 30 mer **DNA**^{Nat}, L-cysteine, glutathione, Cys-containing undecapeptide (TAKSVGGACPA-OH).

Mass Analysis

1. Prepare 30 mer **DNA**^{CA} is prepared by PEX in semi-preparative scale (Sect. 3.6.1) as described above using a 5'-biotinylated template;
2. In a final volume of 20 μL mix **DNA**^{CA} (20 μM) and a solution of L-cysteine or Cys-containing undecapeptide TAKSVGGACPA-OH (50 mM) in TEAA buffer (0.3 M, pH 8.3) (*see* **Note 21**);
3. Incubate the reactions in a thermomixer at 37°C for 18 h at 300 rpm;
4. Purify the reaction by magneto-separation as described above (Sect. 3.7) and analyze it by MALDI-TOF mass spectrometry (*see* **Note 26**) (Fig. 6).

3.8.3 DNA^{ESQ} with Lysine-Containing Peptides [22]

Gel Analysis

Reagents: 19 mer **DNA**^{ESQ}, 19 mer **DNA**^{Nat}, L-cysteine, Ac-Lys-OH, tri- (Ac-AKA-NH_2) or decapeptide (Ac-SGYTAKAQSG-NH_2)

1. Prepare 19 mer **DNA**^{ESQ} and 19 mer **DNA**^{Nat} (*see* **Note 18**) are prepared by PEX as described above (Sect. 3.6.1) using a 5'-6-FAM labeled primer;
2. In a final volume of 20 μL mix **DNA**^{ESQ} (4.3 μM) and Ac-Lys-OH, tri- or decapeptide (11 mM) in borate buffer (0.5 M, pH 9) (*see* **Note 21**); Repeat the same procedure for **DNA**^{Nat} (control);
3. Incubated the reactions in a thermomixer at 37°C for 36 h at 300 rpm. After incubation add to each Eppendorf tube directly 20 μL of PAGE stop solution;
4. Incubate the reactions 5 min at 95°C (*see* **Note 19**) and separate them by 20% PAGE under denaturing conditions (see Sect. 3.1);
5. Visualize the product by fluorescence imaging.

Mass Analysis

1. Prepare 19 mer **DNA^{ESQ}** by PEX in semi-preparative scale as described above (Sect. 3.6.1) using a 5'-biotinylated template;
2. In a final volume of 20 μL mix **DNA^{ESQ}** (10 μM) and Ac-Lys-OH, tri- or decapeptide (22 mM) in TEAA buffer (0.3 M, pH 8.3);
3. Incubate the reactions in a thermomixer at 37°C for 36 h at 300 rpm;
4. Purify the reaction by magneto-separation, as described above (Sect. 3.7), and analyze it by MALDI-TOF mass spectrometry (*see* **Note 26**) (Fig. 6).

3.8.4 DNA^{HDO} with Arginine-Containing Peptides [23]

Gel Analysis

Reagents: 19/31 mer **DNA^{HDO}**, 19/31 mer **DNA^{Nat}** Boc-Arg-OH, tripeptide (Ac-ARA-OH) or decapeptide (Ac-SGYTARAESG-OH)

1. Prepare 19 or 31 mer **DNA^{HDO}** and 19 or 31 mer **DNA^{Nat}** (*see* **Note 18**) were prepared by PEX as described above (Sect. 3.6.1) using a 5′-6-FAM labeled primer;
2. In a final volume of 20 μL mix **DNA^{HDO}** (0.83 μM) and Boc-Arg-OH, tripeptide or decapeptide (250 mM) each in $NaHCO_3$ buffer (0.5 M, pH 10) (*see* **Note 21**); Follow the same procedure for **DNA^{Nat}**;
3. Incubate the reactions in a thermomixer at 37°C for 18 h. After incubation add to each Eppendorf directly 20 μL (the same volume as the reaction mixture) of PAGE stop solution;
4. Incubate the reactions at 95°C for 5 min (*see* **Note 23**) and separate them by 20 % PAGE under denaturing conditions (see Sect. 2.2);

Visualize the product by fluorescence imaging.

Mass Analysis

1. Prepare 19 or 31 mer **DNA^{HDO}** was prepared by PEX in semi-preparative scale (Sect. 3.6.1) as described above using a 5'-biotinylated template;
2. In a final volume of 20 μL mix **DNA^{HDO}** (3.50 μM) and a solution of Boc-Arg-OH, Arg-containing tripeptide or decapeptide (250 mM) in $NaHCO_3$ buffer (0.5 M, pH 10);
3. Incubate the reactions in a thermomixer at 37°C for 18 h at 500 rpm;

4. Purify the reaction by magneto-separation as described above (Sect. 3.7) and analyze it by MALDI-TOF mass spectrometry (*see* **Note 26**) (Fig. 6).

3.8.5 DNA^{GE} with Arginine-Containing Peptides [24]

Gel Analysis

Reagents: 19 or 31 mer **DNA^{GE}**, 19- 31 mer **DNA^{Nat}**, Ac-Arg-OH, Ac-AlaArgAla-OH or Ac-SGYTARAESG-OH.

1. Prepare 19- or 31 mer **DNA^{GE}** and 19 or 31 mer **DNA^{Nat}** (*see* **Note 18**) by PEX as described above (Sect. 3.6.1) using a 5'-6-FAM labeled primer;
2. In a final volume of 10 μL add **DNA^{GE}** (1 μM) and Boc-Arg-OH, tripeptide or decapeptide (22 mM) in HEPES/$KHCO_3$ buffer (0.05 M, pH 8.5) (*see* **Note 21**); Repeat the same procedure for **DNA^{Nat}**;
3. Incubate the reactions in a thermomixer at 37°C for 18 h. After incubation add to each eppendorf directly 10 μL of PAGE stop solution;
4. Incubate the reactions at 95°C for 5 min (*see* **Note 23**) and separated them by 20% PAGE under denaturing conditions (see Sect. 2.2);
5. Visualize the product by fluorescence imaging.

Mass Analysis

1. Prepare 19 or 31 mer **DNA^{GE}** was prepared by PEX in semi-preparative scale as described above (Sect. 3.6.1) using a 5′-biotinylated template;
2. In a final volume of 120 μL add **DNA^{GE}** (1 μM) and a solution of Boc-Arg-OH, Arg-containing tripeptide or decapeptide (250 mM) in HEPES/$KHCO_3$ buffer (0.05 M, pH 8.5);
3. Incubate the reactions in a thermomixer at 37°C for 18 h at 500 rpm;
4. Purify the reaction by magneto-separation as described above (Sect. 3.7) and analyze it by MALDI-TOF mass spectrometry (*see* **Note 26**) (Fig. 6).

3.8.6 DNA^{ABNOH} with Tryptophan-Containing Peptides [25]

Gel Analysis

Reagents: 19/31 mer **DNA^{ABNOH}**, 19/31 mer **DNA^{Nat}**, peptides containing tryptophan tri- (Ac-AWA), nonapeptides (ASKMAWAKT-NH_2, Ac-FGWCGPYVA-NH_2, ANAQSWAAG-NH_2)

1. Prepare modified 19 or 31 mer **DNA^{ABNOH}** and **DNA^{Nat}** (*see* **Note 18**) as described previously, in semipreparative scale (Sect. 3.6.1) using a 5'-6-FAM labeled primer;

2. In the total volume of 50 μL mix in an eppendorf tube **DNA**[ABNOH] (7 μM) and the desired peptide in large excess (250 mM) in $NaHCO_3$ buffer (0.05 M, pH 10) (*see* **Note 21**);
3. Incubate the reactions in a thermomixer at 37°C for 18 h. After incubation quench the reaction by addition of 10 μL of PAGE stop solution to each eppendorf tube;
4. Incubate the reactions at 95°C for 5 min (*see* **Note 23**) and separated them by 20% PAGE under denaturing conditions (see Sect. 2.2);
5. Visualize the product by fluorescence imaging.

Mass Analysis

1. Prepare 19/31 mer **DNA**[ABNOH] by PEX in semipreparative scale as described above (Sect. 3.6.1) using a 5'-biotinylated template;
2. In a final volume of 50 μL mix **DNA**[ABNOH] (7 μM) and a solution of N-Ac-Trp-containing tripeptide or nonapeptide (250 mM) in HEPES/$KHCO_3$ buffer (0.05 M, pH 8.5);
3. Incubate the reactions in a thermomixer at 37°C for 3 h or 18 h at 500 rpm;
4. Purify the reaction by magneto-separation as described above (Sect. 3.7) and analyze it by MALDI-TOF mass spectrometry (*see* **Note 26**) (Fig. 6).

3.9 Cross-Linking with Proteins

3.9.1 DNA[VS] or DNA[CA] with p53

1. Prepare modified 19 mer **DNA**[VS/CA] by PEX as described above in semi-preparative scale (Sect. 3.6.1) using a 5′-biotinylated template;
2. Purify the product on QIAquick Nucleotide Removal Kit and eluted with water;
3. Then, in a final volume of 20 μL mix purified **DNA**[VS/CA] (6 ng/μL, 10 μL), KCl (500 mM, 2 μL), DTT (2 mM, 2 μL), VP buffer and GSTp53CD protein binding stock solution (400 ng/μL in 25 mM HEPES pH 7.6, 200 mM KCl, 10% glycerol, 1 mM DTT, 1 mM benzamidine; 2 μL);
4. Prepare also control samples using correspondent natural DNA;
5. Incubate all samples for 30 min on ice and then 2h at 25°C, then add 2× VPS loading buffer and incubate the sample for 10 min at 65°C;
6. Analyze all samples (including the reaction mixture) by 10% SDS-PAGE at room temperature;
7. Visualize the product by fluorescence imaging.

3.9.2 DNA^{ESQ}, DNA^{HDO} or DNA^{GE} with Histones

DNA^{ESQ} with Histones

Reagents: 19 mer **DNA^{ESQ}**, H2A or H2B or H3.1 or H4

Gel Analysis

1. Prepare natural and modified 19 mer **DNA^{ESQ}** by PEX in semipreparative scale as described above (Sect. 3.6.1) using a 5'-6-FAM labeled primer and then purify on QIAquick Nucleotide Removal Kit (*see* **Note 22**);
2. Incubate natural **DNA^{Nat}** or modified **DNA^{ESQ}** (25 pmol) with 2 or 10 equiv of H3.1 protein in 10 μL of phosphate buffer NaH_2PO_4/Na_2HPO_4 (4.5 mM, pH 7.4) at 37°C 36 h.
3. Then, dilute the reaction mixture with 2× VPS loading buffer and incubate for 2 min at 100°C (*see* **Note 24**);
4. Analyze the reaction mixture by 12% SDS-PAGE (see Sect. 2.2) at room temperature;
5. Visualize the product by fluorescence imaging;

Mass Analysis

1. Prepare the reaction mixture containing 19 mer **DNA^{ESQ}** and histones (H2A or H2B or H3.1 or H4) as described above;
2. Inject 10 μL of reaction mixture into bioZen Intact C4 column (for equipment and gradient of mobile phases (see Sect. 2.2) and characterize by mass analysis [28].

$DNA^{HDO/GE}$ with Histones

Reagents: 19/30 mer **DNA^{HDO} DNA^{GE}**, H2A, H2B, H3.1, H4, and BSA

Gel Analysis

1. Prepare natural and modified 19 or 30 mer **DNA^{HDO} DNA^{GE}** by PEX in semipreparative scale as described above (Sect. 3.6.1) using a 5′-6-FAM-labeled primer and then purify by QIAGEN-QIAquick Nucleotide Removal Kit (Sect. 2.2);
2. Incubate 1 μM of natural or modified DNA with 5 μM of H2A, H2B, H3.1, H4 and BSA using $NaHCO_3$ (0.05 M, pH 10) for **DNA^{HDO}** OR 10 μL of HEPES/$KHCO_3$ buffer (0.05 M, pH 8.5) for **DNA^{GE}** at 37°C 18 h;
3. Then, dilute the reaction mixture with 2× VPS loading buffer and incubate for 2 min at 100°C;
4. Analyze the reaction mixture by 20% SDS-PAGE (see Sect. 2.2) at room temperature. Visualize the product by fluorescence imaging and subsequently stained with PageBlue TM protein staining solution.

Mass Analysis

1. Prepare the cross-linking mixture using 1 μM of modified DNA (**DNA^HDO^** or **DNA^GE^**) and 5 μM of protein (H2A, H2B, H3.1) in 120 μM;
2. Lyophilize the reaction diluted in water to obtain 50 μM theoretical DNA concentration and use it for ESI analysis;
3. Analyze by LC system (I-class, Waters) injecting 5 μM of the reaction mixture into reverse phase column (MassPREP desalting, Waters) and desalted by fast gradient (4 min) of acetonitrile in water with 10 mM ammonium acetate, pH 9 and characterize by mass analysis [28].

3.9.3 DNA^{ABNOH} with bs GntR

Reagents: 19/30 mer **DNA^{ABNOH}**, *bs* GntR (*Bacillus subtilis* gluconate operon transcription factor).

Gel Analysis

1. Prepare modified **DNA^{ABNOH}** as described above in semipreparative scale (Sect. 3.6.1);
2. In a final volume of 10 μL mix **DNA^{ABNOH}** (50 μM, 1 μL) with *bs*GntR (50 μM, 4 μL) in TrisHCl buffer (0.05 M, pH 10), using the ratio 1:4 DNA:protein;
3. Incubate the reaction in the thermal cycler at 37°C for 18 h;
4. Then, dilute the reaction mixture with 2× VPS loading buffer and incubate for 2 min at 100°C (see Sect. 2.2);
5. Analyze the reaction mixture by 17.5% SDS-PAGE (see Sect. 2) at room temperature (200 V, 50 min) using 1×Tris glycin SDS buffer for 1 h;
6. Visualize the product by fluorescence imaging.

4 Notes

1. Distillation under vacuum. Usually, the temperature of the oil bath is around 60°C, but it depends on the vacuum strength. Check pressure-temperature nomographs for proper boiling point at their vacuum strength. For storage use a glass flask with septum. If properly stored it has a shelf life of 6 months.
2. For storage use a glass flask with a glass stopper and not rubber septum. Additionally, it can be kept in the fridge/freezer with PTFE tape. If properly stored it has a shelf life of 1 year. When taking aliquots for reactions use Hamilton syringe and flash argon on the stock solution (creating a "hood" with a funnel connected to argon line).

3. The cation exchange from sodium to tributylammonium is very important since it is affecting the solubility of the pyrophosphate salts and the yield of following triphosphorylation reactions.
4. The indicator of successful preparation of the TEAB is the pH around 7.6 of the solution. If the pH is too basic the TEAB is not ready yet. After successful preparation the TEAB is a dense transparent fluid and has a characteristic smell of almonds quite different from the pungent fishy smell of Et_3N. *p*-anisaldehyde TLC staining preparation: Add 5 mL of concentrated sulfuric acid, 1.5 mL of glacial acetic acid and 3.7 mL of *p*-anisaldehyde to 135 mL of ethanol; Stirr the mixture vigorously to ensure homogeneity; Store the solution in a wide mouth jar covered with aluminum foil to ensure longer shelf life. This staining is sensitive to most functional groups, especially those which are strongly and weakly nucleophilic.
5. Keep the times between washings as written, otherwise the yield is low. Step 3 is done to exchange from Py^+ form to Bu_3NH^+ form. $(NHBu_3)_2H_2P_2O_7$ is unstable at room temperature, even more so at 40°C, so do not overheat it on rotavapors. Minimize the time on rotavapor while in glue-like consistency, it quickly decomposes. It is necessary to have a highly capable pump to be able to evaporate water without excessive heating (for example Vacuubrand PC3000 VARIO series). Pyrophosphate is hydroscopic, store and open it under argon "hood." If the crystallization does not work the first time, evaporate it and try again, using less acetone. If the mixture is biphasic, dilute it a bit more.
6. Structure of ATD is reported, the synthesis is according to [6] (Fig. 9).
7. This step is very important for the yield of dNTP. If you are working with azido-modified nucleosides, be sure to have the compound protected from light (e.g., with aluminum foil).
8. Clean with repetitive washings of water/MeOH the Hamilton syringe immediately after the use to avoid clogging. Dry completely the glass channel of the syringe. Store the syringe always disassembled.

ATD: 2-(2-(2-(2-(2-azidoethoxy)ethoxy)ethoxy)ethoxy)isoindoline-1,3-dione

Fig. 9 Structure of ATD (*see* **Note 6**) [6]

9. The mobile phase for TLC is composed by isopropyl alcohol: NH_3: H_2O (11:7:2) (IPAV). Nucleoside usually is the spot closer to the solvent line (r.t 0.8-0.7), than monophosphate (r.t. 0.5-0.4) and triphosphate is closer to the starting line (r.t. 0.2-0.1). Retention times can variate depending on the starting nucleoside but the order not. Stain the TLC plate after properly drying it with *p*-anisaldehyde staining.
10. The quenching generates CO_2, be sure to insert a needle in the septum before starting quenching to avoid over-pressurization of the vial.
11. The **Purification 1** is an important step in this scale (ca. 200 mg), if the reaction set up is below 50 mg this step can be avoided. The triphosphate elutes usually as the third peak in the chromatogram (50% A–B), after the initial nucleoside and monophosphate. Check via always TLC, MS or ^{31}P-NMR which eluted fraction is the triphosphate.
12. Before running HPLC separation be sure to properly evaporate the crude mixture and dissolve it in sole water. Use HPLC semi preparative columns (see in Sect. 2.1 Miscellaneous for specification of the equipment) with flow 10 mL/min. Triphosphates elute usually between 1 and 20 min run depending on the gradient. Might be necessary to run two separations to get pure compound.
13. Example of the appearance of ^{31}P NMR spectrum of a triphosphate in Fig. 10.
14. This is due to the stability of triphosphate that can be hydrolyzed to the corresponding diphosphate. The yield is also influenced by the presence of the intermediate monophosphate, which is not completely converted to triphosphate. Monophosphate is present in high amounts if the $POCl_3$ is partially hydrolyzed. Therefore, the $POCl_3$ should be as dry as possible (freshly distilled, properly stored, *see* **Note 2**).
15. EDTA is used in order to complex the copper present in the reaction mixture, and then to easily remove the complex by reverse-phase HPLC.
16. **ON1Y** (**O**ligo**N**ucleotide) = single-stranded template containing only 1Y (**Y**: C or T or A or G) according to the template sequence chosen, which is usually 19 or 30 mer long. Example: **ON1C** template 5'-CCC**G**CCCATGCCGCCCATG-3' (19 mer), 3'-CATGGGCGGCATGGG**C**GGG-5' (complementary strand). Primer = single strand primer usually 15 mer = 5'-CATGGGCGGCATGGG-3'.
17. One modified 2'-deoxynucleoside triphosphate (**dN^{X}TP**, **X**: modification) is usually incorporated, enzymatically, by primer extension (PEX) in DNA. For positive control, according to the modified nucleotide **dN^{X}TP** we want to

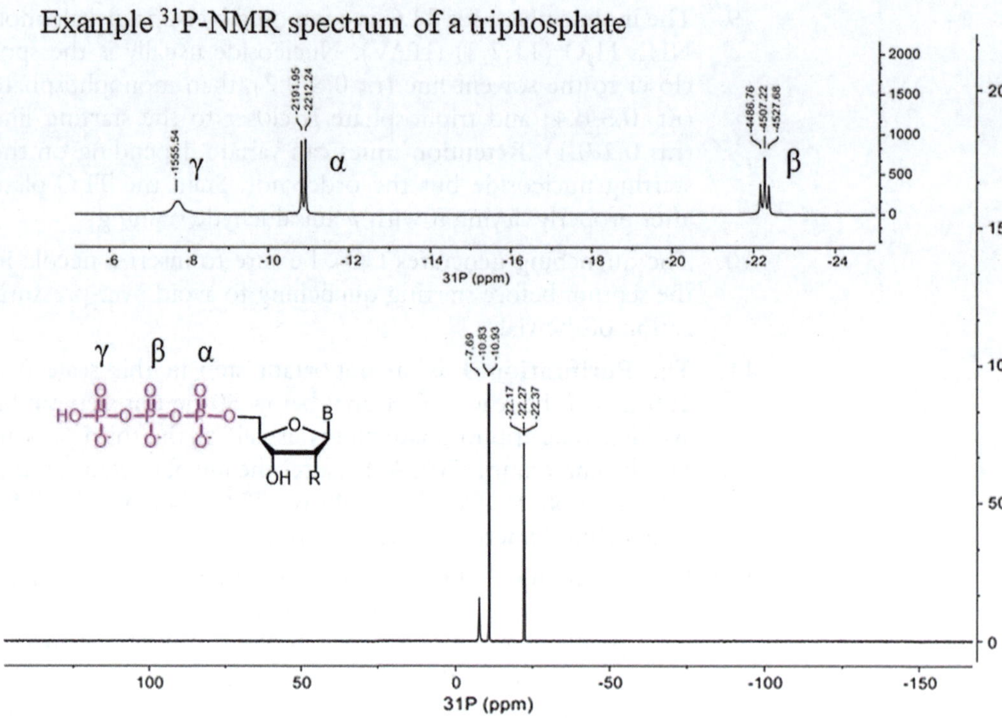

Fig. 10 Example of the appearance of ^{31}P NMR spectrum of a triphosphate

insert (C or A or T or G letter), the correspondent natural (unmodified) dNTP is added instead of modified **dN^{X}TP**. The rest of dNTPs added are chosen according to template sequence (included in MM). For negative control, additional water replaces natural dNTP or modified **dN^{X}TP**. For reaction mixture, modified **dN^{X}TP** is incorporated instead of dNTP and the rest of dNTPs will be natural (unmodified) dGTP. For example, in case of **ON1c**, the letter to insert is C. Therefore, for positive control, natural (unmodified) dCTP is added instead of **dC^{X}TP**. For reaction mixture, modified **dC^{X}TP** is incorporated instead of natural dCTP. For negative control additional water replaces natural (unmodified) dCTP or modified **dC^{X}TP** (Fig. 7).

18. Natural DNA (**DNA^{Nat}**) is used as control in order to check, in comparison with the modified DNA (**DNA^X**), whether the modified **dN^{X}TP** is incorporated. **DNA^{Nat}** and **DNA^X** show different mobility shift on the gel (Fig. 7).
19. The addition of PAGE stop solution is used for DNA denaturation, in order to separate the double-stranded DNA and visualize the FAM labeled single-stranded product by fluorescence imaging (Typhoon FLA 9500).

20. The preparation of a master mix (MM) is useful to minimize the error of pipetting which could be observed while preparing samples one by one.
21. The peptide is used in large excess to mimic the proximity effect between DNA and DNA-binding proteins.
22. QIAquick Nucleotide Removal Kit is also used to purify the double-stranded DNA fluorescently labeled in 5', when is prepared in semi-preparative scale for cross-linking experiments. The purification aims to remove the enzyme and the excess of unreacted dNTPs.
23. The temperature of 95°C is used to denature the double strand of DNA of each sample in order to visualize only the single-stranded product by gel electrophoresis.
24. The temperature of 100°C is used to denature the double-stranded DNA and any non-covalent interactions between DNA and protein.
25. In the gel analysis (PAGE or SDS-PAGE) the visualized band are always single-stranded DNA since the gel is run under denaturing conditions.
26. In MALDI-TOF analysis is detected usually the mass of single-stranded DNAs and rarely double-stranded DNA.

References

1. Rohs R, Jin X et al (1999) Origin and specificity of protein-DNA recognition. Annu Rev Biochem 79:233–269
2. Lambert SA, Jolma A et al (2018) The human transcription factors. Cell 172:650–665
3. Furey TS (2012) hIP–seq and beyond: new and improved methodologies to detect and characterize protein–DNA interactions. Nat Rev Genet 13:840–852
4. Dovgan I, Koniev O et al (2019) Antibody-oligonucleotide conjugates as therapeutic, imaging, and detection agents. Bioconjug Chem 30:2483–2501
5. Shi P, Wang Y (2021) Synthetic DNA for cell-surface engineering. Angew Chem Int Ed 60:11580–11591
6. Trads JB, Tørring T et al (2017) Site-selective conjugation of native proteins with DNA. Acc Chem Res 50:1367–1374
7. Niemeyer CM (2010) Semisynthetic DNA-protein conjugates for biosensing and nanofabrication. Angew Chem Int Ed 49:1200–1216
8. Nielsen TB, Thomsen RP et al (2019) Peptide-directed DNA-templated protein labelling for the assembly of a pseudo-IgM. Angew Chem Int Ed 58:9068–9072
9. Trads JB, Tørring T et al (2017) Building machines with DNA molecules. Acc Chem Res 50:1367–1374
10. Tretyakova NY, Groehler A et al (2015) DNA-protein cross-links: formation, structural identities, and biological outcomes. Acc Chem Res 48:1631–1644
11. Vinkenborg JL, Mayer G (2012) Aptamer-based affinity labeling of proteins. Angew Chem Int Ed 51:9176–9180
12. Rosen CB, Kodal AL et al (2014) Template-directed covalent conjugation of DNA to native antibodies, transferrin and other metal-binding proteins. Nat Chem 6:804–809
13. Ivancová I, Leone DL et al (2019) Reactive modifications of DNA nucleobases for labelling, bioconjugations, and cross-linking. Curr Opin Chem Biol 52:136–144
14. Winnacker M, Breeger S et al (2009) Novel diazirine-containing DNA photoaffinity probes for the investigation of DNA-protein-interactions. ChemBioChem 10:109–118
15. Hottin A, Marx A (2016) Structural insights into the processing of nucleobase-modified nucleotides by DNA polymerases. Acc Chem Res 49:418–427

16. Hocek M (2019) Enzymatic synthesis of base-functionalized nucleic acids for sensing, cross-linking, and modulation of protein-DNA binding and transcription. Acc Chem Res 52:1730–1737
17. Hocek M (2014) Synthesis of base-modified 2'-deoxyribonucleoside triphosphates and their use in enzymatic synthesis of modified DNA for applications in bioanalysis and chemical biology. J Org Chem 79:9914–9921
18. Fantoni NZ, El-Sagheer AH, Brown T (2021) A Hitchhiker's guide to click-chemistry with nucleic acids. Chem Rev 121:7122–7154
19. Daďová J, Vrábel M et al (2015) Azidopropylvinylsulfonamide as a new bifunctional click reagent for bioorthogonal conjugations: application for DNA-protein cross-linking. Chem Eur J 21:16091–16102
20. Daďová J, Orság P et al (2013) Vinylsulfonamide and acrylamide modification of DNA for cross-linking with proteins. Angew Chem Int Ed 52:10515–10518
21. Olszewska A, Pohl R et al (2016) Chloroacetamide-linked nucleotides and DNA for cross-linking with peptides and proteins. Bioconjugate Chem 27:2089–2094
22. Ivancová I, Pohl R et al (2019) Squaramate-modified nucleotides and DNA for specific cross-linking with lysine-containing peptides and proteins. Angew Chem Int Ed 58:13345–13348
23. Leone DL, Hubálek M et al (2021) 1,3-Diketone-modified nucleotides and DNA for cross-linking with arginine-containing peptides and proteins. Angew Chem Int Ed 60:17383–17387
24. Leone DL, Pohl R et al (2022) Glyoxal-linked nucleotides and DNA for bioconjugations and crosslinking with arginine-containing peptides and proteins. Chem Eur J 28:e202104208
25. Spampinato A, Leone DL et al (2024) ABNOH-linked nucleotides and DNA for bioconjugation and cross-linking with tryptophan-containing peptides and proteins. Chem Eur J 30:e202402151
26. Carlson M, Carter JD, Rohloff J (2015) Improved preparation of 2 M triethylammonium bicarbonate. Green Chem Lett Rev 8(3–4):37–39
27. Jäger S, Rasched G et al (2005) A versatile toolbox for variable DNA functionalization at high density. JACS 127:15071–15082
28. Ferrige AG, Seddon MG (1992) Real-time measurement of myosin–nucleotidenoncovalent complexes by electrospray ionization mass spectrometry rapid commun. Mass Spectrom 6:707–711

Chapter 16

Protein Modification via Bioorthogonal HS–BCN Ligation

Ming Fang and Qing Lin

Abstract

Bioorthogonal reactions offer a powerful tool for site-specific labeling of biomolecules in living systems. Among them, the recently reported hydrazonyl sultone (HS)–bicyclo[6.1.0]non-4-yne (BCN) ligation reaction stands out for its fast reaction kinetics and tunable aqueous stability. Herein, we describe the experimental protocols of using HS–BCN ligation for site-specific modification of a recombinant nanobody in vitro and a G protein-coupled receptor (GPCR) on a live mammalian cell surface. These protocols include the genetic encoding of BCN-lysine (BCNK) into the target protein, bioorthogonal modification of the BCNK-encoded proteins, and characterization of the reaction rate and selectivity. Together with the robust genetic encoding of the strained alkyne BCN in any protein structure, the HS–BCN ligation reaction promises to expand the capabilities of bioorthogonal chemistry to enable facile modifications of domain antibodies in vitro for diagnostic applications and selective fluorescent labeling of GPCRs for biophysical studies of receptor dynamics in live cells.

Keywords Bioorthogonal chemistry, Click chemistry, Bioconjugation, Hydrazonyl sultone, Nanobody, G protein-coupled receptor

1 Introduction

Bioorthogonal protein conjugation enables precise chemical modification of proteins and subsequent studies of the labeled proteins in native cellular environments [1]. This two-step bioconjugation approach requires: (1) site-specific incorporation of a bioorthogonal chemical reporter into a biomolecule of interest, and (2) the use of bioorthogonal reactions—organic reactions that proceed in living systems with minimum perturbation to the endogenous biological processes—to selectively react with the chemical reporter to generate the bioconjugate for functional studies in living systems [2]. Specifically, bioorthogonal ligation reactions based on the azide/cyclooctyne [3] and tetrazine/trans-cyclooctene [4] reactant pairs have been demonstrated to be particularly useful in live-cell imaging and in vivo targeting studies. A major challenge in the bioorthogonal approach has been to strike a delicate balance between bioorthogonal reagent reactivity and stability in the intended biological environments [5].

Alexander Deiters (ed.), *Bioconjugation Methods*, Methods in Molecular Biology, vol. 3009,
https://doi.org/10.1007/978-1-0716-5112-4_16,

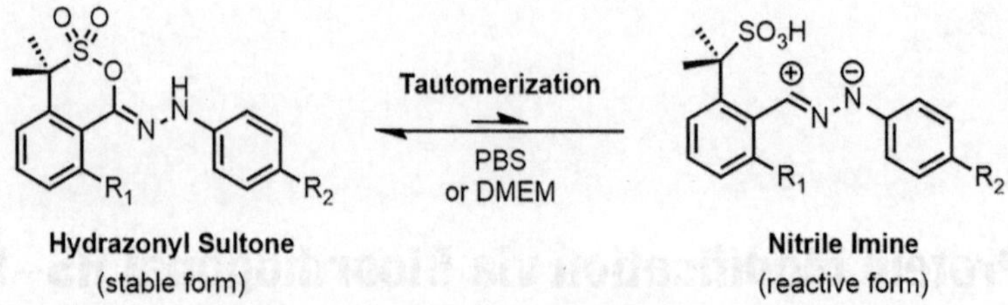

Fig. 1 Spontaneous tautomerization of the stable hydrazonyl sultone form to the reactive nitrile imine form under physiological conditions

We recently reported the design of a new hydrazonyl sultone (HS)–bicyclo[6.1.0]non-4-yne (BCN) reactant pair for bioorthogonal protein conjugations [6]. A prominent feature of the HS–BCN reactant pair is that the reactivity–stability balance can be readily tuned through HS structural modification [7]. BCN has become a popular chemical reporter because of its commercial availability, remarkable stability, fast reaction kinetics, and its ability to be incorporated site-specifically into protein structure [8, 9]. In addition, the BCN-derived lysine (BCNK) can be readily removed from the cytosol through gentle washing, leading to a very low small-molecule-related background reaction [10]. On the other hand, HS exists in a ring-chain tautomeric equilibrium between the stable HS ring form and the reactive nitrile imine chain form (Fig. 1), which allows HS to remain inert before the cycloaddition reaction with BCN and thus minimizes the potential side reactions. A recently reported method of using phosphine-mediated cyclodehydration for HS synthesis has further expanded the structural repertoire of HS [11].

In this chapter, we describe the protocols of applying HS–BCN ligation reaction to selectively modify two recombinant proteins in vitro and in live cells. In one, the nanobody NB1 carrying a genetically encoded BCN-lysine (BCNK) undergoes quantitative modification in minutes after treatment with a HS reagent [6]. In the other, a Cy5-conjugated HS reagent permits selective fluorescent labeling of a BCNK-encoded glucagon receptor without compromising cellular viability [6]. The detailed step-by-step procedures should facilitate a wider adoption of this bioconjugation method in chemical biology research.

2 Materials

2.1 Materials for Bioorthogonal Modification of a Nanobody

2.1.1 Bacterial Strains and Plasmids

1. BL21(DE3) competent cells (New England Biolabs, #C2527I).
2. pET26b(+)-NB1-V4TAG mutant plasmid (kanamycin resistant; obtained from Gene Universal).

3. pEVOL-BCNKRS plasmid (chloramphenicol resistant; derived from pEVOL-*Mm*PylRS, AddGene #223512, with Y306M/L309G/C348A mutations).

2.1.2 Media and Antibiotics

1. Lysogeny broth (LB) liquid medium (20 g/L): Prepared by dissolving 20 g of LB powder in autoclaved ultrapure water, adjusting the final volume to 1 L with autoclaved ultrapure water, then autoclaved. Stored at room temperature.
2. LB Agar plates (stored at 4°C).
3. Terrific broth (TB) liquid medium (50.8 g/L): Prepared by dissolving 50.8 g of TB powder in autoclaved ultrapure water, adjusting the final volume to 1 L with autoclaved ultrapure water, then autoclaved. Stored at 4°C.
4. Kanamycin sulfate stock solution (50 mg/mL): Prepared by dissolving 0.5 g of antibiotic powder in autoclaved ultrapure water, adjusting the final volume to 10 mL with autoclaved ultrapure water, then aliquoted. Stored at −20°C.
5. Chloramphenicol stock solution (34 mg/mL): Prepared by dissolving 0.34 g of antibiotic powder in autoclaved ultrapure water, adjusting the final volume to 10 mL with 95% ethanol, then aliquoted. Stored at −20°C.

2.1.3 Induction and Lysis Reagents

1. L-Arabinose (AmBeed # A395369) stock solution (20% w/v): Prepared by dissolving 20 g of L-arabinose powder in 80 mL of autoclaved ultrapure water, aliquoted, and stored at −20°C.
2. Isopropyl β-D-1-thiogalactopyranoside (IPTG, AmBeed #A193402) stock solution (500 mM): Prepared by dissolving 11.92 g of IPTG powder in autoclaved ultrapure water, adjusting the final volume to 100 mL with autoclaved ultrapure water, then aliquoting the solution and storing at −20°C.
3. Endo-bicyclo[6.1.0]non-4-yne-L-lysine (BCNK, SiChem #SC-8014).
4. Binding buffer: 10 mM imidazole, 300 mM NaCl, 46.5 mM Na_2HPO_4, 3.5 mM NaH_2PO_4, pH 8.0.
5. Washing buffer: 25 mM imidazole, 300 mM NaCl, 38.4 mM Na_2HPO_4, 11.6 mM NaH_2PO_4, pH 7.4.
6. Elution buffer: 250 mM imidazole, 300 mM NaCl, 38.4 mM Na_2HPO_4, 11.6 mM NaH_2PO_4, pH 7.4 (*see* **Note 1**).
7. PMSF protease inhibitor (Thermo Scientific™ #36978).
8. Lysozyme (Thermo Scientific™ #89833).

2.1.4 Purification Matrix and Filter

1. HisPur™ Ni-NTA Superflow Agarose (Thermo Scientific™ #25214).

2. Amicon® Ultra Centrifugal Filter, 3 kDa MWCO (MilliporeSigma™ UFC500308).

2.1.5 Chemical Reagents

1. HS-1 (10 mM stock solution in acetonitrile, stored at −80°C, *see* **Note 2**).
2. 0.5 mM HS-1 solution in PBS as a quenching solution.
3. Phosphate-buffered saline (PBS): 137 mM NaCl, 2.7 mM KCl, 10 mM Na_2HPO_4, 1.8 mM KH_2PO_4, pH 7.4.

2.1.6 Equipment and Labware

1. Eppendorf™ Pipette Pick-a-Pack Set.
2. Corning® LSE™ Vortex Mixer.
3. VWR® Sterile Syringe Filter, PES 25 mm, 0.22 μm.
4. Agilent 6530 LC-MS/Q-TOF system.
5. NanoDrop™ 2000C Spectrophotometer.
6. Sonic Dismembrator (Fisherbrand™ #FB120).
7. Thermo Scientific MaxQ 4000 Benchtop Orbital Shaker.
8. Fisherbrand™ Mini Tube Rotator.
9. Corning® 50 mL centrifuge tubes.
10. Corning® 15 mL centrifuge tubes.
11. VWR® Screw-Cap centrifuge tubes, 5 mL.
12. Avant 1.7 mL microcentrifuge tubes.
13. 0.6 mL Microcentrifuge tubes with attached lid.

2.2 Materials for Bioorthogonal Labeling of GCGR in Live Cells

2.2.1 Biology Materials

1. HEK293T cells (CRL-3216, ATCC).
2. Dulbecco's modified eagle medium (DMEM, Gibco™ #12430054).
3. Fetal bovine serum (FBS, Fisherbrand™ #FB12999102).
4. Dulbecco's phosphate-buffered saline (DPBS, Gibco™ #14190250).
5. 35 mm Glass bottom culture dishes, glass diameter: 14 mm, glass thickness: 1.5 mm (0.16–0.19 mm) (Matsunami Glass #D35-14-1.5-U).
6. Sterile Syringe Filter, PES 25 mm, 0.22 μm (VWR® #514-1263).
7. Gentamicin (50 mg/mL, Gibco™ #15750060).
8. Plasmocin® (2.5 mg/mL, InvivoGen #ant-mpp).

9. 0.05% Trypsin-EDTA (Gibco™ #25300062).
10. Opti-MEM™ I Reduced Serum Medium (Gibco™ #31985070).
11. Polyethylenimine, linear (MW 25,000, transfection grade; PEI, Polysciences #23966-1): A stock solution (1 mg/mL) was prepared by dissolving 50 mg of PEI powder in autoclaved ultrapure water, adjusting the final volume to 50 mL with autoclaved ultrapure water, then aliquoted, and stored at −20°C with strict prohibition of freeze-thaw cycles to preserve the integrity.
12. FluoroBrite™ DMEM (Gibco™ #A1896701).
13. pCMV-BCNKRS-U6-tRNA: The following three mutations, Y306M, L309G, and C348A, were introduced into the pCMV-MmPylRS-U6-tRNA template (Gene Universal).
14. pCMV6-GCGR-372TAG-GFP: Amber codon was introduced at position-372 in pCMV6-GCGR-GFP (OriGene #RG211179) through site-directed mutagenesis.

2.2.2 Chemical Reagents

Cy5-conjugated HS-2 (1 mM stock in DMSO, stored at −20°C protected from light, *see* **Note 3**).

2.2.3 Equipment and Labware

1. 5% CO_2 cell culture incubator (Thermo Electron Corporation).
2. Benchtop centrifuge (Allegra™ X-22R Centrifuge).
3. Biosafety cabinet (The Baker Company).
4. Water bath (VWR).
5. Bright-Line™ Hemacytometer (Sigma #Z359629).
6. Inverted microscope (Olympus-CKX31).
7. Zeiss LSM 710 Confocal Microscope equipped with Plan-Apochromat 20×/0.8 M27 or 40×/1.3 Oil DIC M27 objective.
8. 50 mL Centrifuge tubes (Corning® #430828).
9. 15 mL Centrifuge tubes (Corning® #430790).
10. Screw-cap centrifuge tube, 5 mL (VWR® #10002-738).
11. 1.7 mL Microcentrifuge tubes (Avant #2925).
12. 0.6 mL Microcentrifuge tubes with attached lid (Sigma #T5149).

3 Methods

3.1 Bioorthogonal Modification of a Nanobody

For bioorthogonal protein modification, the overall efficiency depends on the fidelity of chemical reporter incorporation and subsequent bioorthogonal ligation reaction kinetics and selectivity.

The genetic encoding of BCNK via an orthogonal pyrrolysyl-tRNA synthetase/tRNA pair ensures site-specific incorporation, as evidenced by LC/MS characterization of the full-length protein. For modification of the NB1 nanobody, the chemical reporter BCNK was placed at position-4, a solvent-exposed site near the complementarity-determining region (CDR) of the nanobody. The modification reaction by HS reagents was fast, reaching >95% conversion in 2 min in phosphate-buffered saline (Fig. 2). This kinetic performance surpasses classical strain-promoted azide-alkyne cycloadditions (SPAAC) with BCN that typically require 30–60 min for completion under

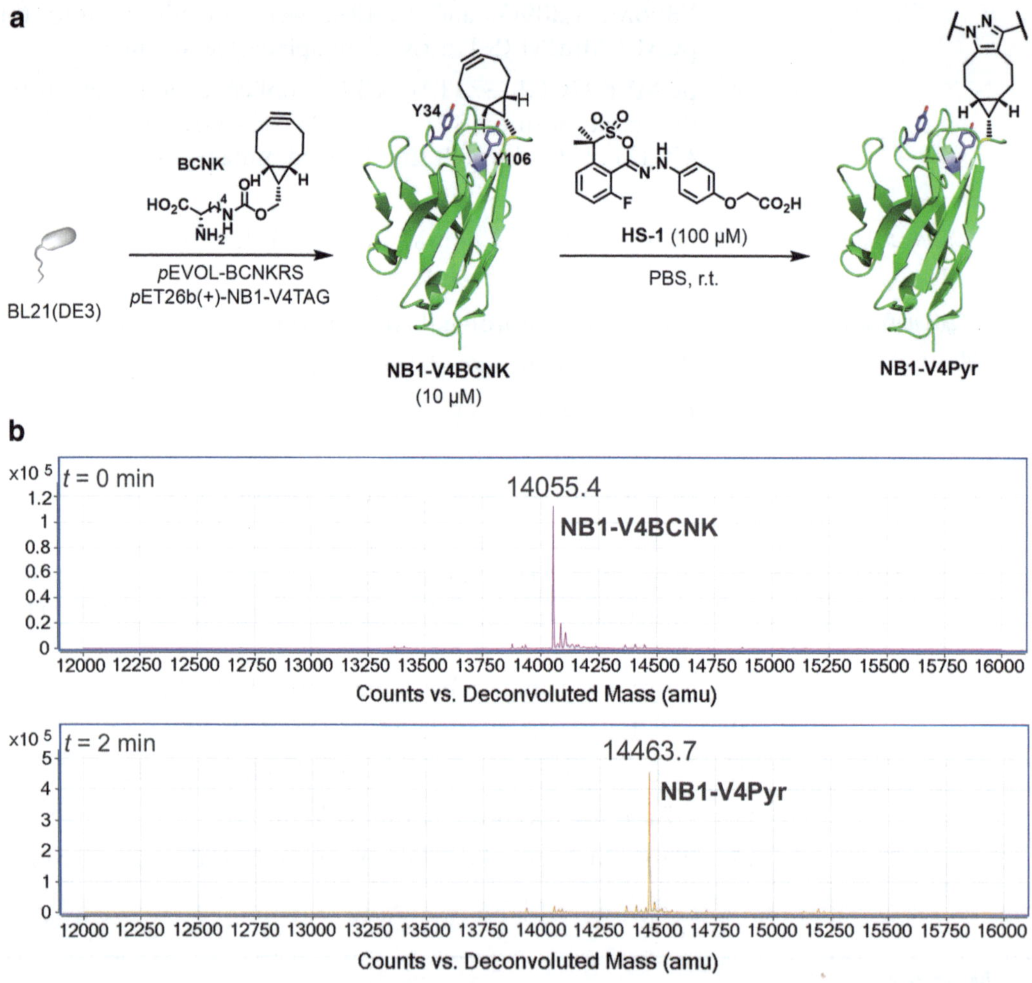

Fig. 2 Bioorthogonal modification of a BCNK-encoded nanobody by a hydrazonyl sultone reagent. (**a**) Reaction scheme. Nanobody NB1 is rendered with a ribbon model (PDB code: 3OGO) showing BCNK at position-4 and adjacent hydrophobic residues Y34 and Y106 (purple sticks) at the complementarity determining region. (**b**) MS analysis of the reaction mixture of NB1-V4BCNK with HS-1 in PBS. Top, NB1-V4BCNK starting materials (calculated m/z 14,056.1 Da); Bottom, full conversion to the pyrazole product (calculated m/z 14,465.2 Da) after 2 min. Adapted with permission from reference [6]: Copyright 2023 American Chemical Society

similar conditions [12], while maintaining comparable selectivity. The microenvironment was crucial as substitutions of adjacent hydrophobic residues Tyr-34 and Tyr-106 with Asp/Glu significantly decreased the reaction rate, a feature distinct from tetrazine-based reactions where electronic effect dictates the reaction rate [13]. Notably, HS-mediated modification exhibited exceptional selectivity for BCNK over endogenous amino acids, as no other adduct masses were detected in the LC/MS analysis (Fig. 2b). The absence of off-target adducts contrasts with the maleimide-thiol conjugation, which requires stringent conditions to suppress the nonspecific reactions. Optimization of the HS structure led to further enhancement in reaction kinetics, e.g., a HS analog bearing *o*-CF_3 group at C-aryl ring and *p*-phenol at N-aryl ring displayed a second-order rate constant of 1500 M^{-1} s^{-1} in PBS–ethanol (9:1) [7]. Importantly, the rate enhancement was due to the protein microenvironment surrounding BCN as the corresponding small-molecule BCN–HS ligation reactions proceeded much slower in aqueous buffer [7]. This microenvironment effect, unique to HS–BCN ligation among bioorthogonal reactions, enables efficient nanobody modification at low HS concentrations (10–100 μM).

3.1.1 Preparation of BCNK Stock Solution

1. For preparation of solution A, 80 mg of NaOH pellets were added into 8.5 mL of autoclaved NanoPure water and 1.5 mL of DMSO in a sterile tube. The mixture was thoroughly vortexed to ensure complete dissolution and subsequently filter-sterilized through a 0.22 μm syringe filter. This yielded a final concentration of 0.2 M NaOH and 15% v/v DMSO.
2. For preparation of solution B, 2.38 g of HEPES was dissolved in 10 mL of autoclaved NanoPure water. The pH of the solution was adjusted to 7.4, and it was then filter-sterilized using a 0.22 μm PES syringe filter.
3. For preparation of 100 mM BCNK stock solution, 96.72 mg of BCNK was weighed and subsequently dissolved in 3 mL of solution A. The mixture was sonicated for 30 s to aid dissolution. Following sonication, the solution was aliquoted into sterile microcentrifuge tubes and stored at −80°C (*see* **Note 4**).

3.1.2 NB1-V4BCNK Expression and Purification

1. For cotransformation, 50 ng each of pET26b(+)-NB1-V4TAG and pEVOL-BCNKRS plasmids were mixed with 50 μL of BL21(DE3) competent cells in a 0.6 mL microcentrifuge tube (*see* **Note 5**). The mixture was incubated on ice for 30 min, followed by a heat shock at 42°C for 45 s. Cells were then placed on ice for 5 min for recovery. Subsequently, 950 μL of TB medium was added to the mixture, which was then incubated at 37°C for 1 h with shaking at 260 rpm.
2. For plating, the cells were centrifuged at 12,000 *g* for 1 min. Following centrifugation, 900 μL of the supernatant was

removed using a pipette, and the cell pellet was resuspended in the remaining 100 µL of media. Finally, 100 µL of the cell suspension was spread onto LB agar plates containing 50 µg/mL kanamycin and 34 µg/mL chloramphenicol (*see* **Note 6**). The plates were incubated overnight at 37°C.

3. To begin culture expansion, a starter culture is inoculated by placing a single colony into 6 mL of LB medium supplemented with antibiotics, specifically 50 µg/mL kanamycin and 34 µg/mL chloramphenicol. This starter culture is then incubated in a shaker incubator at 37°C and 250 rpm for 16 h.
4. Following the initial incubation, the culture is scaled up for expression. This involves transferring 1 mL of the starter culture into 40 mL of fresh LB medium, again containing 50 µg/mL kanamycin and 34 µg/mL chloramphenicol. The culture is then grown at 37°C with shaking at 260 rpm until the optical density at 600 nm (OD_{600}) reaches a value between 0.5 and 0.7 (*see* **Note 7**).
5. Once the OD_{600} of the culture reaches 0.5–0.7, protein induction is initiated. First, 800 µL of a 100 mM BCNK stock, previously stored at −80°C, is thawed. This thawed BCNK stock is then added to 3.2 mL of solution B, resulting in 4× dilution to derive a 20 mM BCNK solution. This solution is subsequently filtered using a 0.22 µm syringe filter
6. Next, 4 mL of the filtered, diluted BCNK solution is added to the culture, achieving a final BCNK concentration of 2 mM. The culture is then incubated at room temperature for 15 min. Following this incubation, 0.4 mL of a 20% (w/v) L-arabinose solution is added to reach a final concentration of 0.2% (w/v), and 80 µL of a 500 mM IPTG solution is added to achieve a final concentration of 1 mM. After the addition of L-arabinose and IPTG, the culture is incubated at 25°C for 18 h with shaking at 260 rpm (*see* **Note 8**).
7. For cell lysis and purification, the process begins with centrifugation at 4000 *g* at 4°C for 30 min. The resulting pellet is then resuspended in 4 mL of ice-cold binding buffer, which is supplemented with 1 mM PMSF and 1 mg/mL lysozyme. This suspension is incubated on ice for 30 min to allow the lysozyme to weaken the cell wall. Following this incubation, the resuspended cells are transferred to an ice-cold tube.
8. The cells are then sonicated on ice using the probe sonicator. The sonication parameters are set to an amplitude of 40%, with pulse cycles of 10 s ON and 20 s OFF to prevent overheating (*see* **Note 9**). This process is repeated for five to ten cycles, or until the lysate becomes translucent and less viscous. Throughout the sonication process, the sample is kept

on ice. After sonication, the lysate is clarified by centrifugation at 12,000*g* for 30 min at 4°C.

9. For purification, HisPur Nickel Superflow Agarose, which is stored as a 50% slurry at 4°C, is gently resuspended. A volume of 80 μL of this slurry, corresponding to 40 μL of packed beads, is transferred to a 5 mL tube. The beads are then equilibrated by adding 1 mL of binding buffer, gently inverting the tube several times, and centrifuged at 700*g* for 2 min at 4°C. The supernatant is carefully removed without disturbing the beads. This equilibration step is repeated twice.
10. The clarified lysate (4 mL) is then added to the equilibrated beads and incubated for 2 h at 4°C with end-over-end mixing (*see* **Note 10**). After incubation, the bead-lysate mixture is centrifuged at 700*g* for 2 min at 4°C, and the supernatant is discarded. The beads are subsequently washed by resuspending them in 0.5 mL of wash buffer. This mixture is gently mixed, centrifuged at 700*g* for 2 min at 4°C, and the supernatant is discarded. This wash step is repeated two more times, for a total of three washes (*see* **Note 11**).
11. To elute the protein, the beads are resuspended in 0.3 mL of elution buffer and incubated for 5 min at room temperature with mixing. The mixture is then centrifuged at 700*g* for 2 min at 4°C, and the supernatant, designated as elute 1, is collected. The elution process is repeated with 0.1 mL of elution buffer to obtain elute 2. The eluates are pooled, resulting in a total volume of 0.4 mL, and the protein concentration is quantified by UV spectrophotometry using a NanoDrop™.
12. Finally, the elute is concentrated using an Amicon Ultra 0.5 mL centrifugal filter with 3 kDa molecular weight cutoff (MWCO). This is achieved by centrifuging at 14,000*g* at 4°C until the final volume is no more than 100 μL. A buffer exchange to PBS (pH 7.4) is then performed by diluting the concentrate to 400 μL with PBS and repeating the concentration cycle three times. The protein concentration is then quantified again by UV spectrophotometry using NanoDrop™.

3.1.3 Bioorthogonal Modification of NB1-V4BCNK by HS

1. To set up the reaction, components are combined in a 0.6 mL microcentrifuge tube. This includes 50 μL of NB1-V4BCNK mutant protein solution (obtained from the last step, with a stock concentration of approximately 50 μM), resulting in a final reaction concentration of 10 μM. Additionally, 200 μL of PBS (pH 7.4) is added. Finally, 2.5 μL of 10 mM HS-1 is added to achieve a final concentration of 100 μM. The mixture is then quickly pipetted up and down three times and incubated at room temperature.

2. For time-course sampling, at specified time points such as 120 s, 25 μL of the reaction mixture is immediately transferred to a fresh tube that contains 50 μL of 0.5 mM BCNK, which serves as a quenching solution (*see* **Note 12**). The tube is vortexed immediately to terminate the reaction, resulting in a final BCNK concentration of 333 μM, representing a greater than 30-fold excess over NB1-V4BCNK. The quenched sample is then centrifuged at 12,000 g for 2 min to precipitate any aggregates that may have formed.
3. For LC/MS analysis, 50 μL of the supernatant from the centrifuged quenched sample is injected into the QTOF-LC/MS system (*see* **Note 13**). The analysis is performed using an Aeris Widepore C4 column (50 × 2.1 mm, 3.6 μm). The mobile phase consists of A: 0.1% formic acid in H_2O, and B: 0.1% formic acid in acetonitrile. A gradient of 5%–95% B over 10 min is employed. Mass spectra are acquired for the unmodified NB1-V4BCNK mutant and the HS-1-labeled product.
4. Conjugation efficiency, expressed as percent conversion, is calculated based on LC/MS peak areas (*see* **Note 14**) using the formula:

$$\text{Conversion (\%)} = \frac{\text{Area}_{\text{labeled}}}{\text{Area}_{\text{labeled}} + \text{Area}_{\text{unlabeled}}} \times 100\%$$

The time-dependent conversion data are then fitted to a pseudo-first-order kinetic model (*see* **Note 15**):

$$\ln(1 - \text{conversion}) = -k_{\text{obs}}t, \text{ where } k_{\text{obs}} = k_2 \times [HS]_0$$

where k_{obs} is the observed pseudo-first-order rate constant, and k_2 is the second-order rate constant. Each labeling reaction is performed in triplicate ($n = 3$) to ensure reproducibility and to enable statistical analysis.

3.2 Bioorthogonal Labeling of GCGR in Live Cells

Traditional live-cell protein labeling involves fusion of a protein fluorophore or an enzymatic tag, which frequently perturb protein function. For example, fluorescent protein fusions may sterically hinder ligand binding or receptor dimerization, and the use of an enzymatic tag such as SNAP-tag [14] requires cumbersome washing steps to eliminate the unreacted probes. To overcome these limitations, bioorthogonal chemistry-driven protein labeling methods have been developed that display minimum cytotoxicity, inertness against endogenous molecules, and rapid reaction kinetics toward the desired targets [5]. The key advantage of a bioorthogonal approach lies in its versatility in introducing a biophysical label at late stage and in native cellular environment through selective bioconjugation with a wild-type-like protein scaffold carrying a small bioorthogonal chemical handle. In particular, the genetically encoded BCNK has emerged as a robust bioorthogonal reporter owing to its high stability in biological systems ($t_{1/2} > 48$ h in mammalian cytosol), compatibility with amber suppression technology for site-specific incorporation (Protocol 3.2.1), and rapid reaction kinetics [8]. On the other hand, hydrazonyl sultones exhibit exceptionally fast reaction kinetics toward the BCNK-encoded proteins, with second-order rate constant, k_2, up to ~1500 $M^{-1}\ s^{-1}$ under physiological conditions [7]. This reaction efficiency exceeds that of strain-promoted azide-alkyne cycloaddition ($k_2 = 0.1–1\ M^{-1}\ s^{-1}$) [15], while requiring no exogenous catalysts or photoirradiation. To illustrate, a C-terminal GFP-fused glucagon receptor, GCGR-GFP, encoding BCNK at position-372 in the extracellular loop 3 can be selectively labeled with a Cy5-conjugated HS-2 within 1 h in HEK293T cells, as indicated by the overlaid fluorescence images in confocal microscopy as well as line profile analysis (Fig. 3) [6]. Protocol 3.2.2 achieves the fluorescent labeling of GCGR with HS-2 at 1.4 μM, a concentration that minimizes nonspecific background reactions while preserving >95% cell viability (Fig. 3). Because BCNK can be incorporated into any position of a GPCR, including the intracellular loop regions [16], using the standard amber codon suppression protocols (Protocol 3.2.1), this HS-BCN ligation reaction offers a precision tool for late-stage fluorophore installation on-demand to generate fluorescence-based biosensors for biophysical studies of GPCR dynamics in live cells [17].

3.2.1 Protocol for Incorporation of BCNK Into GCGR-GFP in HEK293T Cells

1. To ensure proper cell growth and availability for experiments, HEK293T cells are routinely maintained (*see* **Note 16**). This involves culturing them in a 10-cm plate containing 10 mL of DMEM, which is supplemented with 10% (v/v) FBS. These cultures are kept at 37°C in an incubator with a 5% CO_2 atmosphere.
2. For subculturing cells, the existing culture medium is first gently removed. The cells are then washed twice with 3 mL of prewarmed DPBS. Following the washes, 3 mL of 0.05% trypsin-EDTA, prewarmed to 37°C, is added to the cells and spread evenly to cover the entire surface. After a 5-min incubation period, trypsin activity is neutralized by adding 7 mL of complete growth medium containing FBS. The cells are subsequently collected into a 15-mL centrifuge tube and pelleted by centrifugation at 400*g* for 5 min at 22°C. The supernatant is then discarded without disturbing the cell pellet. Finally, the pellet is gently resuspended in 1 mL of fresh growth medium, and the cell number is determined using a hemacytometer.
3. To prepare for transfection, cells are seeded into appropriate culture vessels. Specifically, 6.0×10^5 HEK293T cells are seeded into 35-mm glass bottom culture dishes. Each dish

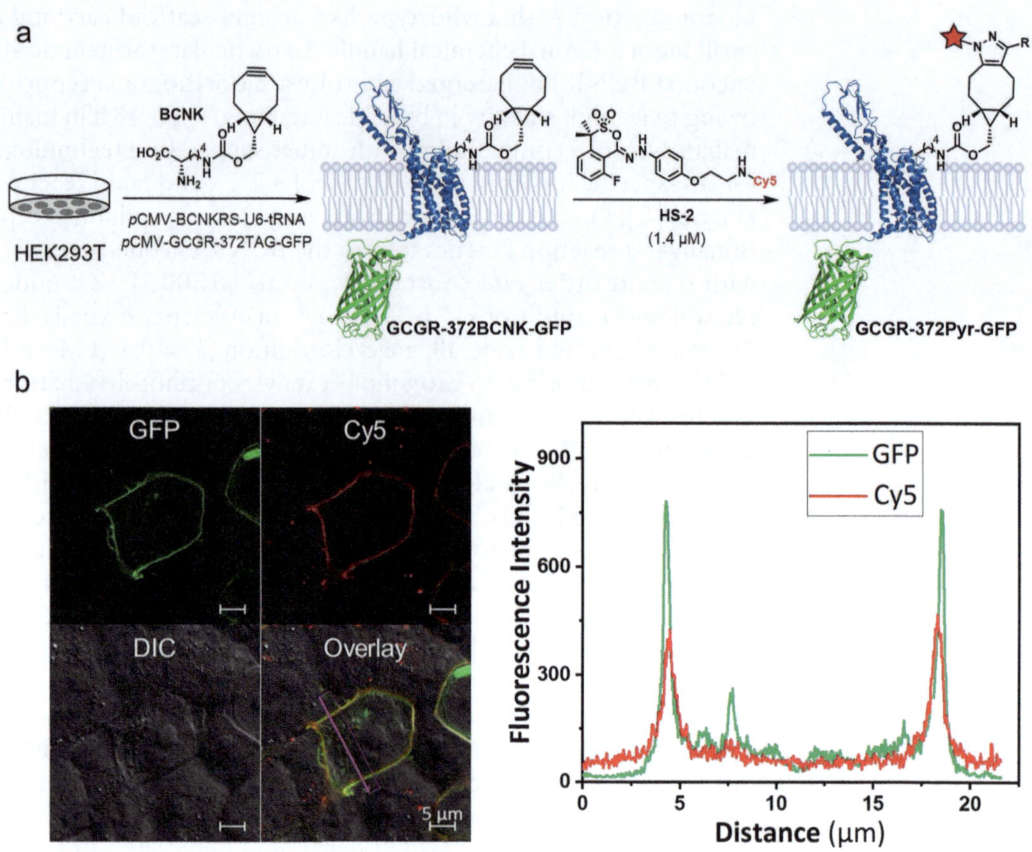

Fig. 3 Bioorthogonal labeling of GCGR-372BCNK-GFP with HS-2 in HEK293T cells. (**a**) Labeling scheme. (**b**) Representative confocal micrographs of HEK293T cells expressing GCGR-H372BCNK-GFP after 1-h incubation with HS-2 in DMEM (left) and line profile analysis of the overlaid image showing specific membrane receptor labeling (right). Adapted with permission from reference [6]: Copyright 2023 American Chemical Society

contains 2 mL of growth medium, which consists of DMEM supplemented with 10% (v/v) FBS, 50 μg/mL gentamicin, and 0.5 μg/mL Plasmocin™. Under these conditions, cells typically reach 70%–80% confluency approximately 48 h after seeding.

4. Prior to transfection, it is crucial to ensure optimal cell confluency. Cell confluency is monitored using a microscope, and transfection is performed when cells reach 70%–80% confluency. At this stage, the existing medium is aspirated and gently replaced with pre-warmed DMEM.

5. The transfection process itself involves preparing a DNA-PEI mixture. For this, a total of 3 μg DNA, consisting of 2.4 μg pCMV-BCNKRS-U6-tRNA and 0.6 μg pCMV6-GCGR-372TAG-GFP, is combined with 9 μg of PEI (taken from a 1 mg/mL stock, i.e., 9 μL). This mixture is then supplemented with Opti-MEM® to achieve a final volume of 200 μL in a 0.6 mL microcentrifuge tube. The tube is vortexed gently and

incubated at room temperature for 15 min to ensure the DNA is complexed with PEI. This transfection mixture is then slowly added to the cell culture while gently swirling the plate.

6. Following the addition of the transfection mixture, the cells require an initial incubation period. The cells are maintained at 37°C in a 5% CO_2 atmosphere for 6–8 h (*see* **Note 17**).
7. After the initial incubation, the medium is exchanged for one containing BCNK. To prepare this, stock solutions of BCNK (100 mM in DMSO) (*see* **Note 18**) are diluted into 2 mL of growth medium in a 5-mL centrifuge tube to obtain a final concentration of 1 mM BCNK (*see* **Note 19**). The resulting BCNK-containing medium is then filtered through a 0.2 μ m PES filter. The transfection medium in each plate is subsequently replaced with 2 mL of this BCNK-containing growth medium.
8. For the final stage of the experiment, the cells undergo a continued incubation. This incubation is conducted at 37°C and 5% CO_2 for a period of 24–48 h.

3.2.2 Protocol for Labeling BCNK-Encoded GCGR-GFP

1. To prepare the labeling agent, an HS-2 working solution is made. First, the HS-2 stock solution (1 mM in DMSO) is thawed on ice while being protected from light. Then, in a 1.7 mL microcentrifuge tube, the HS-2 stock is diluted into prewarmed DMEM to a concentration of 1.4 μM (e.g., by adding 1.4 μL of stock to 1 mL of DMEM). It is critical to ensure that the final DMSO concentration is less than or equal to 1% (v/v). Finally, this diluted solution is filtered through a 0.22 μm PES filter.
2. For the cell labeling procedure, the growth medium is first aspirated from cells previously cultured in glass-bottom dishes (from Step 8 in Sect. 3.2.1). The cells are then gently rinsed twice with 2 mL of DMEM, prewarmed to 37°C, to remove any residual BCNK and FBS. Following the rinses, 1 mL of the filtered HS-2 working solution is added to each well. The cells are then protected from light and incubated at 37°C under a 5% CO_2 atmosphere for 1 h (*see* **Note 20**).
3. After the incubation period with the labeling agent, a washing procedure is performed (*see* **Note 21**). The HS-2 solution is aspirated away from the cells. Subsequently, the cells are rinsed three times, each time with 2 mL of prewarmed DMEM.
4. To prepare the cells for imaging, the DMEM is replaced with 2 mL of FluoroBrite DMEM (*see* **Note 22**) that has been prewarmed to 37°C. It is important to maintain the cells in the dark at 37°C until imaging. Cells should be imaged promptly, within 30 min after labeling, because extended incubation may

lead to a reduction in fluorescence intensity and could alter cell viability.

5. For image acquisition via confocal microscopy, specific parameters are utilized. A 40×/1.3 Oil DIC M27 objective is used. The excitation and emission settings are as follows: for the GFP channel, excitation is at 488 nm with emission collected between 493 nm and 598 nm; for the Cy5 channel, excitation is at 635 nm with emission collected between 640 nm and 759 nm. Laser power should be optimized, typically 1%–3% for GFP and 5%–7% for Cy5, to minimize photobleaching. Scan settings include a 1024 × 1024 resolution, 2× line averaging, and a 12-bit depth. The pixel dwell time should be set to approximately 2 μs/pixel or adjusted as necessary to balance image quality and photobleaching (*see* **Note 23**).

4 Notes

1. Binding, washing, and elution buffers should be prepared with sterile-filtered NanoPure water and sterilized by passing through a 0.22 μm filter prior to use.
2. HS-1 was synthesized by following the reported procedures in reference [6].
3. The Cy5-conjugated HS-2 was synthesized by following the reported procedures in reference [6].
4. The efficiency of BCNK incorporation is highly dependent on the quality and handling of the BCNK stock solution. It is recommended to prepare fresh BCNK stock solutions or use aliquots that have undergone minimal freeze-thaw cycles. Store aliquots at –80°C. Prolonged storage or multiple freeze-thaw cycles can lead to degradation of BCNK, reducing its incorporation efficiency into the target protein.
5. The quality of competent cells BL21(DE3) is critical for efficient transformation. Use highly competent cells and avoid repetitive thaws.
6. Ensure selective pressure by using appropriate concentrations of kanamycin and chloramphenicol on plates and in liquid cultures. Variability in colony size after transformation might indicate issues with plasmid uptake or initial growth.
7. Inducing protein expression at the optimal cell density (OD_{600} of 0.5–0.7) is crucial for maximizing the yield of correctly folded, soluble protein. Premature or late induction can lead to lower yields or increased misfolding.

8. Ensure L-arabinose and IPTG stock solutions are fresh and accurately prepared. The 18-h expression at 25°C is designed to enhance proper protein folding and solubility; higher temperatures or shorter times might increase yield but risk protein misfolding.
9. Efficient cell lysis is key to maximizing protein recovery. Monitor the viscosity and translucency of the lysate to gauge lysis completion. Avoid overheating the sample during sonication as this can lead to protein denaturation and aggregation. Ensure the sonicator probe is cleaned properly and tuned. The "10 s ON/20 s OFF" cycle is critical for temperature control.
10. Ensure the HisPur Nickel Superflow Agarose is thoroughly equilibrated with binding buffer to maximize binding efficiency. The 2-h incubation of lysate with beads at 4°C is generally sufficient; however, for proteins with lower expression levels or weaker His-tag accessibility, this time can be extended, or a larger volume of beads may be used.
11. Wash steps are critical to remove nonspecifically bound proteins; ensure complete resuspension of beads during washes.
12. Rapid and effective quenching of the HS-BCN ligation is crucial for accurate time-course analysis. The >100-fold molar excess of BCNK in the quenching solution should be sufficient. Ensure immediate and thorough mixing upon adding the reaction mixture to the quenching solution. If incomplete quenching is suspected (e.g., continued reaction), the concentration of the quencher or the mixing efficiency may need to be increased.
13. Consistent performance of the LC/MS system is vital. Regularly check column performance and system calibration. Ensure complete dissolution of samples and remove any particulate matter before injection to prevent column clogging.
14. Deconvolution of mass spectra for proteins can sometimes be challenging; ensure appropriate software settings are used for accurate mass determination of both unlabeled and labeled protein.
15. The pseudo-first-order kinetic model assumes that the concentration of HS-1 remains effectively constant throughout the reaction. This is valid when HS-1 is in significant excess (100 μM HS-1 vs 10 μM protein). If this condition is not met, a more complex kinetic model may be required. Accurate peak integration is paramount for calculating conversion percentages. The reproducibility of triplicate measurements provides confidence in the determined rate constants.

16. Maintaining healthy, actively growing HEK293T cells is paramount for successful BCNK incorporation and labeling. Adhere strictly to cell culture protocols, monitor cells for any signs of stress or contamination (e.g., changes in morphology, growth rate, or medium color). Passage cells regularly to avoid over-confluency, which can affect transfection efficiency and physiological responses. The use of gentamicin and Plasmocin™ helps prevent bacterial and mycoplasma contamination, respectively, but aseptic techniques remain critical.
17. Transfection efficiency can vary significantly based on cell passage number, confluency (70%–80% is optimal), DNA quality and quantity, PEI quality, and the DNA:PEI complex formation procedure. Optimize the DNA:PEI ratio if low efficiency is observed. Ensure gentle mixing and even distribution of the transfection complex onto the cells.
18. Ensure high-quality DMSO is used for dissolving BCNK, as impurities can affect cell viability and expression yield of transfected genes.
19. The final concentration of BCNK (1 mM) in the growth medium for 24–48 h is a starting point. This concentration and duration may need optimization for different cell types or target proteins to maximize BCNK incorporation while minimizing potential cytotoxicity. Monitor cell morphology and viability during BCNK incubation. Always filter-sterilize BCNK-containing medium before adding to cells.
20. HS-2 stock solution (in DMSO) should be protected from light and stored appropriately to maintain its reactivity. The final DMSO concentration in the labeling medium must be kept low (≤1% v/v) to avoid adverse effects on cell viability and membrane integrity. Pre-warming all solutions (DMEM, HS-2 working solution) to 37°C helps maintain cell physiology. The 1-h incubation time with HS-2 at 1.4 μM is optimized for GCGR labeling; this may require adjustment for other targets or HS probes to achieve optimal signal-to-background ratios.
21. Thorough washing after labeling is critical to remove unbound HS-2 probe and minimize nonspecific background fluorescence, which can obscure specific signals. Perform wash gently to avoid detaching cells.
22. Switching to FluoroBrite™ DMEM or a similar low-autofluorescence imaging medium significantly improves image quality, especially for dim signals. Imaging cells promptly (within 30 min postlabeling and washing) is recommended, as prolonged incubation can lead to probe internalization, signal redistribution, or decreased cell viability, potentially confounding data interpretation.

23. Laser power settings should be carefully optimized for each channel (GFP and Cy5) to obtain adequate signal without causing significant photobleaching or phototoxicity. Use the lowest laser power that provides a good signal-to-noise ratio. Scan speed, averaging, and resolution should be balanced to achieve desired image quality within a reasonable acquisition time. If spectral bleed-through between GFP and Cy5 channels is suspected, acquire images sequentially or use appropriate control samples (e.g., cells expressing only GFP or labeled only with Cy5) to set up correction parameters if your system supports spectral unmixing.

Acknowledgment

We gratefully acknowledge the National Institutes of Health (R35GM130307) for supporting our bioorthogonal chemistry work.

References

1. Sletten EM, Bertozzi CR (2009) Bioorthogonal chemistry: fishing for selectivity in a sea of functionality. Angew Chem Intl Ed 48:6974–6998
2. Scinto SL et al (2021) Bioorthogonal chemistry. Nat Rev Methods Primers 1:30
3. McKay CS, Finn MG (2014) Click chemistry in complex mixtures: bioorthogonal bioconjugation. Chem Biol 21:1075–1101
4. Blackman ML, Royzen M, Fox JM (2008) Tetrazine ligation: fast bioconjugation based on inverse-electron-demand Diels–Alder reactivity. J Am Chem Soc 130:13518–13519
5. Tian Y, Lin Q (2019) Fitness factors for bioorthogonal chemical probes. ACS Chem Biol 14:2489–2496
6. Fang M et al (2023) Hydrazonyl sultones as stable tautomers of highly reactive nitrile imines for fast bioorthogonal ligation reaction. J Am Chem Soc 145:9959–9964
7. Fang M, Kumar GS, Lin Q (2023) Systematic structural tuning yields hydrazonyl sultones for faster bioorthogonal protein modification. ChemBioChem 24:e202300398
8. Lang K et al (2012) Genetic encoding of bicyclononynes and trans-cyclooctenes for site-specific protein labeling in vitro and in live mammalian cells via rapid fluorogenic Diels–Alder reactions. J Am Chem Soc 134:10317–10320
9. Liu H et al (2016) Ultrafast click chemistry with fluorosydnones. Angew Chem Intl Ed 55:12073–12077
10. Kumar GS, Racioppi S, Zurek E, Lin Q (2022) Superfast tetrazole–BCN cycloaddition reaction for bioorthogonal protein labeling on live cells. J Am Chem Soc 144:57–62
11. Borade BR, Fang M, Lin Q (2024) Synthesis of hydrazonyl sultones via phosphine-mediated cyclodehydration of vicinal sulfo-acyl hydrazides. Org Lett 26:8239–8243
12. Leunissen EHP et al (2014) Copper-free click reactions with polar bicyclononyne derivatives for modulation of cellular imaging. ChemBioChem 15:1446–1451
13. Svatunek D et al (2022) Uncovering the key role of distortion in bioorthogonal tetrazine tools that defy the reactivity/stability trade-off. J Am Chem Soc 144:8171–8177
14. Keppler A et al (2003) A general method for the covalent labeling of fusion proteins with small molecules in vivo. Nat Biotech 21:86–89
15. Luu T, Gristwood K, Knight JC, Jörg M (2024) Click chemistry: reaction rates and their suitability for biomedical applications. Bioconjugate Chem 35:715–731
16. Tian Y, Fang M, Lin Q (2021) Intracellular bioorthogonal labeling of glucagon receptor via tetrazine ligation. Bioorg Med Chem 43:116256
17. Lewandowski TM et al (2022) Dual fluorescent labeling of GLP-1R in live cells via enzymatic tagging and bioorthogonal chemistry. RSC Chem Biol 3:702–706

Subject Index

Alexander Deiters (ed.), *Bioconjugation Methods*, Methods in Molecular Biology, vol. 3009,
https://doi.org/10.1007/978-1-0716-5112-4,

<u>GPSR Compliance</u>

The European Union's (EU) General Product Safety Regulation (GPSR) is a set of rules that requires consumer products to be safe and our obligations to ensure this.

If you have any concerns about our products, you can contact us on ProductSafety@springernature.com

In case Publisher is established outside the EU, the EU authorized representative is:

Springer Nature Customer Service Center GmbH
Europaplatz 3
69115 Heidelberg, Germany

Batch number: 10513764

Printed by Printforce, the Netherlands